Metallurgy of welding

Metallurgy of welding

Sixth Edition

J. F. Lancaster

WOODHEAD PUBLISHING LIMITED
Cambridge, England

Published by Woodhead Publishing Limited,
80 High Street, Sawston, Cambridge CB22 3HJ, UK
www.woodheadpublishing.com
www.woodheadpublishingonline.com

Woodhead Publishing, 1518 Walnut Street, Suite 1100, Philadelphia,
PA 19102-3406, USA

Woodhead Publishing India Private Limited, G-2, Vardaan House, 7/28 Ansari Road,
Daryaganj, New Delhi – 110002, India
www.woodheadpublishingindia.com

First published 1965, George Allen & Unwin
Sixth edition 1999, Abington Publishing, an imprint of Woodhead Publishing Limited
Reprinted 2006, 2007, 2012

© J. F. Lancaster, 1999
The author has asserted his moral rights.

This book contains information obtained from authentic and highly regarded sources. Reprinted material is quoted with permission, and sources are indicated. Reasonable efforts have been made to publish reliable data and information, but the author and the publisher cannot assume responsibility for the validity of all materials. Neither the author nor the publisher, nor anyone else associated with this publication, shall be liable for any loss, damage or liability directly or indirectly caused or alleged to be caused by this book.
   Neither this book nor any part may be reproduced or transmitted in any form or by any means, electronic or mechanical, including photocopying, microfilming and recording, or by any information storage or retrieval system, without permission in writing from Woodhead Publishing Limited.
   The consent of Woodhead Publishing Limited does not extend to copying for general distribution, for promotion, for creating new works, or for resale. Specific permission must be obtained in writing from Woodhead Publishing Limited for such copying.

Trademark notice: Product or corporate names may be trademarks or registered trademarks, and are used only for identification and explanation, without intent to infringe.

British Library Cataloguing in Publication Data
A catalogue record for this book is available from the British Library.

ISBN 978-1-85573-428-9 (print)
ISBN 978-1-84569-486-9 (online)

Contents

Preface to the sixth edition   ix

1	**Introductory**	**1**
1.1	Structural joints	1
1.2	The cohesion of crystalline solids	1
1.3	The plastic behaviour of crystalline solids	6
1.4	Solid–liquid interactions	15
1.5	Fracture	20
	References	38
2	**Solid-phase welding**	**40**
2.1	Physical aspects and metallurgy	40
2.2	Solid-phase welding processes	42
	References	53
	Further reading	53
3	**The use of adhesives for making structural joints**	**54**
3.1	History	54
3.2	Bonding between adhesive and substrate	55
3.3	Polymers	61
3.4	The properties of adhesive polymers	66
3.5	Bonding procedures	72
3.6	Joint design and applications	80
	References	83
	Further reading	84

vi　Contents

4	Soldering and brazing	85
4.1	Physical aspects	85
4.2	Soldering	87
4.3	Brazing	94
	References	104
	Further reading	104

5	The joining of ceramics: microjoining	105
5.1	Scope	105
5.2	The properties of ceramics	105
5.3	Glass–metal seals	107
5.4	Glass-ceramics	111
5.5	Brazing	114
5.6	Other techniques	116
5.7	Microjoining	120
	References	126
	Further reading	127

6	Fusion welding processes and their thermal effects	128
6.1	The development of fusion welding	128
6.2	The nature of fusion welding	130
6.3	Types of fusion-welded joint	145
6.4	Heat flow in fusion welding	147
6.5	Weld defects	166
	References	168
	Further reading	168

7	Metallurgical effects of the weld thermal cycle	169
7.1	Gas–metal equilibria	169
7.2	Gas–metal reactions in arc welding	176
7.3	The mechanism of gas absorption in welding	185
7.4	Porosity	187
7.5	Diffusion	190
7.6	Dilution and uniformity of the weld deposit	192
7.7	Weld pool solidification	193
7.8	Weld cracking	197
7.9	Metallurgical effects in the parent metal and solidified weld metal	201
	References	209
	Further reading	210

8	**Carbon and ferritic alloy steels**	**211**
8.1	Scope	211
8.2	Metallurgy of the liquid weld metal	211
8.3	Transformation and microstructure of steel	225
8.4	The mechanical properties of the welded joint	239
8.5	Stress intensification, embrittlement and cracking of fusion welds below the solidus	248
8.6	Steelmaking	279
8.7	The welding of iron and steel products	287
	References	307
	Further reading	308
9	**Austenitic and high-alloy steels**	**310**
9.1	Scope	310
9.2	The weld pool	310
9.3	Alloy constitution	316
9.4	Mechanical properties	324
9.5	Transformation, embrittlement and cracking	327
9.6	The use of austenitic Cr–Ni alloys for repair welding, cladding and transition joints	337
9.7	Corrosion-resistant steels: alloys and welding procedures	345
9.8	Heat-resisting steels: alloys and welding procedures	348
9.9	Hardenable high-alloy steels	349
	References	350
	Further reading	351
10	**Non-ferrous metals**	**353**
10.1	Aluminium and its alloys	353
10.2	Magnesium and its alloys	371
10.3	Copper and its alloys	373
10.4	Nickel and its alloys	381
10.5	The reactive and refractory metals	388
10.6	The low-melting metals: lead and zinc	396
10.7	The precious metals: silver, gold and platinum	397
	References	397
	Further reading	398
11	**The behaviour of welds in service**	**399**
11.1	General	399
11.2	The initiation and propagation of fast fractures	399
11.3	Slow crack propagation	409

Contents

11.4	Corrosion of welds	420
11.5	Assessing the reliability of welded structures	430
	References	432
	Further reading	432

Appendix 1 Symbols 434

Appendix 2 Conversion factors 437

Index 439

Preface to the sixth edition

Apart from a general revision and updating of the text, some specific alterations and additions have been made to this book. These are as follows:

1. The material on fracture and fracture mechanics originally located in Chapter 11 (behaviour of welds in service) has been moved to a more logical position in Chapter 1. Additional subject-matter includes the brittle and ductile behaviour of solids, ductile fracture, and the velocity of crack propagation.
2. A description of the new and promising welding process, friction stir welding, is given in Chapter 2.
3. A new chapter on adhesive bonding has been written. The treatment includes bonding forces, polymer chemistry, types of adhesive, production technology, quality control and applications.
4. Part of the material on mass and heat flow has been incorporated in the chapter on welding processes. The section on heat flow has been expanded and includes worked examples. A section on weld defects and the evaluation of non-destructive tests has been added.
5. The metallurgy of superaustenitic stainless steel is considered in Chapter 9.
6. A section on the welding metallurgy of aluminium–lithium alloys has been added to Chapter 10.
7. In the final chapter, a new section describes major structural failures, and considers the role of welding in such failures.

The author would like to acknowledge helpful suggestions made by Michael Dunn, who at the time was Director, Engineering and Materials Science, Chapman and Hall. He also wishes to thank the librarians at TWI Abington for their unfailing help, and his wife, Eileen, who devoted much time to the preparation of the manuscript.

1
Introductory

1.1 Structural joints

In general there are two ways in which parts may be fastened together. The first method employs mechanical techniques such as bolting or riveting; in bolting, for example, the joint strength is obtained from frictional forces that keep the nuts in place, and from the shear and tensile strength of the bolt. The second method, with which this book is concerned, is to form a bond between the surfaces to be joined. In welding, brazing and soldering, the objective is to form a continuous metallic bridge between the two surfaces, such that the bonding is of the same character as that which maintains the integrity of the metal itself. The means of accomplishing this end are numerous, and have multiplied rapidly during the past half-century. The greater part of this book is concerned with the metallurgical consequences of such techniques, both in terms of the immediate problems that they may present, and also of their effect on the long-term behaviour of the bond. The joints under consideration will include those between metals and ceramics, where the bonding technique must be adapted to accommodate the brittle character of the non-metallic part. The use of synthetic polymers to make adhesively bonded joints is also considered. Before doing so, however, it is appropriate to say a few words about the nature of chemical and metallic bonds, and how they affect the strength of solids.

1.2 The cohesion of crystalline solids

1.2.1 Types of bond

Crystalline solids are those in which the constituent atoms or ions are arranged in a repetitive geometric pattern known as a **lattice structure**. Most of the solids used in engineering structures are of this character, exceptions being glass, which is a metastable supercooled liquid, and polymers, which are aggregates of large organic molecules.

Crystalline solids obtain their cohesion when a chemical or metallic bond is formed between the constituent atoms. There are two basic types of chemical bond: **ionic** and **covalent**. An ionic bond is formed when a **valence** or **bonding** electron is detached from the outer sheath of one atom and becomes attached to another, to form two oppositely charged **ions**. In the formation of sodium chloride, for example, an electron detaches from the sodium atom, forming a positively charged **cation**, and this electron becomes attached to the chlorine atom, forming a negatively charged **anion**. These ions are arranged in a **lattice structure** of which the basic component is a cube, with sodium and chloride ions alternating at the corners. The bonding force in this instance is the electrostatic attraction between positive and negative ions. Normally the ions occupy an equilibrium position in which the resultant force upon them is zero. When exposed to a tensile force the inter-ionic spacing increases, and the electrostatic attractive force comes into play; under compression, however, the repulsing force between the atomic nuclei is dominant.

In covalent bonding the constituent atoms lose an electron or electrons to form a cluster of positive ions, and the resulting electron cloud is shared by the molecule as a whole. In both ionic and covalent bonding the locations of electrons and ions are constrained relative to one another. Ionic and covalent solids are, in consequence, characteristically **brittle**.

Metallic bonding may be regarded as a type of covalent bonding, but one where the constituent atoms are identical or of the same type and where they do not combine with each other to form a chemical compound. The atoms lose an electron or electrons, forming arrays of positive ions. Electrons are shared by the lattice as a whole, and the electron cloud is therefore **mobile**. This fact accounts for the relatively high thermal and electrical conductivity of metals. It also accounts for their **ductility**, since not only are the electrons free to move, but so, within limits, are the ions. The manner in which this occurs is considered in Section 1.3.2.

The nature of the cohesive force in covalent and metallic crystals is basically similar. When the solid is subject to tensile loading the inter-ionic spacing increases, and there is a corresponding increase in the attractive force due to interaction between the positively charged ions and the negative space charge due to the electron cloud. Compression, on the other hand, is balanced by the mutually repulsive force between the positive ions.

The difference between these two structures is, as already noted, that covalence usually implies brittleness, while the metallic bond allows ductile behaviour. In all but a few exceptional cases, ductility is essential for successful welding. In order to obtain true metallic bonding, the surfaces to be joined must be sufficiently close together for the inter-atomic forces to come into play. There are two ways in which this may be accomplished: either the two surfaces are plastically deformed so as to obtain an intimate contact, or they are melted local to the interface, allowed to run together, and then cooled to make a solidified joint. In the first method the requirement for ductility is self-evident. In the second, melting local to the interface results in strains due to thermal expansion and contraction, and these are invariably high enough to cause fracture in a brittle solid. Exceptional cases are discussed in Chapter 5.

Neither brittleness nor ductility are absolute properties that exist under all circumstances. A normally brittle substance may behave plastically if present in the form of a very thin film. Oxides form such thin films on metal surfaces and, within limits, deform in such a way as to match the metal substrate. At the other end of the scale, ice, a notoriously brittle substance, may behave like an exceedingly viscous liquid when it exists in very large masses: in glaciers, for example. By the same token, metals that are normally ductile may become **embrittled** owing to the presence of impurities, or they may behave in a brittle manner under extreme loading conditions.

1.2.2 The cohesion between metals and non-metals

While it is clear that where two metal surfaces are brought into intimate contact, then bonding will occur, it is by no means self-evident that a similar bond will form between metals and covalent or ionic compounds. In this connection, Nichols (1990) suggests that adequate bonding is more likely to occur with covalent rather than ionic compounds, on the grounds that covalency represents a condition intermediate between metallic and ionic bonding. It is also noted that few solids encountered in metallurgy or engineering practice are either purely ionic or purely covalent. Bonding is mixed, with the degree of covalency increasing from oxides, which are the most ionic, through nitrides and carbides to borides, which are the most covalent. The effectiveness of metal–non-metal bonding would be expected to march in the same direction. Fortunately, the behaviour of non-metallic inclusions in steel provides useful indications as to the effectiveness of such bonding.

Oxide inclusions are normally present in steel in very large numbers. Except for the case of vacuum-degassed steel, oxygen is dissolved in the molten metal, and then precipitated in the form of oxides, silicates and other compounds, as the metal is cast. Similar precipitates are found in weld metal, as illustrated in Fig. 8.3. They are very small, having a diameter of the order of 10^{-3} mm.

When steel is loaded to failure in a normal tensile test, the testpiece deforms plastically at stresses above the yield or flow stress, and a constriction appears. The first stage of fracture is the formation, within this constriction, of a cavity. Microcavities appear at the inclusions, and these join up by ductile tearing to generate the macrocavity. Sometimes the inclusions fracture, but more frequently they decohere at the metal–oxide interface. Here, then, is a measure of the metal–oxide bond strength: it lies somewhere between the yield strength of the steel and its ultimate strength.

The other notable non-metallic constituent of steel is iron carbide. Carbides, in their various manifestations, are largely responsible for the high mechanical strength of steel, but even in the highest tensile grades, plastic strain does not cause decohesion at the carbide–iron interface. Thus it would seem that predictions about the relative strengths of metal–oxide and metal–carbide bonds are correct. It would also seem that where the practical difficulties of obtaining contact between metal and ceramic surfaces can be overcome, a strong bond can be formed. One such case is described in Chapter 5.

1.2.3 The strength of a crystalline solid

Consider the case of a solid consisting of a single crystal which is exposed to a tensile stress acting at right angles to a crystallographic plane. The stress across this plane increases with increasing separation of the ions up to a maximum value, at which point failure occurs. The stress may be represented as a function of the separation x by

$$\sigma = \frac{E\lambda}{\pi a}\sin\left(\frac{\pi x}{\lambda}\right) \tag{1.1}$$

where E is the elastic modulus, a is the spacing of the two planes at zero stress and λ is the wavelength of the interaction (i.e. the effective range of the attractive force). Equation 1.1 conforms to Hooke's law at small displacements.

Since the work done in separating the two planes is equal to the **surface energy** $2\gamma_s$ of the two new surfaces

$$2\gamma_s = \int_0^\lambda \frac{E\lambda}{\pi a}\sin\left(\frac{\pi x}{\lambda}\right)\mathrm{d}x = \frac{2E\lambda^2}{\pi^2 a} \tag{1.2}$$

whence it follows that the maximum stress is

$$\sigma_{max} = (E\gamma_s/a)^{1/2} \tag{1.3}$$

This is the maximum tensile strength of a perfect crystal, and it may be realized in the case of whiskers having a diameter in the region of one micrometre (10^{-3} mm).

Solids that are used in real structures differ from the ideal model in two respects; firstly they contain defects, which in brittle materials take the form of cracks, while the specific energy of fracture is much higher than the surface energy of the solid. The classic work of Griffiths (1920) showed that in the case of a plate containing a central crack of length $2c$, the stress required to initiate an unstable brittle fracture was given by

$$\sigma = \left(\frac{2E\gamma_F}{\pi c}\right)^{1/2} \tag{1.4}$$

which is similar in form to equation 1.3. Note that both equations represent an **instability**, since the stress required to extend the crack decreases as the crack length increases. It is predicted that the velocity of crack growth will increase very rapidly until or unless it is restricted by some physical limitation.

Equation 1.4 differs from 1.3 in the use of the quantity γ_F, which will be designated here the **specific fracture energy**, in place of γ_s, the surface energy. The reason for this is that a real crack does not run between parallel crystallographic planes. Instead, although in the main following a path at right angles to the direction of the principal stress, it makes numerous diversions and excursions, develops transverse cracks, forms splinters and so forth. Such activity requires the expenditure of energy, which generates heat, sometimes sufficiently to cause softening. Whatever the detailed cause, most nominally brittle solids have a significant **fracture toughness**, and this quantity may be measured using standard fracture toughness tests. Such tests, and their technical background, are described in Section 1.5. For the present it is sufficient to note that the fracture toughness K_{IC} is related to the specific fracture energy by the relation

$$G_{IC} = (1 - v^2)K_{IC}^2/E \tag{1.5}$$

where v is Poisson's ratio. G_{IC} or G_C is the usual designation for the specific fracture energy, and will so be used from this point. The relevant properties for zirconia are, from Howatson and Lund (1994)

Elastic modulus, E	2×10^{11} N m^{-2}
Tensile strength, σ	5×10^8 to 8.3×10^8 N m^{-2}
Poisson's ratio, v	0.3
Fracture toughness, K_{IC}	1.2×10^7 N m$^{-3/2}$

Equation 1.3 may now be used to calculate the theoretical maximum tensile strength of a perfect crystal of zirconia. Taking the inter-ionic distance as 2×10^{-10} m, and assuming that the surface energy lies between 0.1 and 1.0 J m^{-2}, the value for σ_{max} lies between 1×10^{10} and 3.2×10^{10} N m^{-2}. This is almost two orders of magnitude greater than the measured value.

Combining equations 1.4 and 1.5 leads to

$$\sigma_u = \left[\frac{2(1-v^2)}{\pi e}\right]^{1/2} K_{IC} \tag{1.6}$$

However, the ultimate (failure) stress of zirconia was measured by means of a bend test, in which failure takes place by propagation of a surface crack, not an interval crack as assumed earlier. The formula for a surface crack of length c gives

$$\sigma_u = \left(\frac{1}{1.2\pi c}\right)^{1/2} K_{IC} \tag{1.7}$$

and

$$c = \frac{1}{1.2}\pi(K_{IC}/\sigma_u)^2 \tag{1.8}$$

whence a crack length of between 0.06 and 0.15 mm may be calculated. Alternatively, where the crack length is known, the fracture stress may be calculated. This fact is of great importance in assessing the failure risk when metals behave in a brittle fashion, as will be seen in Chapter 11.

1.3 The plastic behaviour of crystalline solids

1.3.1 Crystal structure and slip

The three types of crystal lattice most commonly encountered in metals are illustrated in Fig. 1.1. These are body-centred cubic, face-centred cubic and close-packed hexagonal. Ferritic steel has a body-centred cubic structure, while the face-centred cube is found in austenitic steel and the non-ferrous metals aluminium, copper and nickel. The diagrams show the arrangement of ions in a **unit cell**; extending this arrangement in three dimensions would produce the relevant crystal lattice structure.

Plastic flow in a crystalline solid occurs as the result of **slip**. Blocks of the solid above and below the crystallographic plane move laterally relative to each other when a shear force is applied. Figure 1.2 illustrates such movement in the case of a specimen consisting of a **single crystal**. The motion takes place across a **preferred crystallographic plane**, which as a rule is a plane in which the density of ions is greatest. In each type of crystal lattice there are a number of such planes, and in a specimen loaded in tension the slip will take place across the plane where the resolved shear stress is greatest. For a testpiece loaded in uniaxial tension, as illustrated in Fig. 1.2, the component of force acting tangentially across the slip plane is $F\sin\phi$, while the area of the plane is $A/\cos\phi$, where A is the cross-sectional area. Putting $F/A = \sigma$, the longitudinal

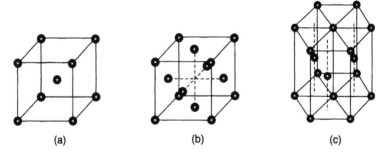

1.1 Elementary unit cell of (a) body-centred cubic, (b) face-centred cubic and (c) close-packed hexagonal structures.

Introductory 7

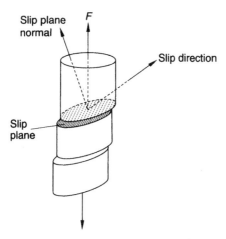

1.2 Slip in a crystalline solid having a uniform lattice orientation.

stress, the shear stress across the slip plane is $\tau = \sigma \sin\phi \cos\phi = \frac{1}{2}\sigma \sin 2\phi$. The shear stress is greatest, therefore, across planes lying at an angle of 45° to the longitudinal axis. There is also a **preferred direction** for slip, and this is also determined by the lattice structure.

Metals used in engineering structures are polycrystalline, and the individual crystals are very small and **randomly oriented**. In a sample of such metals there is no preferred slip plane, and no preferred slip direction; when subject to tensile loading of sufficient magnitude there will be slip along crystallographic planes in individual grains, but the overall result will be elongation in the tensile direction accompanied by a reduction in cross-sectional area: in other words, there is a generalized flow whose direction is not constrained by crystal geometry. Such flow has some features in common with that of a very viscous fluid.

The continuum plastic flow in a cylindrical tensile testpiece may be regarded as the result of slip across internal conical surfaces. The shear stress across such a conical region is the same as that across an inclined plane, namely $\frac{1}{2}\sigma \sin 2\phi$ where in the case of the cone ϕ is the half-angle at the conical tip. The most active slip will therefore take place at an angle of 45° to the axis. The supposed cones point both upwards and downwards so that **cross slip** occurs. The net result is a combination of axial elongation with radially inward flow.

Metals flow in a plastic manner when the applied stress exceeds a limiting value peculiar to the metal or alloy in question. For example, in the uniaxial tensile test, this stress is defined as that where the stress–strain curve departs significantly from a straight line. As the amount of **plastic strain** increases, so does the stress, until eventually rupture occurs. It has been stated above that plasticity results from shear across certain crystallographic planes. It remains to be explained how, if such an apparent discontinuity as a slip plane exists, it is nevertheless possible for the metal to retain its coherence. Such an explanation is one of the major achievements of the theory of **dislocations**.

1.3.2 Dislocations

It is very exceptional for the crystal lattice of any solid, be it metallic or non-metallic, to achieve geometric perfection. Most real substances contain impurities, which may be present as atoms or ions in solution, or as chemical compounds in the form of inclusions. A site in the lattice normally occupied by an ion may be empty: this is a **point defect** known as a **vacancy**. All such defects result in local lattice strains. Much more significant, however, is the presence of dislocations. These occur where **planes** of ions are partly missing. Figure 1.3, for example, is a diagram in which the lattice planes are represented as lines. One half-plane is absent, and the remainder are correspondingly distorted. Dislocations may take various geometric forms, and may in practice be complex and may also be distorted. The manner in which they behave when exposed to a shear stress in the plastic range may, however, be quite adequately illustrated by reference to a simple edge dislocation, as shown in Fig. 1.4. Lines of ions along the slip plane relocate progressively and stepwise from the right-hand side of the dislocation to the left, so that the solid below the slip line moves in the direction of the shear while the dislocation moves in the opposite direction. In this way the two blocks of metal move relatively to each other without losing cohesion. This, as noted earlier, is an essential requirement for plastic flow in a crystalline solid if it is to retain both its integrity and its crystalline form.

1.3 Edge dislocation.

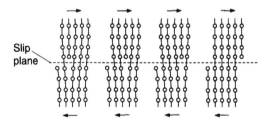

1.4 The mechanism of dislocation movement. The solid below the slip plane moves from right to left while the dislocation moves in the opposite direction.

1.3.3 Dislocation velocity and the strain rate

It will be evident from Fig. 1.4 that dislocations must possess one property that will influence the rate of plastic flow, namely, the **dislocation velocity**. In certain substances the presence of dislocations may be disclosed by attacking the polished surfaces of a sample with an etching agent. Exposing samples of such materials to a stress pulse of known intensity and duration, followed by treatment with a suitable etch, makes it possible to plot dislocation velocity as a function of shear stress. Figure 1.5 shows such a plot for $3\frac{1}{2}\%$ silicon iron at various temperatures. It will be evident that the dislocation velocity varies very rapidly with applied stress:

$$V_D = k(\tau)^n \tag{1.9}$$

where V_D is dislocation velocity, τ is applied shear stress and k is a constant. Hence

$$\frac{V_D}{V_{Dy}} = \left(\frac{\tau}{\tau_y}\right)^n \tag{1.10}$$

where V_{Dy} is the dislocation velocity at the yield stress.

Now the axial strain rate $d\varepsilon/dt$ must be related to the dislocation velocity. Suppose that the volume density of mobile dislocations is ρ_D and the mean area displacement associated with each emergent dislocation is δA resolved in the axial direction, then

$$\frac{d\varepsilon}{dt} = \dot{\varepsilon} = \rho_D V_D \delta A \tag{1.11}$$

Thus, for any single batch of a particular metal, where ρ_D and δA are constant,

$$\frac{\dot{\varepsilon}}{\dot{\varepsilon}_y} = \frac{V_D}{V_{Dy}} \tag{1.12}$$

and from equation 1.10

$$\frac{\dot{\varepsilon}}{\dot{\varepsilon}_y} = \left(\frac{\tau}{\tau_y}\right)^n = \left(\frac{\sigma}{\sigma_y}\right)^n \tag{1.13}$$

since shear stress is proportional to axial stress.

This relationship may be interpreted as follows. Suppose that a series of tensile testpieces are prepared from a single batch of material, and suppose that they are tested at successively higher strain rates. Then an individual specimen would be expected to behave elastically up to the stress at which the dislocation velocity is high enough to match the relevant strain rate. Such tests have been carried out on a high-tensile structural steel and the results are shown in Fig. 1.6(a). In Fig. 1.6(b) the same data are presented in log–log form with strain rates as ordinates in order to allow comparison with Fig. 1.5. The two curves are in fact very similar in

10 Metallurgy of welding

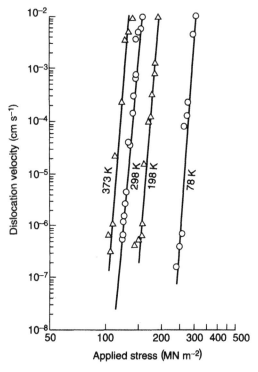

1.5 Dislocation velocity as a function of applied shear stress at various temperatures for $3\frac{1}{2}$% silicon iron (Hull, 1997).

character, and their slopes (the exponent n in equations 1.10 and 1.13) are 35 and 37 respectively. It would seem that the proposed relationship between strain rate and dislocation velocity is probably correct.

Other tests have shown that there is an upper limit to the dislocation velocity which is equal to the propagation rate of transverse oscillations in the medium concerned. For steel this is about 3000 m s^{-1}. Taking the dislocation velocity at yield as 3×10^{-10} m s^{-1}, the limiting strain rate may be estimated to be of the order 10^9.

The need (other than in a few exceptional cases) for plastic deformation during welding operations has already been emphasized. However, there is one welding process that is of particular interest in connection with plastic flow; namely **explosive welding**. The way in which one variant of this process works is illustrated in Fig. 1.7. This diagram shows the welding of a **cladding** plate on to a **backing** plate. The two plates are set at a specified distance apart, and the explosive is placed so as to cover the cladding plate. It is then detonated from one end. The upper plate deforms in the manner shown, colliding progressively with the lower plate and welding to it. The interface has a characteristically wavy profile, as indicated in the diagram.

Introductory 11

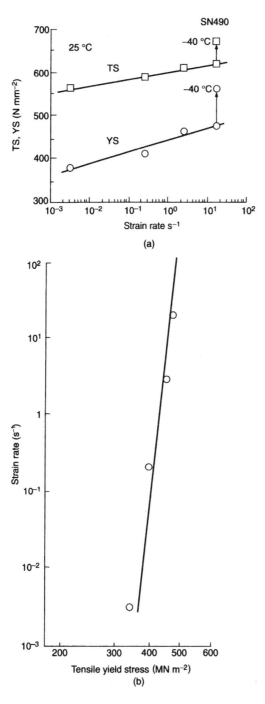

1.6 Effect of strain rate on yield strength: (a) original plot (Toyoda, 1995); (b) replot of same data in the format of Fig. 1.5.

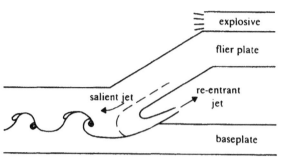

1.7 Explosive welding.

The explosive commonly used in the explosive cladding of plate is based on ammonium nitrate, and its detonation velocity lies in the range 2×10^3 to 3×10^3 m s^{-1}. Assuming that the deformed portion of the cladding plate makes an angle of 45° with the backing plate, the rate of shear in this region will be about half the detonation rate. Supposing the thickness of the cladding to be 3 mm, the strain rate would be of the order 10^5 s^{-1}, which is well within the limit of 10^9 s^{-1} estimated earlier. Regardless of the accuracy of such estimates, experience demonstrates that metals, and steel in particular, withstand the very high strain rates imposed by explosive welding, and by other modes of explosively induced deformation. Moreover, as far as steel is concerned, high strain rates cause a relatively modest increase in yield strength, and do not by themselves induce premature failure.

The considerations set out above are applicable to those instances where plastic flow is predominantly in one direction; where, for example, the applied stress is uniaxial. In regions of stress concentration this is not the case; stresses are multiaxial. In the extreme, when stresses in the three principal directions are equal, plastic flow is impossible, and if the stresses are high enough, the metal will fail without prior deformation. A similar condition applies to liquids, where a sub-atmospheric (negative) pressure may be interpreted as causing a triaxial tensile stress. When such stresses are high, the liquid fractures. Failure is nucleated (in water for example) at microscopic particles or bubbles, and propagates for a short distance to form macroscopic bubbles. This phenomenon is known as **cavitation**. Metals may be affected in a similar way, but the three principal stresses are rarely equal. In these circumstances two factors affect the risk of failure: the degree of triaxiality and the strain rate. Experience in the 1995 earthquake in Japan, where large structures were subject to high strain rates, indicated that with a severe stress concentration, such as that associated with a partial penetration weld, brittle fractures would initiate and propagate in good-quality steel when the strain rate was 0.1 s^{-1}. This figure is orders of magnitude lower than that suggested earlier as a limiting value for uniaxial strain, but it is probably realistic, and reflects the severe limitation that self-restraint around discontinuities can impose on ductile behaviour.

In commercial practice, tensile tests are carried out at a constant specified strain rate; for example, British Standard EN 10002-1990 requires that for steel, the strain rate must be limited as follows:

- In the elastic range: between $3 \times 10^{-5}\,\text{s}^{-1}$ and $1.5 \times 10^{-4}\,\text{s}^{-1}$.
- Between upper and lower yield points: between $2.5 \times 10^{-4}\,\text{s}^{-1}$ and $2.5 \times 10^{-3}\,\text{s}^{-1}$.
- In the plastic range: below $8 \times 10^{-3}\,\text{s}^{-1}$.

The yield strength of a metal as quoted in the literature is that determined under such standard testing conditions.

The way in which plastic yielding occurs depends in part on the type of lattice structure. Figure 1.8 shows a **stress–strain curve** typical of a polycrystalline face-centred cubic metal. The transition from elastic to plastic behaviour is gradual, such that it would be impossible to pinpoint the yield stress with any certainty. A similar test carried out on a single-crystal specimen of the same metal would, however, show a sudden fall in the gradient of the curve when plastic yielding starts. The behaviour of the polycrystalline specimen is due no doubt to the random orientation of the grains; those that are favourably oriented start to slip at a lower stress than others, so that there is a transition region in which the metal comprises a mixture of plastic and elastic grains.

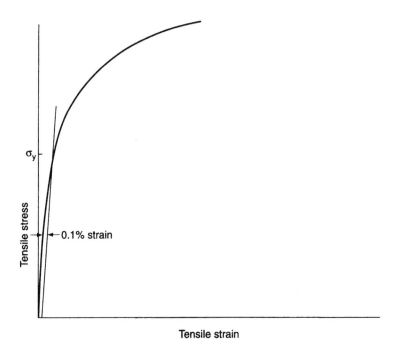

1.8 Typical stress–strain curve for a polycrystalline face-centred cubic metal.

As a rule, body-centred cubic metals, whether in the form of single crystal or polycrystalline specimens, show a discontinuity in the stress–strain curve when tested in tension. Figure 1.9 is a typical example of a polycrystalline sample. In this diagram the stress–strain curve for the plastic region has been projected backwards to meet that for the elastic region. The point of juncture is a notional yield point where, according to previous conjectures, the dislocation velocity and consequent rate of slip match the shear strain rate produced by the testing machine. However, at this point the dislocations remain fixed, possibly owing to pinning by second-phase particles, or because of a rigidity inherent to the lattice structure. It requires a higher stress to initiate movement, but at this stress the dislocation velocity is higher, probably very much higher, than that appropriate to the fixed strain rate. Therefore the test specimen relaxes, and the stress does not start to rise again until the machine has taken up the slack. The stress required to produce dislocation movement and yielding is known as the **upper yield stress**, while the lowest value obtained before the load increases again is the **lower yield stress**. Both these quantities may vary with the strain rate.

In this book no distinction is made between the two modes of yielding described above; both are designated 'yield point' with a corresponding 'yield stress'. Where it is not feasible to ascertain the yield point with certainty, as in tests giving the type of stress–strain curve shown in Fig. 1.8, it is customary to

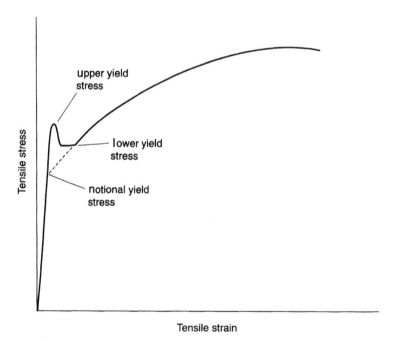

1.9 Typical stress–strain curve for a body-centred cubic metal.

designate a **proof stress** as being the stress required to produce, say, 0.1% plastic elongation or **permanent set** (as indicated in the figure).

1.4 Solid–liquid interactions

1.4.1 Bonding at the solid–liquid interface

An individual ion in a crystalline lattice is not fixed in its equilibrium position: it is in a state of oscillatory motion with the equilibrium position as the neutral point. With increasing temperature the amplitude of the oscillation increases. At the melting temperature this amplitude is such that coherence is lost; the lattice disperses and the ions become mobile. The nature of the bond between ions, however, remains essentially the same. This is equally true of the bond between liquid and solid metal when the two are in perfect contact. The art of fusion welding is that of obtaining such perfect contact.

In considering this matter, it is necessary to distinguish between two cases; that in which solid and liquid are of the same or closely similar composition, and that (as in brazing and soldering) where there is a substantial difference in composition and in melting temperature.

In the first case, that of similar metals, perfect contact is obtained quite simply by partially melting the solid and allowing the melt so formed to mix with any **filler metal** that may be present. When the liquid metal solidifies, the only discontinuity is that between the metallurgical structure of the original solid or **parent metal** and that of the newly cast **weld metal**. Conditions may be such, however, that the solid is not melted, and the interface between liquid and solid remains at a temperature lower than the melting point. The liquid local to the interface then solidifies in such a manner that no bond is formed. Areas of **lack of fusion** that form in this way in fusion welds constitute serious defects, since they may propagate in service.

In brazing and soldering (the dissimilar metals case) there is no question of melting the parts that are to be joined. It is, however, a necessary condition for bonding that the temperature of the solid at the solid–liquid interface be equal to or higher than the melting point of the liquid. In brazing this is accomplished by heating with a flame or in a furnace, but in soldering the required heat is transmitted by conduction and convection through the solder. If such heating is inadequate, a **dry joint** (lack of fusion) will result.

Given that the temperature of the solid metal is high enough, it remains necessary to expose a clean, smooth surface to which the liquid may bond. The first step is to remove the oxide and other surface films. These may be removed by solution or dispersal using a flux; alternatively, oxides may be decomposed by exposure to a reducing gas or to a sufficient degree of vacuum. Secondly, a small amount of the original solid surface is removed, generally by solution in the liquid metal, but sometimes by more complex interactions such as electro-deposition.

16 Metallurgy of welding

This type of process has the merit of forming a relatively strong joint, while minimizing undesirable dimensional and metallurgical changes.

When metals are joined by means of adhesives the surfaces must be free from contaminants such as dust or grease, but the oxide film remains; indeed aluminium alloys may be anodized, thus increasing the thickness of the oxide film prior to joining. So bonding occurs across the sandwich polymer–oxide–metal. Possible bonding mechanisms are discussed in Chapter 3, but here it is sufficient to note that when metal-to-metal adhesive joints are tested under dry conditions, failure takes place in the polymer, and not at the interface.

1.4.2 The effect of surface forces: static conditions

In an earlier section the surface energy of a solid was equated to the work done in separating two adjacent crystallographic planes. The **surface tension** is numerically equal to the surface energy but is expressed in force per unit length instead of energy per unit area. The surface energy/tension of metals is relatively high, and decreases with increasing temperature. Figures for liquid metals are given in Table 1.1 together with those for some common substances. This surface force is common to both solids and liquids, and has various manifestations. For example, if a liquid surface is curved, there is an internal pressure given by

$$p = \gamma \left(\frac{1}{R_1} + \frac{1}{R_2} \right) \qquad (1.14)$$

where R_1 and R_2 are the principal radii of curvature. In the case of a liquid drop resting on a plane solid surface (the **sessile drop**) the balance of forces, as illustrated in Fig. 1.10 and 1.11 is

$$\gamma_{SV} = \gamma_{SL} + \gamma_{LV} \cos \theta \qquad (1.15)$$

where γ_{SV} is the surface tension of the solid in equilibrium with its vapour, γ_{SL} is the surface tension of the solid–liquid interface, and γ_{LV} is the surface tension of the liquid in equilibrium with its vapour. This is an idealized equation in which

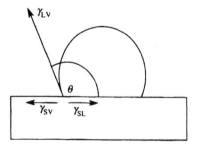

1.10 Sessile drop with contact angle greater than 90°.

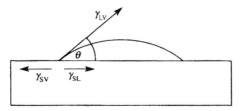

1.11 Sessile drop with low contact angle.

the mass of the liquid drop is ignored. It is not often realized in welding, brazing and soldering because, amongst other things, there is usually some interaction, such as alloying, between the liquid metal and the substrate. Nevertheless, sessile drop tests are still sometimes used to compare the properties of different brazing alloys.

If a liquid metal is to wet and spread over another then the condition

$$\gamma_{SV} > \gamma_{SL} + \gamma_{LV} \tag{1.16}$$

must obtain. This is the desirable state of affairs where, for example, it is required to coat a metal, as in galvanizing steel or tinning copper. At the other end of the scale, when $\gamma_{SV} < \gamma_{SL} + \gamma_{LV}$ we have the situation illustrated in Fig. 1.10. Such a condition may apply to ferrous metals that are heavily contaminated with sulphur. This element is **surface-active**, and reduces the surface tension of iron and steel. However, for any given sulphur content this reduction is much greater for the solid metal than for the liquid or the liquid–solid interface. Hence, in fusion welding material so affected, it is possible for the weld bead to draw inwards away from the fusion boundary, resulting in **undercut**. More frequently, this defect is due to irregular manipulation of the electrode. Undercut may also be caused by excessive **arc force**, outlined in the next section.

1.4.3 Dynamic effects: Maragoni flow

Consider the situation illustrated in Fig. 1.12. A liquid covers a solid substrate to a uniform depth d. The liquid has a uniform surface tension gradient $d\gamma/dx$ which generates a surface velocity v, and which sets up a shear stress τ. Within the liquid there is a velocity gradient dv/dy giving rise to an equal and opposite shear stress $\eta dv/dy$ where η is the viscosity of the liquid. In this arrangement $dv/dy = v/d$ so that

$$v = \frac{d}{\eta}\frac{d\gamma}{dx} \tag{1.17}$$

Now a surface tension gradient may be due to a gradient of temperature or to a change in the concentration of a solute.

18 Metallurgy of welding

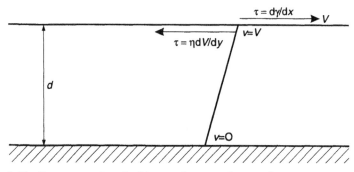

1.12 Stress associated with a surface tension gradient.

Suppose that there is a temperature gradient, then $d\gamma/dx = d\gamma/aT \cdot dT/dx$. Values of $b = d\gamma/dT$ for liquid metals are given in Table 1.1.

To estimate a value for v (the velocity of **surface tension streaming**) consider a hemispherical weld pool of radius R. Ignoring the difference in geometry, $d = R$ and $dT/dx = \Delta T/R$ where ΔT is the difference in temperature between the centre of the weld pool and that at the boundary. Then

$$v \simeq \left(\frac{d\gamma}{dT}\right) \cdot \frac{\Delta T}{\eta} \qquad (1.18)$$

The viscosity of liquid iron is about 4×10^{-3} N m^{-2} s, ΔT for a gas tungsten arc weld pool is about 400 K, and from Table 1.1 $d\gamma/dT$ is about 1×10^{-4} N m^{-1} K^{-1}, which leads to

$$v \simeq 10 \, \text{m s}^{-1} \qquad (1.19)$$

In this rough estimate, no account is taken of the weld pool shape or of inertia and electromagnetic forces, and the result is much too high. Nevertheless, more sophisticated calculations also indicate that where surface tension gradients are present in gas tungsten arc weld pools, they may cause rapid and undesirable streaming flow. Fortunately such gradients are exceptional.

It has already been noted that sulphur has a surface-active effect in steel. Oxygen behaves in the same way, and when present in liquid steel will reduce the surface tension quite sharply. In addition, however, the proportion of the total oxygen content that segregates to the surface falls with increasing temperature. Where this takes place, the surface tension, instead of decreasing with increased temperature, may actually increase. In practice, the presence of oxygen simply nullifies surface tension forces. For example, if the centre of a pool of mercury is heated in a vacuum, a strong outward flow results; if air is then introduced, the flow is arrested.

Surface tension-induced flow is not therefore a common problem. Exceptions occur in the case of low-sulphur nickel and iron–base alloys containing sub-

Table 1.1. Surface tension

(a) Metals. Because of potential contamination measured values tend to err on the low side. The following are based on the mean of the upper third of available data, and are shown in the form

$$\gamma = a + b\,(T - T_m)$$

where T is temperature and T_m is melting point, both in °C (Keene, 1991)

Element	a (N m^{-1})	$b\,(T - T_m)$(N m^{-1})
Aluminium*	0.89	$-1.82 \times 10^{-4}\,(T - 660)$
Cobalt	1.928	$-4.4 \times 10^{-4}\,(T - 1500)$
Copper	1.374	$-2.6 \times 10^{-4}\,(T - 1085)$
Calcium	0.724	$-7.2 \times 10^{-5}\,(T - 30)$
Gold	1.162	$-1.8 \times 10^{-4}\,(T - 1065)$
Indium	0.561	$-9.5 \times 10^{-5}\,(T - 157)$
Iron	1.909	$-5.2 \times 10^{-4}\,(T - 1530)$
Lead	0.471	$-1.56 \times 10^{-4}\,(T - 327)$
Mercury	0.498	$-2.31 \times 10^{-4}\,(T - 38)$
Nickel	1.834	$-3.76 \times 10^{-4}\,(T - 1455)$
Silver	0.955	$-3.1 \times 10^{-4}\,(T - 960)$
Tin	0.586	$-1.24 \times 10^{-4}\,(T - 232)$

*These figures are for liquid aluminium saturated with oxygen. The surface tension of oxygen-free aluminium is probably between 1.05 and 1.10 N m^{-1} at the melting point.

(b) Non-metals

Subsidue	Temperature (°C)	Surface tension (N m^{-1})
Ammonia	34.1	0.018
Argon	-188	0.013
Benzene	20	0.029
Ethyl alcohol	20	0.023
Helium	-269	1.2×10^{-4}
Phenol	20	0.041
Water	0	0.0756
Water	20	0.073
Water	100	0.059

stantial amounts of chromium, austenitic chromium–nickel steels, for example. In such alloys, the free oxygen content of a melt is kept to a very low level due to the presence of chromium, and a surface tension gradient may develop. This matter is discussed further in Chapter 10.

The phenomenon outlined above is known as **Maragoni flow** after the nineteenth-century scientist who first described it. The surface tension gradient which drives such flows may result from either a temperature gradient or a gradient in solute content; in the former case the term **thermocapillary flow** may be used. Maragoni flow is responsible for some phenomena observed in everyday life: the camphor boat, for example. This is a toy boat with a block of camphor attached to its stern; the boat is driven by a surface tension gradient caused by solution of camphor. In welding, apart from the streaming which may affect gas tungsten arc welds in stainless steel, it may be responsible for the fusion of spatter drops to plate. It is probable that thermocapillary flow will be set up when such drops come into contact with the relatively cold plate, and this in turn could generate sufficient convectional heat flow to cause fusion at the interface.

1.5 Fracture

1.5.1 General

The need for ductility in order to achieve bonding in a welded joint has been emphasized in the earlier parts of this chapter. It has been noted that ductility plays an important part in maintaining the integrity of structures by enabling them to accommodate local stress concentrations without fracture. It is also an essential feature of structures that may be exposed to shock loading. An obvious example is the motorcar which, so far as is practicable, must be able to absorb the shock of a collision without the occupants suffering serious injury. Likewise, steel-framed buildings in areas subject to earthquakes should bend rather than break when a tremor occurs.

Against this must be set the need for adequate strength. In the aerospace industry, for example, there is a pressing requirement for a high strength-to-weight ratio, and this in turn means that the materials used should have a high yield strength. In general, as the yield strength of any given alloy increases, the reduction of area in a tensile test (which is a good measure of ductility) falls, and there is a corresponding reduction in the fracture toughness. For any given project, it is necessary to consider how far such factors could prejudice safety, and one of the objects of this book is to describe the relevant hazards insofar as they may result from welding operations.

A loss of fracture toughness may also stem from the embrittling effect of impurities or, in the case of body-centred cubic metals, from a reduction of temperature. Such metals, and steel in particular, have a **transition temperature range** above which the notch-ductility is good or at least acceptable, and below which brittle behaviour is likely. Within the transition range itself behaviour is somewhat unpredictable, and there may be a wide scatter in test results. Nevertheless, it is possible to operate safely within the range provided that material properties are adequately controlled.

The fall in fracture toughness through the transition temperature range is accompanied by a change in **fracture appearance**. Above the transition, fracture is mainly due to **micro-void coalescence**; that is to say, the process of decohesion at inclusion–metal interfaces followed by ductile failure of the intervening ligaments, as described earlier in connection with cohesion. The appearance of such fractures is **fibrous**. As the temperature falls, however, an increasing proportion of the fracture surface is occupied by **cleavage facets**. These are flat, shiny areas where the solid (steel in particular) has cracked without any significant amount of plasticity. Other fracture morphologies may be visible under the microscope, but these two are distinguishable by the naked eye or with a hand lens. The proportion of fibrous fracture may be used as a control measure, as described in Section 1.5.4 on Charpy V-notch testing.

Brittle fractures in steel plate may also show **chevron markings**. These are macroscopic arrow-like ridges on the crack surface. They point towards the origin of the fracture, and are therefore useful to failure investigators.

The remainder of this chapter is concerned with the fracture process, starting with brittle fractures and the quasi-brittle, fast, unstable fracture of solids that are normally ductile. The more frequently used tests for assessing susceptibility to such fractures are described. After a section on crack velocity, ductile fracture and the occurrence of fast, unstable failure during this process are considered briefly.

1.5.2 Brittle fracture in the presence of a crack

The solids considered in this section are those which are normally brittle, together with those that are normally ductile, but which under certain circumstances fail in a brittle manner; that is to say failure takes place without significant plastic deformation, and takes the form of a running crack whose face is more or less normal to the plate surface. The work of Griffiths (1920) on the extension of cracks in a brittle substance was mentioned briefly in Section 1.2.3. The derivation of equation 1.4 is as follows. Consider an infinite plate of a brittle solid and having unit thickness. It contains an internal through-thickness crack of length $2a$, and a uniform tensile stress σ is applied remote from the crack (Fig. 1.13). In the absence of a crack the plate would have a uniform strain-energy content of $\frac{1}{2}\sigma^2/E$ per unit volume. The presence of a crack, however, reduces the strain-energy content by an amount $\pi a^2 \sigma^2/E$. At the same time there is a positive contribution due to the surface energy of the crack, which is $4a\gamma$. It may be postulated that the crack will extend if thereby the total energy of the system is reduced:

$$\frac{d}{da}\left(\frac{-\pi a^2 \sigma^2}{E} + 4a\gamma\right) < 0 \qquad (1.20)$$

or

$$\left(\frac{-2\pi a \sigma^2}{E} + 4\gamma\right) < 0 \qquad (1.21)$$

22 Metallurgy of welding

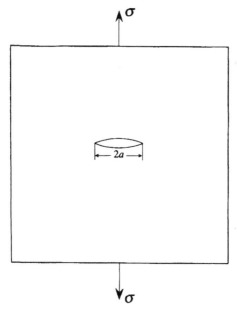

1.13 A through crack of length 2*a* subject to a uniform tensile stress σ.

This leads to

$$\sigma_c = \left(\frac{2\gamma E}{\pi a}\right)^{1/2} \quad (1.22)$$

where σ_c is the critical stress required to initiate a running crack, and γ may be interpreted as the surface energy or the fracture energy, as the case may be. Equation 1.22 may be rearranged to give

$$\sigma_c^2 a = \text{constant} \quad (1.23)$$

and tests carried out with centre-cracked specimens have confirmed this result, even with relatively small testpieces, very far from the infinite plate assumed earlier.

In these, and in other fracture toughness tests described later, it is necessary to form a sharp crack tip. This is normally done by first machining a notch of the appropriate form, and then extending it by exposure to axial fatigue loading in a suitable machine.

1.5.3 Fracture mechanics

The Griffiths theory of brittle fracture is based on the postulate that a pre-existing crack will extend when the rate of strain-energy release from the stress field around the crack is at least equal to the rate at which it is absorbed by crack extension. In fracture mechanics, on the other hand, it is supposed that the crack will start to run when the stress field around its tip reaches a certain intensity. The solid is assumed to be elastic at all stress levels, and stresses ($\sigma_x, \sigma_y, \sigma_z$) and displacements ($u, v, z$) are expressed as a function of location relative to the crack tip:

$$\sigma_x = K \int_x (r, \theta) \tag{1.24}$$

etc. (see Fig. 1.14).

K is the **stress intensity factor**, which varies with the depth of the crack and the applied stress, and r and θ are the spherical polar coordinates of the point in question. Three types of loading are envisaged: tensile, designated as mode I, shear, which is mode II, and torsion, mode III. For any given K, the stress distribution differs with the type of loading. Other things being equal, there is a critical value of the stress intensity factor at which the crack will extend; in the case of tensile loading (mode I) for example, this is designated K_{IC}.

There are two basic conditions to which a plate specimen may be subject: **plane strain** or **plane stress**. When displacements in the z direction (at right angles to the plate surface) are zero, plane strain conditions obtain; alternatively in plane stress the stress at right angles to the plate surface is zero and there are substantial inward displacements. The first condition is that normally associated with brittle fracture, while the second would be classified as a ductile failure. In

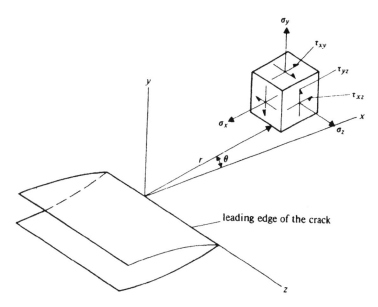

1.14 Coordinates measured from the leading edge of a crack and the stress components in the crack-tip stress field (from Paris and Sih, 1965).

both cases, however, the crack may propagate at high speed, possibly resulting in the catastrophic failure of a structure. Plane strain conditions are more likely to prevail with higher yield strength, greater plate thickness, and lower values of the fracture toughness K_{IC}. Plane strain conditions also increase the likelihood of failure at stresses below the yield point.

The relationship between the fracture toughness and the specific fracture energy G_I (also known as the **energy release rate** or the **crack extension force**) is obtained by calculating the work done to close a short length δ of the crack, as indicated in Fig. 1.15. The required value is obtained by integrating the product of stress and displacement (the strain energy), over the length δ

$$G_1 = \lim_{\delta \to 0} \frac{2}{\delta} \int_0^\delta \left(\frac{\sigma_y v}{2} + \frac{\tau_{yx} u}{2} \right) dx \tag{1.25}$$

whence

$$G_1 = \frac{1 - v^2}{E} K_1^2 \tag{1.26}$$

This is the same relationship as that given in equation 1.5.

Fracture toughness is measured using notched and pre-cracked rectangular specimens, the dimensions of which are standardized in the relevant national standards, or by notched slow bend tests. Figure 1.16 shows the general configuration of a specimen according to ASTM E-24; pre-cracked bend specimens similar to those used for crack-tip opening displacement tests may also be employed. An important limitation is that for plane strain conditions both a and B (plate thickness) must be equal to or greater than 2.5 $(K_{IC}/\sigma_{ys})^2$, and test results not conforming to this criterion are considered invalid. For unembrittled carbon and carbon–manganese steels, the specimen dimensions required to meet the ASTM E-24 criterion are impracticably large; even for a steel with a yield stress of 500 MN m^{-2} a specimen thickness of 300 mm is required. For lower-

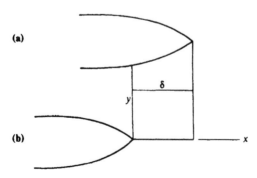

1.15 (a) The tip of a crack that has been pulled closed (b) along a segment adjacent to the tip (from Paris and Sih, 1965).

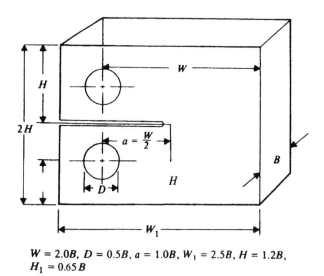

$W = 2.0B$, $D = 0.5B$, $a = 1.0B$, $W_1 = 2.5B$, $H = 1.2B$, $H_1 = 0.65B$

1.16 An ASTM compact tensile specimen for the measurement of K_{IC}.

strength steels, and for cases where general yielding precedes fracture, other methods are used to determine fracture toughness, as discussed below.

1.5.4 Alternative means of estimating or measuring fracture toughness

The CTOD test

Consider a centre crack of length $2a$ in an infinite plate and apply uniform tensile stress at right angles to the crack. Under plastic conditions the crack opens, and it is assumed that plastic wedges are formed at each end that prevent crack extension until a critical value of the plastic displacement at the crack tip is reached (Fig. 1.17). It may be shown that the displacement δ is given by

$$\delta = 8\frac{\sigma_{ys}}{\pi E} a \ln \sec\left(\frac{\pi \sigma}{2\sigma_{ys}}\right) \tag{1.27}$$

Expanding the ln sec term gives

$$\delta = \frac{8\sigma_{ys}}{\pi E} a \left[\frac{1}{2}\left(\frac{\pi \sigma}{2\sigma_{ys}}\right)^2 + \frac{1}{12}\left(\frac{\pi \sigma}{2\sigma_{ys}}\right)^4 + \frac{1}{45}\left(\frac{\pi \sigma}{2\sigma_{ys}}\right)^6 \ldots\right] \tag{1.28}$$

26   Metallurgy of welding

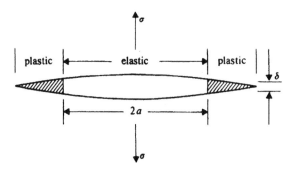

1.17 The assumed form of plastic deformation at the tip of a centre crack of length 2a.

Using the first term of the expansion only

$$\delta \simeq \frac{\pi\sigma^2 a}{E\sigma_{ys}} \qquad (1.29)$$

From equation 1.26

$$G_I \simeq \frac{\pi\sigma^2 a}{E}$$

Therefore, at the point of instability

$$G_{IC} \simeq \sigma_{ys}\delta_c \qquad (1.30)$$

where δ_c is the critical crack-tip opening displacement (critical CTOD).

Crack-tip opening displacement tests are made using a full-plate thickness specimen containing a machined notch which is extended by fatigue cracking (Fig. 1.18). The testpiece is normally loaded in three-point bending, but a four-point bend may also be used. The testing method is specified in BS 5762. Displacement is measured as a function of load using a clip gauge, and the critical crack-tip opening displacement is that at which a cleavage crack initiates and leads either to complete failure of the specimen or to an arrested brittle crack (the second condition being known as pop-in). The critical value of CTOD corresponding to the transition between ductile and brittle behaviour is typically 0.1 or 0.2 mm.

The J integral test

A method of obtaining an equivalent to the stress intensity factor for elastic–plastic conditions is to calculate the stresses and displacements for the configuration in question, and then integrate strain energy around a closed path that includes the crack tip. The result (designated *J*) is theoretically independent

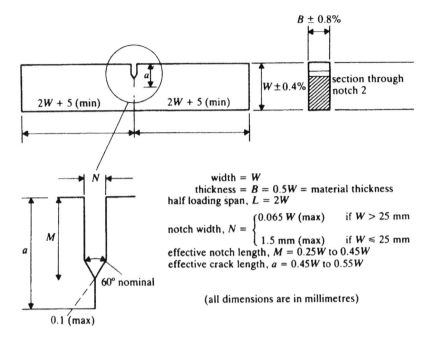

1.18 A CTOD test specimen. The testpiece is bent in three-point loading and the displacement at the crack tip is measured continuously to failure. There are no restrictions on the validity of the test results.

of the path taken, and it is related to the stress intensity factor in the same way as G:

$$J = K^2/E' \tag{1.31}$$

where $E' = E$ (plane stress) and $E' = E(1 - v^2)$ (plane strain).

An alternative and equivalent definition of J is that it is equal to the rate of change of strain energy U with crack extension

$$J = \frac{\delta U}{\delta a} \tag{1.32}$$

There are two ways in which this expression may be used to obtain a measurement of J. In the first technique a number of standard ASTM bend test specimens are loaded to different levels and the crack length obtained by heat tinting and breaking the specimen. The area under the load–deflection curve, which is the quantity U in equation 1.37, is thus obtained as a function of crack length. A least-squares fit of log u versus log a is used to replot a normalized U/a curve. This is extrapolated back to intersect a 'blunting line' as shown in Fig. 1.19. This gives a tentative value for J at zero crack length, J_Q. If this quantity

28 Metallurgy of welding

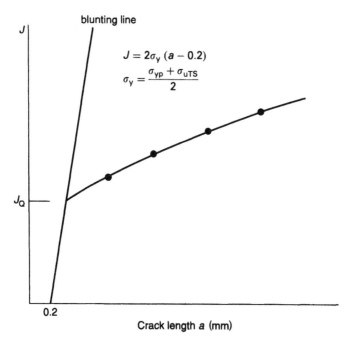

1.19 Evaluating J_Q from J integral tests.

passes a validity test similar to that used in K_{IC} testing, then it is accepted as being correct.

The second method is to use a single specimen and load it to successively higher levels, measuring the crack length at each level. J is then determined by the procedure used for multiple specimens. The crack length is measured either by electrical resistance or by its compliance. Compliance is the ratio between displacement and load, and it is measured on specimens cracked to a known depth. The second method avoids the scatter associated with a number of different samples, but there are uncertainties about the measurement of crack depth. J integral testing is standardized in ASTM E-813, to which the reader is referred for greater detail.

The Charpy impact test

The standard Charpy test uses a 10 mm × 10 mm specimen with a 2 mm deep V notch, placed upon an anvil and broken by a pendulum weight. The energy absorbed is measured by the height of the swing of the pendulum after fracture. In the UK and USA this is recorded simply as energy but in continental Europe it is not uncommon to divide the energy by the cracked area of 800 mm² to give results in J m^{-2}. The fracture appearance is also used as a criterion. The proportion of the fracture surface occupied by fibrous or shear fracture may be measured, and one means of defining a transition temperature is where the fracture is 50 % fibrous. This is the fracture appearance transition temperature FATT 50.

Alternatively the transition temperature is defined in terms of impact energy, the 27 J (equivalent to 20 foot-lb force) or 40 J transition for example. Such figures are of course arbitrary but nevertheless may provide a useful datum for control purposes.

A third measure provided by impact specimens is the lateral contraction at the root of the notch. For certain materials this may provide a more sensitive quality criterion than energy; more often it is used as additional information.

To provide for material thinner than 10 mm there are substandard thicknesses of Charpy specimen, and where specifications require a minimum Charpy energy, lower values are specified for the thinner specimens. It is customary to require three specimens for each test and to specify a minimum value for the average of the three, plus a minimum for any single specimen.

There is a correlation between Charpy V-notch energy and K_{IC} for the upper shelf region:

$$K_{IC} \simeq \sigma_{yp}\left(\frac{0.646 C_V}{\sigma_{yp}} - 0.00635\right)^{1/2} \qquad (1.33)$$

where C_V is Charpy energy in joules and σ_{yp} is in MN m^{-2}.

In the transition region the situation is more complex and a number of investigators have concluded that there is no relationship between C_V and K_{IC}. However, according to Barsom and Rolfe (1970):

$$\frac{K_{IC}}{E} = 0.221 C_V^{1.5} \qquad (1.34)$$

where K_{IC} is in N mm$^{-3/2}$, E is in N mm^{-2} and C_V is in J.

Dolby (1981) found a correlation between the transition temperatures measured by Charpy V and COD for individual types of weld deposit. Figure 1.20 shows the relationship for a manual metal arc (MMA) multipass deposit. However, lower-strength metal has a higher strain rate sensitivity, and this shifts the impact transition to a higher temperature. Therefore the slope and position of the correlation curve depends on the yield strength of the weld metal (Fig. 1.21).

The Pellini drop-weight test

The drop-weight test was developed by the US Naval Research Laboratories in 1952. At that time fracture toughness testing had not yet evolved, and the Navy wished to obtain a more precise evaluation of the ductile–brittle transition temperature than was possible using Charpy impact specimens. The object of the test is to determine the temperature (the nil-ductility temperature) at which a plate subject to impact will fail in a brittle manner. A typical sample size is

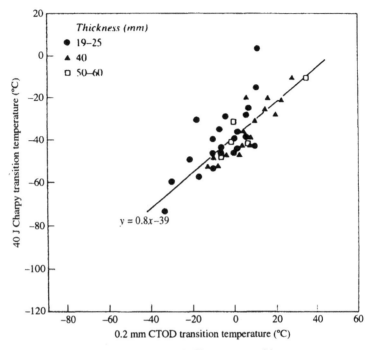

1.20 The relation between 40 J Charpy transition temperature and 0.2 mm COD transition temperature for as-welded MMA multipass deposits (Dolby, 1981).

125 mm × 50 mm (5 inches by 2 inches). A brittle weld (made using a hard-facing electrode) 65 mm (2½ inches) long is laid centrally on one side of the sample and then notched at its midlength. This is laid with the brittle weld face down across two supports, and a tup is released so as to drop and strike the plate in its centre. The sample is cooled to its expected nil-ductility temperature and, depending on the result, repeated at 3 °C (5 °F) intervals up or down until the marginal condition is obtained. Here the plate either cracks in two or cracks across a half-width. This test, which is standardized in ASTM E 208, is used, for example, for assessing the embrittlement of steel by neutron irradiation. A variant, the Pellini explosion bulge test, is employed for testing quenched and tempered steel used in submarine construction.

1.5.5 Fracture velocity

Most of the work on this subject was carried out during the period 1935–1960 in Germany and France, and nearly all is concerned with the fracture of glass. This material was reviewed by Schardin (1958).

Introductory 31

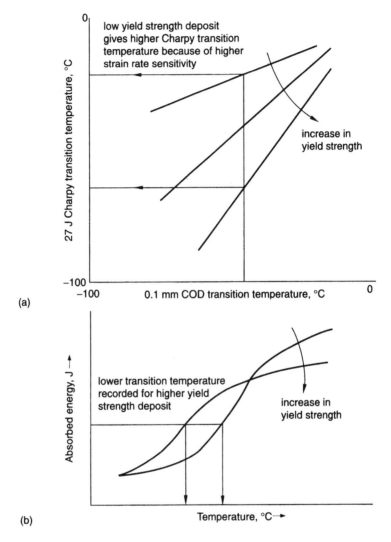

1.21 Schematic diagrams showing (a) Charpy COD correlations and influence of yield strength; and (b) effect of yield strength on Charpy-V data (Dolby, 1981).

Fracture velocity of glass has been studied using high-speed cameras capable of a frame rate of 10^5 to 10^6 per second, combined with ultrasonic vibration of the specimen at a frequency typically of 5×10^6 Hz. The ultrasonic vibrator generated ripples on the fracture surface which were observable under the microscope. This technique was of particular value in studying the velocity change after the initiation of fracture.

A typical result is shown in Fig. 1.22. The crack accelerates from zero velocity up to a constant maximum value, which is about half the velocity of transverse (shear) waves in the glass concerned. Figure 1.22 shows the initial phase of the velocity curve, while Fig. 1.23 is a more extended plot showing distance travelled versus time; the straight-line portion of the graph indicates a velocity of about 1500 m s^{-1}.

Tests were made at various temperatures and an approximate relationship with the maximum constant velocity was found to be

$$\frac{1}{V_f}\frac{dv_f}{dT} \simeq 10^{-4} \text{ K}^{-1} \tag{1.35}$$

This means that the velocity change corresponding to a temperature change of 100 K is about 1%.

Tests were also made under various loading conditions, and it emerged that the constant maximum velocity was **independent of the applied stress**. This was the case not only for uniform applied stress, but also for safety glass, which has a relatively high level of internal stress.

Plexiglass, which is a transparent synthetic polymer and which has some plasticity, was found to behave somewhat differently. The fracture velocity built up to a constant maximum as with glass, but the level of this constant maximum increased with higher applied stress. Robertson (1953) investigated the brittle fracture of steel using a technique that was, in principle, similar to that used for glass in the tests reported by Schardin. In both cases a tensile stress was imposed on the specimen, and fracture was initiated by an impact load. Robertson found that the fracture velocity close to the point of initiation was greater with a higher imposed stress; with a load of 10 tons the velocity was 6000 ft s^{-1} (1829 m s^{-1}), and at 6 tons it was 4000 ft s^{-1} (1219 m s^{-1}). Schardin gives the velocity for

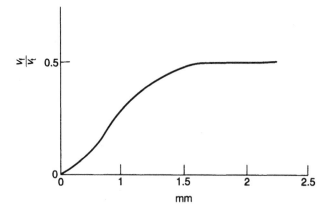

1.22 The increase in velocity in the initial phase of a fracture up to a constant maximum value. v_f is the fracture velocity, v_t is the velocity of the transverse wave (Schardin, 1958).

Introductory 33

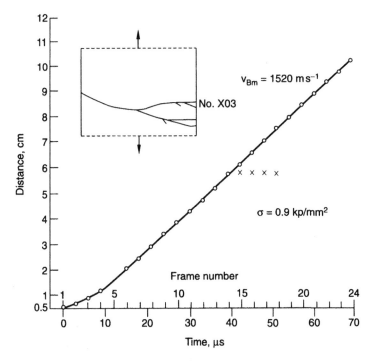

1.23 A typical distance–time plot for crack propagation in a glass test specimen (Schardin, 1958).

steel as 963–1090 m s^{-1}. The implication of these observations is that for glass (and possibly other brittle substances) the crack propagation rate is independent of the rate of energy release from the surroundings, and must be governed by other factors. In ductile solids, on the other hand, the fracture velocity increases with increased strain energy.

In recent years interest in this subject has declined, and little or no experimental work has been reported. The main concern has been with the conditions for fracture initiation, the theoretical basis for which is well established. There is no such established theory for fracture propagation, however. It has been proposed that fracture velocity should be related to that of elastic waves. Figure 1.24 is a plot of crack velocities versus the velocity of sound for a number of different types of glass. These have been grouped into two categories: flint glass, which contains silica with potassium, sodium and lead oxides; and borosilicate types, in which the lead oxide is replaced by B_2O_3 and other oxides. The velocity of sound, or compression wave velocity, v_s, was calculated from the expression

$$v_s = \left(\frac{E}{\rho}\right)^{1/2} \tag{1.36}$$

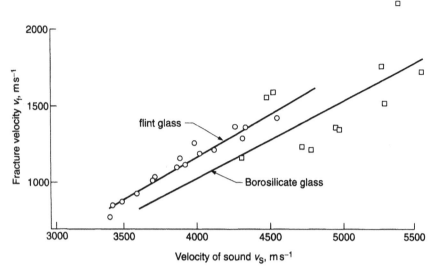

1.24 Relationship between fracture velocity and the velocity of sound in various flint and borosilicate glasses (from Schardin, 1958).

which holds for elastic solids generally. In the case of flint glass there is a good correlation and the least-squares fit gives

$$v_f = 0.55v_s + 105b \tag{1.37}$$

in m s^{-1}. For borosilicate glasses the scatter is greater but the trend is similar.

Figure 1.24 does not indicate any simple relationship between fracture velocity and the speed of sound; indeed it shows that the ratio v_f/v_s is not constant. It would, however, be reasonable to conclude that fracture propagation rates in brittle substances are influenced by their physical and mechanical properties, in particular density and elastic modulus.

Glass may also crack slowly when exposed to a moist atmosphere. This process is associated with the diffusion of alkali metal ions to the crack surface, and the rate of growth may be very low indeed; such cracking is not uncommon in window-panes.

1.5.6 Ductile fracture

It is convenient, as a start, to consider the failure of a cylindrical tensile test-bar, since this will serve as a model for some other fracture morphologies, and also because it has been the subject of a number of useful studies.

Initially the cylindrical gauge-length portion of the test-bar extends more or less uniformly, as described earlier in Section 1.3.1. The force required to extend the specimen increases up to the point of **maximum load**, when a constriction or neck appears, and the force required for further extension **falls**. As the neck contracts, a negative hydrostatic pressure builds up within it, leading to rupture and the formation of a lenticular cavity. The rupture is the result, as described previously, of decohesion at the interfaces between metal and non-metallic inclusions, followed by ductile failure of the residual ligament. Alternatively the inclusions crack, or cavities form at grain boundaries to initiate a similar process. The cavity then grows to a point where shear bands form between the outer circumference of the cavity and the surface of the metal. Shortly after the formation of the slip bands, they suffer a sudden catastrophic failure. The final result is the formation of the familiar cup-and-cone fracture.

The interesting feature about this type of fracture is that the length of the cone (and the depth of the cup) depend upon the compliance of the testing machine. With a stiff machine the cone is relatively short, and vice versa. This situation may be illustrated by representing that part of the machine to which one of the grips is attached as a rectangular block of depth L and area A, this being in turn attached to a firm inelastic base. Assume further that the other grip is fixed directly to an inelastic body. When, in the course of a tensile test, a force F is applied to the testpiece, the corresponding strain-energy content of the block representing the machine is

$$U = \frac{1}{2}\frac{F^2}{A^2E} \times AL = \frac{F^2 L}{2AE} \tag{1.38}$$

When there is an instability leading to fracture, this energy is released at a rate

$$\frac{dU}{dt} = \frac{dU}{dL}v = \frac{1}{2}\frac{F^2 v}{AE} \tag{1.39}$$

where v is the rate at which the machine relaxes. A soft testing machine is here represented as one with a smaller cross-sectional area A, so its rate of energy release under such circumstances is higher than that of a stiff machine.

Now it may reasonably be supposed that fracture will be initiated when the shear stress in the slip bands reaches a critical value. If the conical area is A_c, the applied force is $\tau_c A_c$ and the rate at which energy is absorbed during the shear failure is $A_c v/\cos\theta$ where v, as before, is the relaxation rate and θ is the half-angle of the cone. Hence, equating the two energy rates

$$A_c = \frac{F^2 \cos\theta}{AE\tau_c} \tag{1.40}$$

Thus, it is predicted that the area of a cone in a cup-and-cone fracture will increase as A decreases; that is to say, it will be larger with a soft testing machine, as has been observed.

1.5.7 Material properties relating to fracture

As discussed in previous sections, there are many ways of measuring the notch-ductility of a solid substance. Pre-eminent among such measurements is that of fracture toughness and in particular K_{IC}, since from this quantity it is possible to calculate the defect size that will put any given structure at risk. It is, of course, the fracture toughness of steel which is of primary concern. This quantity may vary over a large range, depending on temperature, prior heat treatment and exposure to embrittling mechanisms. However, if values of K_{IC} for steel in a good notch-ductile condition tested at about 20 °C are plotted against the yield or 0.2% proof stress, a good correlation is obtained, giving

$$K_{IC} = 225 - 0.1\sigma_y \tag{1.41}$$

where fracture toughness is in MN$^{-3/2}$ and yield strength σ_y is in MN m^{-2}. This relationship is shown in Fig. 1.25. In fracture toughness testing, the thickness of the testpiece must be increased as the yield strength decreases in order to maintain plane-strain conditions. Consequently, the lowest yield point for which valid K_{IC} results are available (at 10 °C) is 500 MN m^{-2}. Extrapolating down to

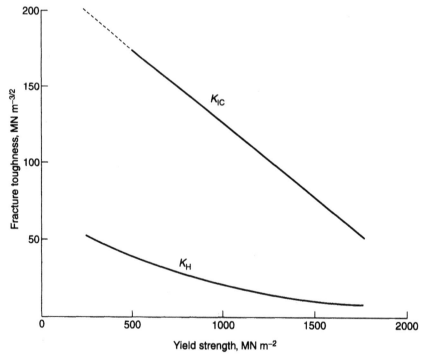

1.25 Fracture toughness of steel (K_{IC} and K_H) as a function of yield strength.

Table 1.2. Metal properties relevant to fracture

Substance	Density ρ (kg dm^{-3})	Elastic modulus E (GN m^{-2})	Poisson's ratio ν	Yield stress σ_y (MN m^{-2})	Ultimate or fracture stress σ_v, σ_f (MN m^{-2})	Specific fracture energy G_{IC} (J m^{-2})	Fracture toughness K_{IC} (MN m$^{-3/2}$)
Alumina	3.9	380	0.24		350	2.4–6.8	3–5
Al 3½ Ca 1 mg age-hardened	2.8	72	0.33	300	390	3×10^4	50
Brick	1.4–2.2	10–50					1–2
Carbon fibre reinforced plastic	1.5–2.0	150–250	0.1–1.2		1000–1500	2×10^4	60
Concrete	2.4	10–20	0.4				0.3
Epoxy resin	1.2–1.4	2–5			40–85	220	0.6–1.0
Glass fibre reinforced plastic	1.5–2.5	80–100			1000	9×10^4	90
Glass (soda)	2.5	74	0.23		50	6.3*	0.7
Ice	0.92	9.1			1.7		0.12
Steel, mild	7.85	210	0.3	240	450	1.7×10^5	200**
Steel, medium tensile	7.85	210	0.3	450	600	1.4×10^5	180**
Steel, high tensile	7.85	210	0.3	1600	2000	1.8×10^4	65**
Zirconia	5.6	200	0.3		500–830		12

*Surface energy γ is 1.2 J m^{-2}.
**Unembrittled at 20 °C.

240 MN m^{-2}, which is a typical yield strength for mild steel, gives a fracture toughness of 200 MN m$^{-3/2}$.

The line given by equation 1.46 represents an upper bound for the fracture toughness of plain carbon steel in the as-rolled condition and high tensile steel in the quenched and tempered condition. Also shown in Fig. 1.25 is a lower bound curve for similar steels in the hydrogen-embrittled state, where the fracture toughness is designated K_H. The derivation of this curve is given in Section 8.5.3, where it is stated that because of the nature of hydrogen embrittlement, measured values of K_H scatter widely above this curve. Steels that are embrittled by low temperature or by some metallurgical process would also be expected to have fracture toughness values falling between these two limits.

K_{IC} figures for non-ferrous metals are rarely reported, since normally they are notch-ductile. However, Wells (1955) carried out tests on age-hardened aluminium alloys of the Duralumin type, and found that for centre-cracked sheet metal specimens a brittle fracture occurred when (in imperial units):

$$a_c \sigma_c^2 = \text{constant} = 250 \text{ inch} - (\text{tons/sq. in.})^2 \qquad (1.42)$$

where a_c was the critical crack length and σ_c the critical stress. For an infinite sheet with a central crack of length $2a$ the stress intensity factor is

$$K_I = \left(\frac{\pi a}{2}\right)^{1/2} \sigma \qquad (1.43)$$

so for the aluminium alloy in question

$$K_{IC} = \left(\frac{\pi}{2} a_c \sigma_c^2\right)^{1/2} \qquad (1.44)$$

which, translating into SI units, gives $K_{IC} = 48.8$ MN m$^{-3/2}$. This figure is for an aluminium–copper–magnesium alloy precipitation-hardened to a yield strength of 300 MN m^{-2}. The fracture toughness would be expected to increase with lower yield strength.

Other properties that may be relevant to the fracture process include density, elastic modulus and Poisson's ratio. These and other relevant properties are listed in Table 1.2.

References

Barsom, J.M. and Rolfe, S.T. (1970) 'Correlations between K_{IC} and Charpy V-notch in the transition range', ASTM STP **466**, 281–302.

Dolby, R.E. (1981) *Metal Construction*, **13**, 43–51.

Griffiths, A.A. (1920) *Philosophical Transactions of the Royal Society of London Series A*, **221**, 163–198.

Howatson, P.G. and Lund, J.D. (1994) *Engineering Tables and Data*, Chapman and Hall, London.

Hull, D. and Bacon, D.J. (1997) *An Introduction to Dislocations* (3rd edn 1984, reprinted 1997), Butterworth Heinemann, Oxford.

Keene, B.J. (1991) *Surface tension of pure metals*, NPL Report DMA(A) 39, National Physical Laboratory, Teddington, UK.

Nichols, M.G. (1990) 'Overview' in *Joining of Ceramics*, Ed. M.G. Nichols, Chapman and Hall, London.

Paris, P.C. and Sih, G.C. (1965) in *Fracture Toughness Testing*, ASTM Special Technical Publications STP 381.

Robertson, T.S. (1953) *Journal of Iron and Steel Institute*, December, 361–374.

Schardin, H. (1958) 'Velocity effects in fracture' in *Fracture*, eds Averbach, B.L. *et al.*, Wiley, New York, and Chapman & Hall, London.

Toyada, M. (1995) *Welding Journal*, **74** (12), 31–44.

Wells, A.A. (1955) 'The conditions for fast fracture in aluminium alloys with particular reference to the Comet failures', BWRA Research Report RB 129, April 1955.

2
Solid-phase welding

2.1 Physical aspects and metallurgy

2.1.1 General

In solid-phase welding the object is to make a welded joint between two solid pieces of metal by bringing their surfaces into sufficiently close proximity for a metallic bond to be formed. In some processes, flash-butt welding for example, a liquid phase is formed between the two surfaces at an intermediate stage of the welding operation, but in the final stage the joint is upset to extrude the liquid and form the weld between the solid surfaces. Compared with fusion welding such a process has a number of advantages. In the first place, there is no cast metal in the joint, except in so far as it may be accidentally included. Secondly, the joint is made under compression, so that the risk of cracking is small; and thirdly the embrittling element hydrogen is absent. Unfortunately, solid-phase processes are only applicable to components of modest dimensions, and although the size of structures made in this way is increasing, they are never likely to be used for the construction of, say, an oil rig.

2.1.2 Bond formation

The primary requirement in making a solid-phase weld is to bring two clean surfaces close together. The barriers to obtaining perfect contact are twofold: the presence of non-metallic films, including chemisorbed gases, on the surface, and the physical difficulty of obtaining an exact fit over the whole of the two surfaces to be joined. Films of water, oil or grease are obvious hazards; being fluid they have good contact with the metal surfaces and are strongly bonded thereto, but since they have no strength, the strength of any joint so contaminated is greatly reduced. Oxide films are commonly brittle, and a joint consisting of the sandwich metal–oxide–metal, even if bonded, will in general be brittle and have low strength. Liquid films are removed by heating or, if welding is to be accomplished cold, by scratch-brushing. Oxide films are also removed by scratch-brushing, by lateral movement of the two surfaces, or by fluxing and melting.

Obtaining good contact by direct pressure between two metal surfaces is difficult and, except in diffusion bonding, joints made in this way have very low strength. For effective mating in other solid-phase welding processes, lateral movement is necessary. When the surfaces are forced together with relatively light normal pressure, the proportion of surface that is brought within bonding range is small. However, the asperities of each surface penetrate the other and, if the two are moved laterally relative to each other, some of the asperities shear, and the clean metal surfaces so produced bond together. Repetition of this process naturally increases the bonded area. Alternatively, if the two surfaces are made to flow laterally in the same direction (**conjoint flow**), areas of unfilmed metal are formed in close proximity and therefore bonding takes place. **Relative lateral flow** occurs in ultrasonic welding and friction welding, and conjoint flow in butt and flash-butt welding, cold and hot pressure welding, and explosive welding.

2.1.3 Surface films

A metal is normally coated with a film of oxide, sulphide or carbonate, whose thickness lies within the range 10^{-3} to 10^{-1} µm. It may also have a layer of adsorbed gas, and it may be contaminated by oil, grease or other non-metallic substances.

Oil films inhibit solid-phase welding either partially or completely when the temperature is such that they can persist. Oxide films hinder, but do not prevent, pressure welding. It is generally considered that, during deformation of the surface, the oxide film (or hardened surface layer produced by scratch-brushing) fractures and exposes areas of clean metal, which bond to the opposite surface wherever two clean areas come into contact.

At elevated temperatures, oxide is dispersed by deformation, by solution or by agglomeration, or by a combination of these processes. Excessive amounts of dissolved oxygen cause brittle welds. Oxide inclusions, which may be present in massive form in welds made at elevated temperature, are also damaging to the mechanical properties. Oxide inclusions, dissolved oxygen and voids may be dispersed from the junction zone of carbon-steel pressure welds by soaking above 1000 °C for an adequate period, although if a proper welding technique is used no significant oxide contamination will be present.

2.1.4 Recrystallization

In welds made at room temperature, recrystallization of a surface zone occurs with low-melting point metals such as tin and lead, but most engineering metals must be welded at elevated temperature if recrystallization is to occur during the welding process. Recrystallization such that grains grow and coalesce across the original interface is not essential to welding, nor does it ensure that the best achievable properties have been obtained. Steel pressure welds made at temperatures above the upper critical temperature show continuity of grains across the interface but may still lack ductility due to oxide inclusions or other causes. However, increasing the welding temperature, which favours recrystallization, also favours the elimination of other defects, and improves the room-temperature ductility of the completed joint.

2.1.5 Diffusion

Gross macroscopic voids at the interface of pressure welds increase in size and diminish in number if the joint is soaked at elevated temperature, indicating that a process akin to that which causes the increase of density of metal powder compacts during sintering may be at work. Increasing temperature would favour such a mechanism, and it does in fact improve the ductility of pressure-welded joints.

Diffusion may be important in removing contaminants from the weld zone, particularly oxygen in reactive metals such as titanium. For such material a postwelding solution treatment is required for optimum joint ductility. Diffusion is an essential feature of **diffusion bonding**, which is described in Section 2.2.5.

2.2 Solid-phase welding processes

2.2.1 Pressure welding at elevated temperature

Forge, **hammer**, **butt** and **oxyacetylene pressure welding** are all techniques designed to make solid-phase welds at elevated metal temperature. In butt welding and oxyacetylene welding, the metal is simply heated to a high temperature (1200–1250 °C in the case of carbon steel) while the joint is subject to axial compression. When the metal in the region of the interface reaches this temperature, it deforms under the axial load and there is a lateral spread which disrupts the surface films and permits welding to take place. Essential controls are the applied pressure and the amount of shortening of the parts being joined. Pressure may be constant or may be increased at the end of the welding cycle.

In resistance butt welding, heating is accomplished by passing an electric current across the joint. This process is applied to welding bar and rod, to end-to-end welding of strip, and to the manufacture of longitudinally welded tube. The tube is formed from strip by a series of rolls, whence it passes between two copper rollers through which the welding current is applied, then through a pair of forging rolls, which force together the heated edges and make the weld.

High-frequency resistance welding (Fig. 2.1) is the process most often used for the manufacture of ERW (electrically resistance welded) pipe. In principle it is

Solid-phase welding 43

similar to resistance butt welding but a high-frequency current is used as the power source. High-frequency current flows preferentially through the surface layers of a conductor, and in welding this effect minimizes the degree of upset and distortion. Downstream of the welding station knives remove the flash and further downstream a high-frequency coil reheats the weld to normalizing temperature. Flying saws cut the pipe to standard lengths (about 12.5 m). This technique is applied to line pipe (that is, pipe used for oil and gas transmission lines) in the medium diameter range and, being a relatively economical route, is used for a large tonnage of this product.

A relatively new method of pressure welding, suitable for butt joints in hollow sections and pipe, is **magnetically impelled arc butt welding**. The square-edged tube ends are separated by a small gap and an arc is struck between them. This arc is rotated around the tube or section by means of a radial magnetic field, and when the surfaces are heated sufficiently they are brought together under pressure to form a solid-phase weld. Liquid metal and impurities are extruded into the flash, which is removed by cutting tools. Suitable machines are commercially available and are used in the automotive industry for welding such items as car axles, drive shafts and shock absorbers, while a portable machine suitable for welding pipe up to 300 mm in diameter has been developed for the oil and gas pipeline industry.

The productivity achievable with hot pressure welding techniques is high, but if defects are present they may penetrate the complete wall thickness. Such welds may also suffer from low impact properties, reputedly owing to the presence of non-metallic films. Ultrasonic testing will eliminate gross defects; otherwise fabricators rely on control of the welding variables to maintain quality.

Flash welding is essentially different in that the two parts are first brought together under light pressure so that contact is localized. On passing a current, metal at the points of contact first melts, and is then violently expelled in the form of globules through the joint gap, producing the phenomenon known as **flashing**

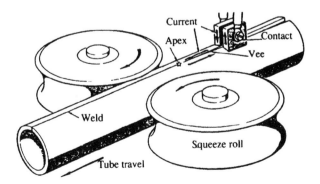

2.1 High-frequency resistance welding of pipe or tube (from Houldcroft, 1979).

(flashing is due to intermittent arcs, the voltage being insufficient to produce a continuous arc). At the end of the flashing period, the surfaces have been heated to a high temperature, and they are brought together under an axial pressure. This pressure extrudes the liquid metal and oxide present at the interface, and welds the underlying clean metal surfaces.

In forge welding steel, the two parts to be welded are heated in a forge fire or furnace to between 1200 and 1400 °C. The surfaces are fluxed with borax for high-carbon steel, or sand for medium-carbon steel. Wrought iron and low-carbon steel do not require flux, because the melting point of the metal is above that of the oxide. The two parts are then hammered together in order to extrude the molten oxide or slag and make the weld.

The **roll bonding** of plate material is an important application of pressure welding. In principle, two slabs of the materials to be bonded are placed in contact and welded around the edges or otherwise treated to exclude air. They are then heated and rolled until the required thickness is obtained. Carbon steel is clad with (for example) austenitic chromium–nickel steel in this way. A carbon steel weld deposit is laid on the stainless steel slab so that the roll bond is carbon steel to carbon steel, or a nickel interlayer may be used. Alclad sheet is produced in a similar way, but here no interlayer is required. The process is applied to high-strength aluminium alloys that are sensitive to corrosion in atmospheric exposure: a thin pure aluminium or 1% zinc alloy layer is roll bonded to both surfaces. These surface layers are anodic to the underlying alloy and are protective in the same way as galvanizing on steel.

2.2.2 Friction welding

Friction welding is not a new process; it was used for attaching bar to plate in the early years of the twentieth century. There was renewed interest in the technique, however, after World War II, and in recent years its applications have widened considerably. The original version of the process, which is applicable to the welding of pipe and bar, will be described first.

In principle, friction welding is one of the simplest welding processes (Fig. 2.2). One part is rotated relative to the other with the surfaces in contact, so that frictional heating of the interface occurs. There are two major variants of the process. In the first, the rotated component is driven **continuously** by an electric motor, and the energy supply to the joint is controlled by rotational speed, axial pressure and time. The second type, called **stored energy** or **inertia friction welding**, employs a flywheel to drive the rotating component. The flywheel is brought up to speed and then disengaged from the driving motor before applying thrust. The speed, torque, axial thrust and burn-off cycles for these two variants are illustrated diagrammatically in Fig. 2.3. With continuous drive, the cold parts are initially subject to dry friction, but as the temperature rises there are

Solid-phase welding 45

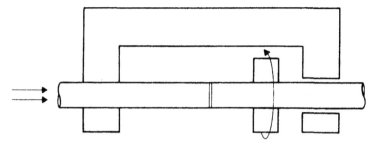

2.2 The principle of friction welding.

local seizures at the interface requiring an increased torque from the machine. Eventually a stage of plastic deformation is reached, when the torque falls and becomes steady. At the end of the plastic phase, the rotation is stopped by a brake, and axial pressure is applied to make the weld. The cycle for inertia

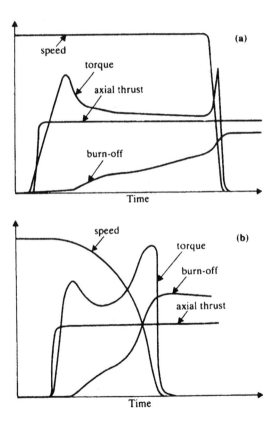

2.3 A schematic diagram of speed, torque, burn-off and axial thrust variation during friction welding: (a) continuous drive; (b) stored energy (from Elliot and Wallach, 1981).

46 Metallurgy of welding

friction welding is generally similar, but the speed decreases with time and the period of steady torque is absent.

The welding machine usually resembles a lathe, but there are a number of possible geometric arrangements by which friction welds can be made. No melting takes place in friction welding. Any incipient melting reduces the frictional force and heating effect, so that the process is self-regulating. In general, heating and cooling rates are low enough to avoid the metallurgical problems that sometimes result from thermal cycling. Power requirements are moderate compared with flash-butt welding, for example, and the equipment is relatively simple and rugged.

Friction welding has been adapted for the butt welding of pipes that are too long to be rotated by using the technique illustrated in Fig. 2.4(a). A ring-shaped insert is rotated between the pipe ends and when the heating phase is complete, axial pressure is applied to complete the weld. An internal expanding mandrel prevents collapse and intrusions into the bore. A similar method may be employed to apply a collar to a pipe. Friction welding may be used to clad a plate with, say, a corrosion-resistant alloy by traversing a rotating bar across the surface in the manner illustrated in Fig. 2.4(b). Butt welds in plate may be made in a similar way by running the rotating bar along a groove formed between the plates. Such operations are all carried out with sections having circular symmetry; rectangular or irregular bars may be joined by using a reciprocating or circular motion relative to each other to generate the required heat (Fig. 2.4c).

Friction welding requires relatively heavy equipment and is best suited to operations in a fixed location. Typical applications are automobile exhaust pipes, steering columns and rear axle assemblies.

Friction welding has the characteristic advantages of a solid-phase welding process, and is applicable to a wide range of ferrous and non-ferrous metals. It is well suited to the joining of dissimilar metals, including that particularly difficult combination, aluminium to carbon steel.

A contemporary development that has considerable potential is **friction stir welding** (Dawes and Thomas, 1996). The principle of this process is illustrated in Fig. 2.4(d). Sheet or plate is butted edge-to-edge and clamped to prevent upward or lateral movement. A tool, which consists of a cylinder terminating in a specially shaped pin, is rotated and traversed along the joint. The metal below the cylinder and around the pin is heated and softened by friction, and flows from front to rear of the pin during its traverse. The result is a solid-phase weld made between two plates with the same geometry as a fusion weld. Such a process has the potential advantages enumerated in Section 2.1.1. In particular, it may be applied to alloys such as the age-hardenable aluminium–copper types which are difficult to fusion weld because of solidification and liquation cracking. As with other friction welding processes, the welding equipment resembles a machine tool, and is controlled almost to the same degree as a machining operation.

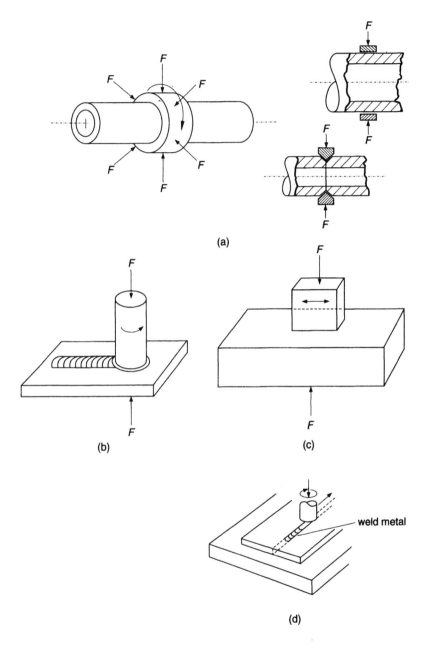

2.4 Developments in friction welding: (a) butt welding of pipe; (b) cladding; (c) welding of non-circular components; (d) friction stir welding.

48   Metallurgy of welding

Friction stir welding was invented and is being developed by TWI (formerly The Welding Institute), Cambridge. Its possible use for welding aluminium alloy sheet, plate and sections is described in Chapter 10.

2.2.3 Explosive welding

The principle of explosive welding and some of its characteristics are discussed in Chapter 1 in relation to the deformability of metals (see Section 1.3.3 and Fig. 1.7). Figure 2.5 is a photomicrograph showing the typically wavy interface of an explosive weld.

Most of the metal is hardened by shock waves, and there may be anomalous slip and twinning and increased dislocation density. In the bond zone, in addition to the considerable plastic deformation visible in Fig. 2.5, there may be some recrystallization due to local heating.

Explosive welding is of special interest for the bonding of one plate to the surface of another. Normally this is carried out by roll bonding, as discussed in Section 2.2.1. However, as the thickness of the backing plate increases, the interfacial bond becomes less consistent when made using the roll bonding technique. Therefore the cladding of thick plate, such as is used for the tubesheet

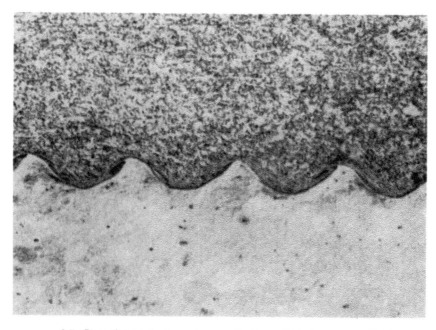

2.5 Part of the interface of an explosive weld between two C–Mn steel components. Locally fused regions are evident, in some cases containing small pores. Etched by 2% nital; ×50, reduced by a quarter in reproduction (photograph courtesy of TWI).

of a high-pressure heat exchanger, is carried out either by means of a fusion welding process or by explosive welding. Explosive welding also offers the possibility of making clad plate in two metals that cannot be joined satisfactorily by roll bonding.

Explosive techniques may also be used to bond tube to tubesheets. A detonator is placed centrally in the tube and exploded. If the tubehole is parallel, the tube will actually be expanded without welding, but if the tubehole is flared welding can be achieved. When using an explosive to clad a tubesheet or to make a tube-to-tubesheet joint it is necessary to consider the possibility of distortion or brittle fracture of the tubesheet.

2.2.4 Ultrasonic welding

The components to be joined are clamped together on an anvil, as shown in Fig. 2.6, and vibrated laterally by an ultrasonic transducer at a frequency of 10–75 kHz. Shear forces at the interface disrupt the oxide film and expose areas of clean metal, which weld under the applied force. Spot, ring, line and continuous seam welds can be produced. As might be expected, a spot weld consists of a central unwelded region, an annular welded zone and, around this, a region of unbonded metal that has an undulating surface.

Ultrasonic welding is important as a means of joining fine wires to themselves and to aluminium-coated surfaces of microcircuit devices, for splicing aluminium foil, and for a limited number of secondary structural applications in the aircraft industry.

2.2.5 Diffusion bonding

Diffusion bonding is one of the most ancient methods of joining metals, and some very early pieces of jewellery were manufactured in this way. Its industrial use is fairly recent, and has received additional impetus in recent years because of applications to relatively large structures in the aircraft industry.

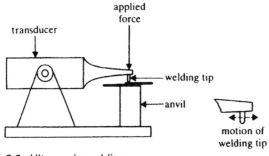

2.6 Ultrasonic welding.

50   Metallurgy of welding

Figure 2.7 shows some of the combinations of metals and non-metals that can be bonded by the diffusion process. For the present only metallic bonds will be considered; metal/ceramic diffusion bonded joints are dealt with in Chapter 5. The process may be divided into two subcategories:

1. **diffusion welding**, in which the two parts are held together under moderate pressure and at elevated temperature without melting, with or without an interlay of dissimilar metal;
2. **diffusion brazing**, in which an interlayer foil or coating produces a liquid at the bond interface (pressure may or may not be used).

Diffusion welding is carried out in an inert atmosphere or vacuum, using a pressure typically 0.7–$10\,\text{N}\,\text{mm}^{-2}$ and a temperature above half the melting point. Some bonding conditions are shown in Table 2.1. Bonding takes place in two phases. In the first phase, asperities collapse and a partial bond is formed, interspersed with cavities. The bond area then increases due to creep, and the cavities diminish in size and eventually disappear through surface and volume diffusion (Fig. 2.8).

The surface finish is important. Machining must give a roughness not greater than 0.4 mm CLA (centre-line average), and degreasing with acetone or

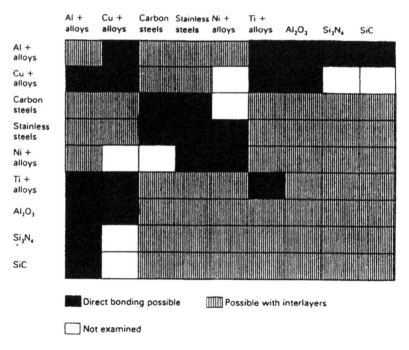

2.7 Materials that can be joined by diffusion bonding (from Dunkerton, 1991).

Solid-phase welding 51

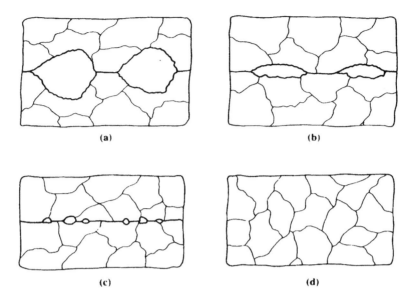

2.8 A schematic illustration of stages of diffusion bonding: (a) stage 0, instantaneous plastic deformation of contacting asperities; (b) stage I, shrinkage of elliptical voids; (c) stage II, shrinkage of spherical voids; (d) final parent metal microstructure (from Wallach, 1984).

Table 2.1. Diffusion bonding variables and strength of joints (Partridge and Ward-Close, 1989)

Materials	Technique	Temperature (°C)	Pressure (N mm^{-2})	Time (h)	Joint strength (N mm^{-2})
Similar metals					
Ti-6A1-4V	Solid phase	930	2–3	1.5	575 (shear)
T-6A1-4V (Cu and Ni interlayers)	Liquid phase	950	nil	1–4	950 (tensile)
Al alloy 7475-T6	Solid phase	516	0.7	4	331 (shear)
Al-Li alloy 8090-T6	Solid phase	560	1.5	4	190 (shear)
Dissimilar metals					
Ti-6A1-4V to stainless steel (V and Cu interlayers)	Solid phase	850	10	1	450 (tensile)

52 Metallurgy of welding

petroleum ether is necessary. Oxide films may be a problem. For titanium alloys the oxide is soluble in the parent metal at elevated temperature but stable oxides such as Al_2O_3 and Cr_2O_3 may need to be removed by sputter cleaning (ion bombardment). Such operations may be too costly and the use of a liquid-phase interlayer may be preferable.

The liquid-phase process, or brazing, is mostly used for dissimilar metal combinations. An interlayer is used; alternatively, if the two metals form a low-melting eutectic phase the bonding temperature is held within the appropriate range until a film of liquid has spread over the surface. The temperature is then reduced below the solidus and in some instances the eutectic may be diffused into the surrounding metal. In liquid-phase bonding the surface finish requirement is less critical and a finish of 0.8 mm CLA may be acceptable.

2.2.6 Superplastic forming combined with diffusion bonding

Some titanium alloys have superplastic properties at elevated temperature, and the necessary pressures and temperatures coincide with those required for diffusion bonding. In the aircraft industry this combination of properties is used for the

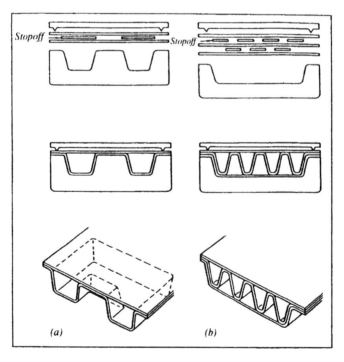

2.9 Superplastic forming combined with diffusion bonding for: (a) an integrally stiffened structure; (b) a sandwich structure (from Dunkerton, 1991).

fabrication of airframe structures, heat exchanger ducts, fan blades and other applications. The technique for making a stiffened airframe part is illustrated in Fig. 2.9. The operation is conducted in a hot isostatic press, which is a furnace contained within a pressure vessel. The part is enclosed, typically in an evacuated steel can, and pressure is applied in such a way that the gas does not flow into the bonding area. Typical conditions are, for titanium, 900 °C for 3 h at 100 N mm^{-2} and, for aluminium, 500 °C for 2 h at 75 N mm^{-2}. The pressures are higher than for uniaxial bonding and the requirements for surface finish are less stringent.

Since the first application of the process to relatively small components the capacity of hot isostatic press units has increased substantially, such that in 1991 British Aerospace were able to bond sheets of up to 2.5 m by 1.8 m. Other applications of diffusion bonding include the joining of Zircaloy-2 to depleted uranium for the encapsulation of nuclear fuel elements, hardfacing small-bore valves, and miniaturized heat exchangers. Diffusion bonding is also used for metal–non-metal joining as described in Chapter 4.

References

Dawes, C.J. and Thomas, W.M. (1996) *Welding Journal*, **75**, 41–45.
Dunkerton, S.B. (1991) *Welding and Metal Fabrication*, **59**, 132–136.
Elliot, S. and Wallach, E.R. (1981) *Metal Construction*, **13**, 221–225.
Houldcroft, P.T. (1979) *Welding Process Technology*, Cambridge University Press, Cambridge, UK.
Partridge, P.G. and Ward-Close, C.M. (1989) *Metals and Materials*, **5**, 334–339.

Further reading

Welding Handbook Vol. 2, *Welding Processes*, American Welding Society, Miami (1991).

3
The use of adhesives for making structural joints

3.1 History

Glues have been in use since the beginning of civilization. Traditional types consist of a colloidal solution in water, which sets after drying into a hard adhesive film. Animal glues consist largely of the protein collagen, which is extracted by boiling bones, hides or horns. Vegetable adhesives include starch, gums such as gum arabic, and latex rubber. There are also inorganic natural glues, examples being waterglass (once used on a wide scale in the manufacture of coated welding electrodes) and pitch, derived from coal tar.

These natural products have now been almost completely superseded by adhesives based on synthetic polymers. Such polymers mostly fall into one of two classes, **thermoplastic** and **thermosetting**. Thermoplastic adhesives are generally water-soluble and melt on heating. The thermosetting type, on the other hand, is set by a polymerization reaction which produces a hard film that is insoluble in water and does not melt on heating. Thermoplastic adhesives are used where flexibility is required; for joining fabrics, or for adhesive tape. Thermosetting polymers, on the other hand, are suitable for making relatively strong joints between parts of an industrial product. This chapter is concerned with the use of such polymers to make load-bearing structural joints.

It is normal practice when bonding sheet metal material with adhesives to make lap joints or variants thereof; butt joints are avoided. For metals in particular, where the tensile properties of the polymer are much lower than those of the substrate, the lap joint allows the development of adequate shear strength. Figure 3.1 is a diagrammatic representation of such a joint. The parts to be joined are known as **adherends** and the adhesive part of the sandwich is called the **glue line** or **bond line**. The bond is made to the oxide layer, which is itself strongly bonded to the metal surface. The natural oxide layer may be only a few nanometres (10^{-9} m) in thickness but aluminium alloys are often anodized, which increases the oxide thickness as indicated in the diagram. A layer of primer is shown; this is required with some adhesive/adherend combinations but not with all. Such primers serve much the same purpose as those used in

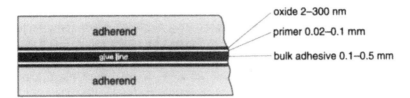

3.1 A typical adhesively bonded lap joint showing nomenclature and the thickness of various layers.

paintwork: they penetrate irregularities in the surface, provide temporary protection until the final joint is made, and adhere well to the bulk adhesive.

In designing adhesively bonded joints it is customary to allow a generous safety factor. A typical working shear stress is 20% of the failure stress obtained in a standard lap-shear test described in Section 3.5.5. Of course it is not possible to calculate stress in adhesively bonded joints as precisely as in the case of a pressure vessel, for example. Experience may provide essential guidance under such circumstances.

3.2 Bonding between adhesive and substrate

3.2.1 General

If a lap-shear test is made under dry conditions on a properly made joint between two metal sheets, failure will take place through the adhesive. In other words, the strength of the bond between adhesive and metal is greater than the cohesive forces between the polymer molecules. If a similar test is carried out after the adhesive is saturated with water the shear strength is much reduced and failure is at the interface. The bonding force in dry conditions is strong therefore, but it is much reduced by the presence of water.

There is no general agreement as to the nature of the forces that maintain the integrity of an adhesively bonded joint. There are, however, forces that are known to act between molecules regardless of their species. These are the **van der Waals forces**, which are considered below.

3.2.2 Van der Waals forces

There are two main types of intermolecular force that are covered by this generic title: **polar** and **dispersion** forces. Such forces result from the fact that in covalent chemical compounds or radicals the centre of gravity of the electrons, and that of the positive ions, do not necessarily coincide. If one of the ions is more **electronegative** than the others, then electrons will be attracted towards it. Hence there will be a gradient of space charge and an electric **dipole** will be set up. Associated with such dipoles are forces which attract similar dipoles that are either present or are induced in a neighbouring molecule. These form the polar component of the van der Waals forces. Dispersion forces result from a similar distortion in the space charge associated with the constituent ions of the molecule. In both cases these forces act only at very short range and are inversely proportional to the distance separating the molecules. They are also reduced if the relative permittivity (the dielectric constant) of the substance is increased. This fact may be relevant to the damaging effect of water on adhesive bonds, because water has a very high dielectric constant: 80, as compared with about 5 for a typical polymer.

The strength, or moment, of the dipoles present at the surface of a polymer affects its surface energy and its ability to bond with an adhesive. C–C and C–H bonds have low polar moments, while C–N, C–O, C–Cl, C–F and O–H are polar. Consequently polyethylene, where C–C and C–H bonds predominate, is difficult to bond in the untreated condition. Treatment with a flame or corona discharge results in the formation of C–O, C–OH and carboxylic acid–COOH groups. Adhesive bonding is then possible. Alternatively, such polymers may be bonded using a cyanoacrylate adhesive, possibly in conjunction with a primer.

The strength of the polar and dispersion forces is reflected in the surface energy which, for a solid, is represented by γ_{SV}. This quantity may be resolved into two parts, that due to polar forces and that due to dispersion forces:

$$\gamma_{SV} = \gamma_S^p + \gamma_S^d \tag{3.1}$$

and for a liquid

$$\gamma_{LV} = \gamma_L^p + \gamma_L^d \tag{3.2}$$

3.2.3 Measurement of surface energy

In Chapter 1 it was suggested that because of interactions between liquid and substrate, the sessile drop technique was not suitable for quantitative measurements in the case of liquid metals. This objection does not hold for organic liquids at room temperature. Distortion due to gravity is minimized by using very small drops that are observed using a microscope. The balance of forces is given by Young's equation:

$$\gamma_{SV} = \gamma_{SL} + \gamma_{LV}\cos\theta \tag{1.15}$$

and it is possible to obtain the surface tension of a liquid by measuring its contact angle with a particular substrate and comparing this with that of a liquid of known properties. For values of the surface energy of the solid, γ_{SV}, it is necessary to make assumptions about the interfacial tension, γ_{SL}. One such assumption (Adams et al., 1997) leads to a formula that be may put in the form

$$\gamma_{SL} = \left(\gamma_S^{1/2} - \gamma_L^{1/2}\right)^2 \tag{3.3}$$

For a sessile drop with a contact angle of 90° or less, this expression meets the boundary conditions, namely that when $\gamma_S = 0$, $\gamma_{SL} = \gamma_L$ when $\gamma_L = 0$, $\gamma_{SL} = \gamma_S$ and when $\gamma_S = \gamma_L$, $\gamma_{SL} = 0$. In these equations γ_L or γ_{LV} is the surface energy of the liquid which is numerically equal to surface tension, but expressed as $J\,m^{-2}$ instead of $N\,m^{-1}$.

The surface energies are now resolved into their polar and dispersion components, to give

$$\gamma_{SL} = \gamma_S + \gamma_L - 2(\gamma_S^p \gamma_L^p)^{1/2} - 2(\gamma_S^d \gamma_L^d)^{1/2} \tag{3.4}$$

Combining this with the equation 1.21 and rearranging

$$\gamma_L \frac{1 + \cos\theta}{2(\gamma_L^d)^{1/2}} = (\gamma_S^d)^{1/2} + \left(\frac{\gamma_S^p \gamma_L^p}{\gamma_L^d}\right)^{1/2} \tag{3.5}$$

Then if contact angles are measured for various liquids, and the left-hand side of equation 3.5 is plotted against $(\gamma_L^p/\gamma_L^d)^{1/2}$, the result should be a straight line with slope $(\gamma_S^p)^{1/2}$ and intercept $(\gamma_S^d)^{1/2}$. So, using liquids of known properties and measuring contact angles makes possible a determination of the components of surface energy for the substrate. Examples are shown in Fig. 3.2 and 3.3, from which it will be seen that the data do indeed plot in linear fashion, thus providing justification for the assumptions made concerning the interfacial free energy. Figure 3.2 relates to an adhesive showing considerable polarity; Figure 3.3, on the other hand, is for an almost completely non-polar substance, zinc stearate. Because of this property zinc stearate is used as a parting compound in moulds for plastics.

Table 3.1 lists data for the surface energies of some polymers and oxides. The first three items are difficult to bond in the untreated condition, whereas epoxides bond readily, as do oxidized metal surfaces. From these and other data it would seem that bonding requires the presence of polar forces in the adherend. Oxides have a very high polar compound of surface energy when perfectly clean, but this very fact means that they rapidly absorb a film of water when exposed to the atmosphere, and easily become contaminated by hydrocarbons. The surface energy of austenitic chromium–nickel steel has been measured after various cleaning operations, and varies from $40\,mJ\,m^{-2}$ in the as-received condition to about $120\,mJ\,m^{-2}$ after plasma treatment.

58 Metallurgy of welding

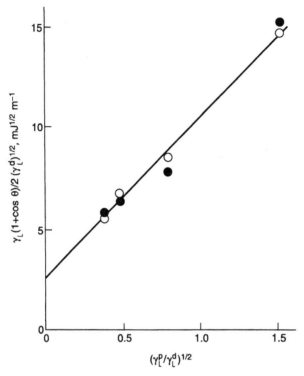

3.2 Plot based on equation 3.5 for liquids on a film of adhesive. $\gamma_S^p = 69.2\,\text{mJ m}^{-2}$ and $\gamma_S^d = 4.8\,\text{mJ m}^{-2}$ (Comyn *et al.*, 1993).

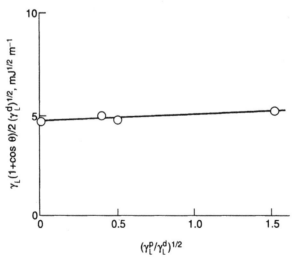

3.3 Plot based on equation 3.5 for liquids on zinc stearate. $\gamma_S^p = 0.06\,\text{mJ m}^{-2}$ and $\gamma_s^d = 22.4\,\text{mJ m}^{-2}$ (Comyn *et al.*, 1993).

Table 3.1. The surface energy of some solids

Solid	Surface energy γ_S (mJ m^{-2})		
	Polar component γ_S^p	Dispersive component γ_S^d	Total γ_S
Polytetrafluorethylene	0.5	18.6	19.1
Polypropylene	0.0	30.2	30.2
Polyethylene	0.0	33.2	33.2
Rubber-modified epoxide	8.3	37.2	45.5
Amine-cured epoxide	5.0	41.2	46.2
Silica SiO$_2$	209	78	287
Alumina Al$_2$O$_3$	538	100	638
Ferric oxide Fe$_2$O$_3$	1250	107	1357

Source: from Adams *et al.* (1997).

3.2.4 Work of adhesion

The work of adhesion is the amount of work that is done when two adherent surfaces are parted and removed to a large distance from each other. This quantity, W_{adh} may be either positive or negative; if positive the two surfaces will bond and the higher the value of W_{adh} the stronger the bond. Conversely, if it is negative, no bonding will occur. The work of adhesion is given by

$$W_{adh} = \gamma_A + \gamma_B - \gamma_{AB} \qquad (3.6)$$

where γ_A and γ_B are the surface energies of the two substances being bonded and γ_{AB} is their interfacial energy. Using equation 1.21, the work of adhesion is obtained in terms of the contact angle

$$W_{adh} = \gamma_A(1 + \cos\theta) \qquad (3.7)$$

where the substrate is substance B. Measurements have been made of these various quantities for a number of metal/oxide combinations. The principle of the technique is illustrated in Fig. 3.4. A small cube or particle of substance A is placed on a flat surface of substance B and the whole is heated in vacuum to a temperature of 500–1000 °C. Provided that γ_A is known, measurement of the contact angle gives W_{adh}. Figure 3.5 shows some results (Hondros, 1985). The regression curve for these data is a straight line with a slope very close to unity and nearly equal intercepts of 2.65 J m^{-2}. The lower the value of the interfacial tension, the higher the value of W_{adh}; when $\gamma_{AB} = 0$, $(\gamma_A + \gamma_B) = 2.65$ J m^{-2}. Under these conditions $\gamma_A = \gamma_B$, giving a value of 1.325 J m^{-2} for the surface energies of both metal and oxide. This is close to the figure for Fe$_2$O$_3$ given in Table 1.1.

In the case of polymers, useful work has been done on adhesion in the presence of liquids. As already noted (and see also Section 3.4), the strength of adhesive

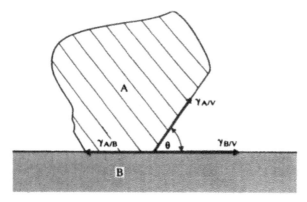

3.4 Equilibrium between a solid particle and a solid surface.

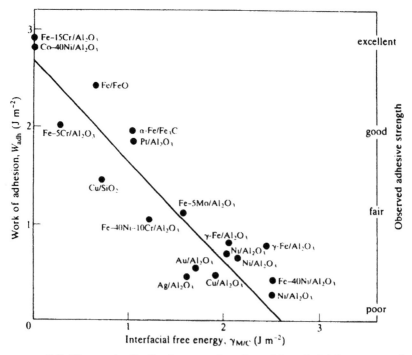

3.5 The work of adhesion as a function of interfacial free energy for various metal–ceramic interfaces (from Hondros, 1985).

bonds may be reduced if water is taken up by the polymer. Table 3.2 lists work of adhesions for various interfaces in air, water and organic liquids. Disbonding was observed for all cases where the work of adhesion was negative. Water caused disbonding for all metal–adhesive combinations, but none of the polymer–adhesive joints was so affected.

Table 3.2. The effects of liquids on the work of adhesion for various adhesive/adherend combinations

Adhesive–adherend combination	Environment	Work of adhesion (mJ m^{-2})	Disbonding in liquid
Epoxide–steel	Air	291	
	Ethanol	22	No
	Formaldehyde	−166	Yes
	Water	−255	Yes
Epoxide–aluminium	Air	252	
	Water	−157	Yes
Epoxide–silica	Air	178	
	Water	−57	Yes
Epoxide–carbon	Air	80–90	
Epoxide–fibre composite	Water	22–44	No
Vinylidene chloride–methyl acrylate	Air	88	
	Water	37	No

Source: from Adams *et al.* (1997).

3.3 Polymers

3.3.1 The chemistry of polymerization

In making an adhesively bonded structural joint there are three essential steps. The first is to prepare the surfaces to be joined in such a manner that optimum bond strength is obtained. The second is to apply the adhesive in the form of a liquid, so that intimate contact is obtained between adhesive and adherend surfaces. The third is to cure or harden the adhesive so as to give it the necessary strength and rigidity. The final step requires the evaporation of any solvent that may have been used, polymerization, and the formation of cross-linkages between the polymer molecules.

Polymerization is the process whereby **monomers**, which are small individual molecules, combine chemically with each other. It occurs by the release of valence bonds due either to the break-up of double bonds between two carbon atoms, or by the rupture of a ring structure. The simplest case is that of polyethylene:

$$\left[\begin{array}{c} H \quad H \\ | \quad | \\ C=C \\ | \quad | \\ H \quad H \end{array} \right]_n = \cdots -\begin{array}{ccc} H & H & H \\ | & | & | \\ -C-C-C- \\ | & | & | \\ H & H & H \end{array}\text{ etc.}$$

where the double bond in the ethylene molecule is broken to permit the formation of a long-chain polymer.

This type of reaction is known as **addition** polymerization. Alternatively, polymers may form by **condensation**. In this process, a chemical reaction takes place between groups of atoms which form part of the monomer, sometimes with the elimination of a simple molecule such as water. There must be two such reactive groups to each monomer molecule in order to form the polymer.

This type of reaction may take place between two like molecules, or between two different species. For example, epoxides are polymerized by mixing with a **hardener**, usually an amine:

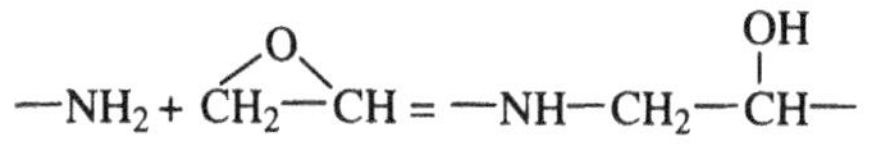

$$-NH_2 + \overset{O}{CH_2{-}CH} = -NH-CH_2-\underset{|}{\overset{OH}{CH}}-$$

In this instance the chemical bond between the monomers is made available by breaking the three-membered **epoxide ring** shown on the left-hand side of the equation. It is one of the advantages of epoxide adhesives that no by-product results from this reaction. Where water is generated during the cure, it is necessary to apply pressure in order to suppress porosity.

Addition polymerization is not quite so simple as was indicated earlier for polyethylene. The chain reaction requires an **initiator**, which may be a **free radical**, an anion or a cation. Free radicals are molecular fragments that have an independent existence for periods up to about one second, during which time they are chemically very active. In adhesive bonding the free radical is obtained by the reaction of an organic peroxide with ferrous ions:

$$Fe^+ + ROOH = Fe^{++} + RO^{\cdot} + OH^-$$

where ROOH represents a hydroperoxide and the dot indicates a free radical.

The free radical so formed then reacts with a monomer:

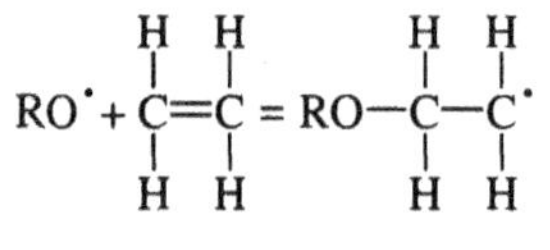

$$RO^{\cdot} + \underset{\underset{H}{|}}{\overset{\overset{H}{|}}{C}}=\underset{\underset{H}{|}}{\overset{\overset{H}{|}}{C}} = RO-\underset{\underset{H}{|}}{\overset{\overset{H}{|}}{C}}-\underset{\underset{H}{|}}{\overset{\overset{H}{|}}{C}}^{\cdot}$$

for example. A rapid chain reaction follows, and is arrested when two chains join together.

An adhesive of the addition polymer type therefore has three components, polymer, peroxide and ferrous-organic compound; if all are mixed together they may explode. Normal practice is then to apply a component to one adherend, and a mixture of the other two to the other. The polymerization reaction is initiated when the joint is closed. Other types of addition polymerization reactions proceed without the explosion hazard noted above, and in such cases the components can be pre-mixed.

3.3.2 Polymers for structural adhesives

Structural adhesives are usually proprietary formulations, and while the manufacturers will supply details of properties, method of use and generic type, details of composition are not necessarily disclosed. Some resins, notably epoxides and phenol–formaldehyde, tend to be brittle after curing, and their properties may be improved by the addition of rubbery compounds. Quite frequently two types of monomer will be mixed, and the hybrid polymer that results may have better properties than its parents. Apart from such complexities, adhesives are extremely varied in their form, chemistry and mode of use. They are therefore not easily lumped into categories, and are considered individually below.

3.3.3 Epoxides

Epoxide adhesives are extremely versatile: users range from the home carpentry enthusiast to the aerospace industry. The resin is a complex organic compound that contains the three-member epoxide ring discussed in Section 3.3.1. They bond to a wide range of adherends, including metals, but not untreated non-polar polymers such as polyethylene. Epoxides may be modified by mixing with a number of polymers. The pure resin is solid at room temperature, but mixed epoxide may be liquid. In either case the monomer melts at the curing temperature.

Polymerization is by condensation, and the curing agent is normally an amine. When aliphatic **straight-chain** amines are used the cure will take place at room temperature, but higher strength is obtained by a post-cure treatment at 80 °C for 4 h. Aromatic amines (those containing benzene rings) must be cured at elevated temperature; typically 150 °C for 2 h. Where epoxides are supplied as a two-part liquid, the composition of the hardener is usually adjusted so that optimum properties are obtained with a simple volumetric ratio between resin and hardener. Epoxides are also supplied in the form of film.

Two-component epoxide adhesives are not easy to handle on the production line and in particular do not lend themselves to automated operations. An alternative form is a single-component paste. This must be cured at elevated temperature, typically for 15 min at 180 °C or 5 min at 200 °C.

3.3.4 Phenol–formaldehyde

This was one of the earliest polymers to be developed. The reaction between phenol and formaldehyde takes place in two stages; in the first, polymerization occurs, then raising the temperature results in cross-linking with the elimination of water. In the unmodified condition, the polymer so formed is unsuitable for the bonding of metals, firstly because one of the components must be taken up in a solvent, secondly because steam is formed in the second stage of the reaction, and thirdly because the final product is brittle. Nevertheless, some useful structural adhesives are based on phenol–formaldehyde systems modified by the addition of elastomers. In this way the three problems noted above are reduced to manageable proportions. It is still necessary to form the bond under pressure to avoid porosity, and to cure at elevated temperature. The adhesive is taken up in a solvent and is applied to both the surfaces to be joined. After the solvent has evaporated the two parts are clamped together under pressure and heated. A typical curing pressure is 0.3 MN m^{-2} (45 psi), with a temperature of, say, 200 °C for 10 min. Phenolics are used for attaching items such as brake shoes and clutch pads.

3.3.5 Acrylics

Acrylic adhesives polymerize by a free-radical addition process: the basic monomer is methyl methacrylate:

$$CH_2 = \overset{\overset{\displaystyle CH_3}{|}}{C} - COOCH_3$$

but other components such as methacrylic acid may be added to improve bonding. A typical initiator is cumene hydroperoxide. If the two components are coated on to opposite surfaces and the joint is brought together under light pressure it will adhere until polymerization is complete and the full joint strength is obtained. Cure takes place at room temperature. Alternatively the two components are mixed immediately upstream of the application nozzle, or overlapping beads of adhesive and hardener may be applied to one component so that they mix when the joint is closed. Acrylic adhesives have self-evident virtues: they cure at room temperature without the need for pressure, they are applicable to metals, and have good strength (20–35 MN m^{-2} in the lap-shear test, similar to values quoted for epoxies).

Anaerobic adhesives are one-part acrylics that remain monomeric in the presence of air but polymerize in its absence. They are supplied in air-permeable containers. Since polymerization depends on the exclusion of air, anaerobics are best suited to the bonding of machined components. They have good resistance to solvents, and for these two reasons are used in automotive engine construction: for example in sealing between castings, in the retention of gear trains on their shafts. In aircraft they are used for fixing roller bearings in instruments.

Cyanocrylates are one-component adhesives that usually take the form of thin, colourless liquids, although they may also be obtained as a gel. The liquid is retailed as 'superglue' and is notorious for bonding together the fingers of unwary home enthusiasts. Polymerization is catalyzed by the water adsorbed on the surface of adherends. Initial hardening takes place in 1–10 s, and full strength is achieved in about 12 h at room temperature. Polymerization is inhibited by the presence of air, so it does not start until the joint is closed.

Cyanocrylates are capable of bonding difficult substances such as polyethylene without pretreatment, although for load-bearing joints adhesive manufacturers recommend the use of a primer. A typical application is the joining of plastic mouldings, such as for example may be used in making a pump for handling pharmaceutical products. Cyanocrylates are also used to bond components to printed circuit boards. The structure of ethyl cyanocrylates is shown below:

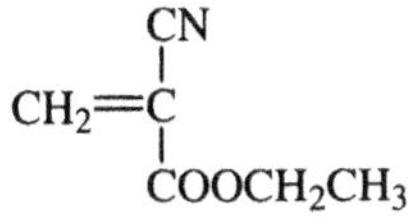

$$CH_2=C\begin{array}{c}CN\\|\\|\\COOCH_2CH_3\end{array}$$

3.3.6 Other structural adhesives

Polyurethane adhesives are available either as two components which harden rapidly after mixing, or as a single component that hardens through contact with atmospheric moisture. They are subject to loss of strength with prolonged exposure to moist conditions, but this defect may be mitigated by the use of a primer. Polyurethanes have a useful degree of flexibility when cured. They are used for bonding windscreens in the automobile industry, and for bonding large panels in coaches and caravans.

High-temperature adhesives based on bismalemide and polyimides are available for service temperatures up to 290 °C. They are cured under pressure at 175–200 °C, and provide useful strength (15 MN m^{-2} in the lap-shear test) up to the maximum allowable temperature. Such adhesives are expensive and not for general use.

In the present context, adhesive joints in timber would be required to resist all forms of outdoor exposure, including humidity and sunlight, and to give a useful life of, say, 50 years. Such requirements impose severe limitations on the number of polymers that may be used. On the other hand, wood is capable of absorbing water, so a wider range of condensation polymers is applicable. The main types are:

- resorcinol–formaldehyde;
- phenol–formaldehyde;
- melamine–formaldehyde;
- melamine–urea–formaldehyde.

The condensation reaction may be initiated by the application of heat, or by a hardener; for example, phenol–formaldehyde polymerizes at room temperature in the presence of an acid catalyst. In all cases it is necessary to hold the joint together under some pressure until completion of the cure. Hot-cured phenol–

formaldehyde adhesives require high pressure because of steam formation, but most others need only moderate force.

The adhesive may be in the form of film, or as a water-based liquid. Sometimes an organic solvent is used, and in such formations the solvent is allowed to evaporate before closing the joint.

3.4 The properties of adhesive polymers

3.4.1 General

If a basic physical property of a polymer, the molar volume for example, is plotted against temperature, it is found that there is a temperature above which the slope of the curve starts to increase. This point is known as the **glass transition temperature**, and is designated T_g. Other properties are similarly affected. Below this transition the polymer is elastic and may be somewhat brittle; above it the behaviour is increasingly rubber-like or plastic. There is an analogue here with the brittle–ductile transition temperature in steel. Both transitions represent the divide between relatively rigid and relatively plastic conditions.

It will be evident that if they are to withstand significant loads, then adhesive joints must operate at temperatures below T_g. Table 3.3 lists figures for various adhesive types together with an approximate guide to upper service temperatures, taken from Adams *et al.* (1997). The two sets of figures are quite close. Composition is the factor that determines T_g. Methacrylic acid has a high glass transition, and when added to acrylic adhesives it increases both polar activity and T_g.

Thermoplastic polymers are partly crystalline and, like crystalline solids generally, they have a melting point which is, of course, above the glass

Table 3.3. Glass transition temperatures and upper service temperatures for structural adhesives

Adhesive type	Glass transition	Upper temperature limit for service (°C)
Modified phenol–formaldehyde		90
Acrylic	20–120	up to 120
Epoxide	70–160	65–150
Epoxide, aliphatic, amine-cured	100	100
Epoxide, aromatic, amine-cured	160	150
High-temperature, bismalemide, polyimide	210–375	220–260
Rubber	−125 to −50	−

Source: from Adams *et al.* (1997).

transition. The thermosetting polymers used in structural adhesive joints do not melt; when heated to a sufficiently high temperature they decompose.

3.4.2 Mechanical properties

When tested in bulk form at room temperature and in dry conditions, the polymers used in structural adhesives behave in more or less the same way as metals. In a tensile test the material first extends elastically, then yields and deforms plastically until failure. It is possible to establish an elastic modulus, a yield strength and an ultimate strength. Figure 3.6 shows stress–strain curves for epoxide and polyurethane adhesives, and Fig. 3.7 gives elastic modulus as a function of temperature. So far as mechanical properties are concerned, unmodified epoxides and polyurethanes occupy opposite ends of the range, epoxides being relatively strong and brittle, whilst polyurethanes are relatively weak and flexible. Thus the latter type find applications such as the fixing of windscreens, as mentioned earlier, and for non-structural work such as the bonding of footwear.

The tensile and shear properties of adhesive polymers are an order of magnitude or more lower than those of metals. This is not of much consequence since the lap joint provides a relatively large area of contact. Under axial fatigue loading an adhesively bonded lap-shear joint in aluminium alloy sheet behaves in much the same way as unbonded material; that is to say, there is a linear relationship between stress and log (cycles to failure). At higher stresses failure is through the adhesive, but at lower stress it occurs through the metal. The stress required to cause failure through the adhesive is, however, substantially higher

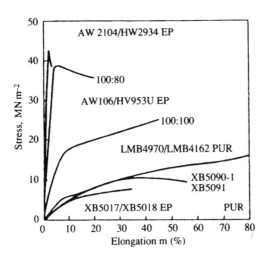

3.6 Tensile properties of polyurethane (PUR) and epoxy (EP) adhesive formulations (from Simon and Lehmann, 1991).

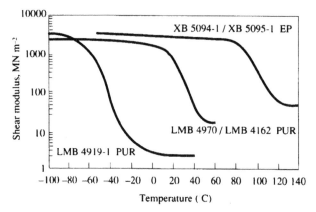

3.7 Shear modulus as a function of temperature for polyurethane (PUR) and epoxy (EP) adhesive formulations (from Simon and Lehmann, 1991).

than that normally used for such joints, so fatigue failure would not be expected in adhesive joints. There was evidence of disbonding of such joints in tests simulating the failure of the early 'Comet' aircraft, but this was associated with severe strain concentration and fatigue failure of the structure.

3.4.3 Effect of water

The disbonding of various adhesive joints in water has been reported earlier in Table 3.2. All the joints with metal adherends failed, while the two polymer-to-polymer testpieces remained intact. These results reflect general experience; it is adhesive joints with one or two metal adherends that are adversely affected by water. It would appear that when water molecules arrive at the adhesive–metal oxide interface some are adsorbed in preference to the polymer, causing a reduction in bond strength.

In the laboratory the effect of water on the adhesive bond may be measured by holding a specimen in moist air under constant or variable load, and determining the time to failure or, more commonly, exposing the specimen to moisture for a given period and then measuring the proportional reduction in strength. Figure 3.8 is an example of the second type of test. During the initial period of exposure there is a relatively rapid fall in strength; then it stabilizes at a constant level. Moreover, if the humidity is reduced, there is some recovery of strength, indicating that the damage process is to some degree reversible. As pointed out earlier, failure after exposure to moist conditions takes place partially or wholly across the interface; the effect of water appears to be that of weakening the polymer–oxide bond rather than causing disbonding.

Time-dependent failure may also occur under dry conditions; that is to say, if an adhesively bonded lap-shear joint between metal adherends is held under

The use of adhesives for making structural joints   69

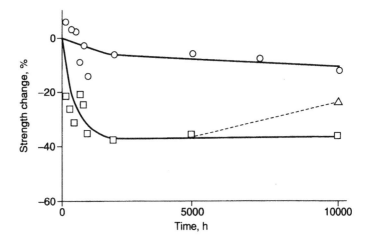

3.8 Strength of joints between aluminium alloy adherends made with a nitrile phenolic adhesive. Exposed to air at 50 °C: ○ with 50% relative humidity; □ with 100% relative humidity; △ after a further 5000 h in dry air (Brewis et al., 1987).

constant load it may fail after an interval of time. On carrying out such tests with different adhesives, it has been found that there is a straight-line relationship between time to failure and the specific fracture energy of the adhesive G_{IC} (a brief description of fracture energy testing is given in Section 3.5.5). When similar tests are carried out under wet conditions a line of the same slope is found, but displaced to shorter times and correspondingly lower values of G_{IC}, the gap between the two curves being about $50\,J\,m^{-2}$.

It is reasonable to suppose that the time-dependent character of bond failure in the presence of a humid atmosphere results from the fact that in an adhesively bonded metal-to-metal joint, water must diffuse into the adhesive from the exposed surface. For one-dimensional flow, the diffusion of a solute (in this case water dissolved in a polymer) is given by

$$\frac{\partial c}{\partial t} = D\frac{\partial^2 c}{\partial x^2} \tag{3.8}$$

This equation has precisely the same form as that for the diffusion of heat in a solid, except that temperature is replaced by concentration (c) and the coefficient of heat diffusivity is replaced by the volume diffusion coefficient D. Therefore, for any given boundary conditions, solutions to the equation for heat diffusion, which are to be found in, for example, Carslaw and Jaeger (1959) may be adapted for the diffusion of solutes.

Figure 3.9(a) illustrates the case of an infinite bar or strip of adhesive whose initial water content is zero, exposed at time $t = 0$ to a humid atmosphere giving an initial equilibrium water concentration at the surface equal to c_e. Then after

70 Metallurgy of welding

time t the average water content c_{av} of the adhesive relative to its equilibrium content is obtained from

$$1 - \frac{c_{av}}{c_e} = \frac{8}{\pi^2} \sum_{h=0}^{\infty} \frac{1}{(2n+1)^2} \exp\left[-(2n+1)^2 \pi^2 Dt/4x^2\right] \qquad (3.9)$$

In practice testpieces have the form shown in plan view by Fig. 3.9(b), with a block of adhesive exposed on all four sides. In such a case the water content is obtained by multiplying that due to diffusion along the x direction with that due to flow along the y-axis:

$$\left[1 - \left(\frac{c_{av}}{c_e}\right)_{x,y}\right] = \left[1 - \left(\frac{c_{av}}{c_e}\right)_x\right] \times \left[1 - \left(\frac{c_{av}}{c_e}\right)_y\right] \qquad (3.10)$$

Some correlation would be expected between the water uptake of an adhesive joint and its loss of strength. In Fig. 3.10 the strength loss is plotted against time of exposure to moist air. The continuous line on this diagram is the water uptake calculated from equations 3.9 and 3.10, with the scales adjusted to match the strength loss obtained. The correlation appears to be very good indeed.

Values of the diffusion coefficient and equilibrium solubility may be obtained by immersing thin films of adhesive in water and measuring the uptake as a function of time. Some figures that have been obtained in this way are listed in Table 3.4.

The mechanism by which water affects the bond strength in adhesive joints remains a matter for speculation; the fact that strength is recovered, at least to some degree, when the joint is exposed to drier conditions (Fig. 3.8) means that disbonding cannot be the sole cause. On the other hand, failure shifts from the

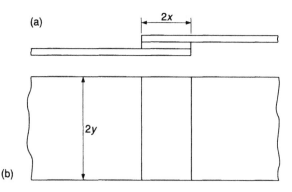

3.9 (a) Section of infinite bar of adhesive polymer width 2x protected top and bottom but exposed laterally to humid atmosphere. (b) Rectangular block of adhesive polymer dimensions 2x × 2y, otherwise similar.

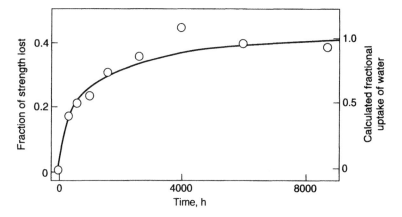

3.10 Comparison between percentage strength loss on wet exposure with calculated water uptake for a joint made with epoxide adhesive (Comyn *et al.*, 1979).

Table 3.4. Coefficient of diffusion and equilibrium solubility of water in adhesive polymers

Adhesive	Temperature	Coefficient of diffusion (10^{-2} m^{-2} s^{-1})	Equilibrium solubility c_e (%)
Nitrile phenolic	25	3.3	1.5
Vinyl phenolic	25	1.8	3.5
Toughened acrylic	23	0.19	1.7
Urea formaldehyde (wood adhesive)	25	0.25	7.4
Epoxide	20	0.24	1.4
	40	0.65	1.6
	60	1.8	2.2
	90	6.1	6.9

Source: from Adams *et al.* (1997).

adhesive to the interface as the water content increases, so the problem is due to a weakening of the adhesive–adherent bond. One possible reason for such a reduction has already been suggested; namely that water increases the permittivity of the polymer. Theoretical expressions for both ionic and van der Waals forces indicate that they are reduced by increased permittivity. Alternatively, water molecules may be preferentially adsorbed at the interface, and may likewise be desorbed when the adhesive dries, allowing bonds between polymer and oxide to re-form.

3.5 Bonding procedures

3.5.1 General

The usual sequence for making an adhesive joint is:

1 surface cleaning;
2 surface preparation;
3 application of adhesive;
4 closing the joint;
5 curing.

Initial cleaning is required to remove loose material, scale, oil or other contaminant that could interfere with bonding. Some adhesives are compatible with oil, and may, for example, be applied to steel in the as-received condition. In general, however, this is not the case, and initial cleaning is required.

Surface preparation of metals consists of etching or, for aluminium and titanium, anodizing. The sole purpose of such treatment is to improve resistance to degradation by water. In many cases degradation by water is not a problem, either because (as in some automobile applications) water is not present, or because the operating stress is low, or for some other reason. Therefore this step may not be required. In the case of polymers of low surface energy, corona discharge or similar treatment may be applied, alternatively the joint is made using a cyanoacrylate adhesive.

A means of dispensing the correct quantity of adhesive in the right place is a primary essential. The systems employed may be adapted for manual or automatic operations, and these may take a number of forms, as will be described later.

Components are normally held in a jig, or press, so that when the joint is closed, alignment is correct. Where pressure must be used curing will take place in the press; otherwise, when required, it may be done in a paint oven.

3.5.2 Surface cleaning and surface preparation

Details of cleaning methods and of etching and anodizing, should these be required, are set out below.

Aluminium

The adherends will normally be a precipitation-hardened alloy in sheet form. Etching may follow the following procedures:

1 Degrease in an organic vapour, or alternatively in an aqueous solution of proprietary detergent.

The use of adhesives for making structural joints 73

2   Immerse in an aqueous solution containing 5% by weight chromium trioxide and 15% by volume concentrated sulphuric acid. Hold for 30 min at 60 °C.
3   Spray rinse in cold tap water.
4   Dry in warm air.

Anodizing may be carried out in an aqueous solution containing 5% chromium trioxide with a wetting agent. The metal is first degreased, as described for etching. A typical anodizing procedure in chromic acid solution at 40 °C is to raise the voltage to 40 V in 10 min, hold for 20 min, then raise to 50 V in 5 min and hold for another 5 min.

Alternatively anodizing is carried out in a phosphoric acid bath; this process is generally favoured in the USA.

Anodizing thickens the oxide film and develops a porous or honeycomb structure, sometimes with whiskers (Fig. 3.11). Adhesives penetrate the pores provided that the oxide layer is not more than 1 μm in thickness. Anodized films may be attacked by water; this was presumably the case in the test for which the result is shown in Fig. 3.12. Such cases are exceptional, and would show up in control tests such as the Boeing wedge test (see Section 3.5.5).

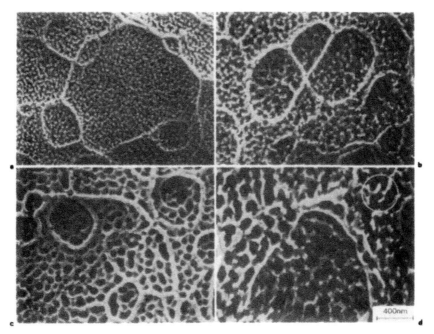

3.11 Scanning electron microscope pictures of aluminium surface anodized for various times in 10% phosphoric acid: (a) 1 min; (b) 5 min; (c) 10 min; (d) 15 min (circled area shows residual spikes of oxide) (from Kosma and Olefjord, 1987).

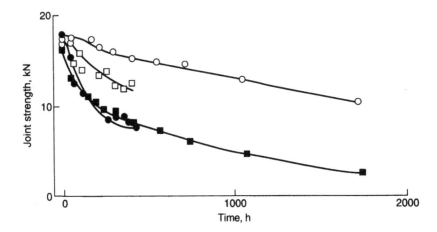

3.12 Strength of aluminium joints bonded with an epoxide–nylon adhesive after exposure to air at 97% relative humidity and at 43 °C. Surface treatment: ○ chromic–sulphuric acid etch; □ alkaline etch; ● anodized in phosphoric acid; ■ solvent-degreased (Adams et al., 1997).

Titanium

Titanium and its alloys (commonly Ti–6Al–4Va) may suffer hydrogen embrittlement during acid etching, so the operation is best performed under oxidizing conditions. Anodizing may also be carried out in a chromic acid bath.

A typical alkaline etch procedure is:

1 Vapour degrease.
2 Hold for 20 min in an aqueous solution containing 2% by weight sodium hydroxide and 2% by volume hydrogen peroxide, added immediately before immersing the metal parts. Hold temperature at 70 °C.
3 Wash for 10 min in hot water.
4 Dry in hot air.

Alternatively, the metal is pre-etched in 30% nitric acid plus $2\frac{1}{2}$% hydrofluoric acid, then anodized in a chromic acid bath.

Steel

Included under this heading are low carbon (mild) steel, ferritic stainless steel and austenitic chromium–nickel steels. The surface treatment is complicated by two problems. Firstly, after acid etching the surface is covered by a layer of carbon, which must be removed in a 'desmutting' bath. Secondly, in the case of mild steel, a non-adherent rust will form when clean wet surfaces are exposed to air, so

The use of adhesives for making structural joints 75

the metal must be de-watered in isopropanol. The following is a typical procedure for mild steel. Stainless steel is treated in a generally similar way, but details must be in accordance with manufacturers' recommendations.

1 Air blast or brush to remove loose deposits.
2 Grit blast.
3 Etch in 4% hydrofluoric acid at room temperature.
4 De-smut in a bath containing 6% sulphuric acid and 10% chromic acid.
5 Wash in tap water.
6 Transfer rapidly to an isopropanol bath.
7 Treat in dry isopropanol.
8 Dry in warm air.
9 Prime immediately, or maintain in dry conditions.

Composites

The surface of glass fibre or carbon fibre reinforced polymers may be coated with a mould release compound, which must be removed by means of a solvent prior to further treatment. Such treatment may be carried out using a flame, corona discharge, excimer laser or other excitation source. Equally good results are obtained by simple abrasion, for example by rubbing with an abrasive paper.

An alternative method is to use a **peel ply**. Fibre reinforced composite shapes are made by laying down resin-impregnated plies of fibre inside a mould (or on a mandrel in the case of pipe) until the required thickness is obtained. The body is then removed from the mould and cured. In the peel ply technique the first one or two plies are impregnated to a lesser degree, and preferably with a different type of resin. After curing the outer ply may be stripped off, leaving a surface to which adhesive will bond readily. The stripping operation is performed immediately prior to bonding.

3.5.3 Dispensing systems for adhesives

Most dispensers comprise the following components:

1 A pressurized reservoir containing the adhesive.
2 A tube leading from the reservoir to a control valve.
3 A tube leading from the control valve to the dispensing nozzle.

Pressure may be applied by means of compressed air, or by a piston, or other mechanical device. For two-component adhesives, two reservoirs are required, and these may lead to a mixer upstream of the valve, or overlapping beads may be laid on one adherend, as indicated earlier for acrylics. The amount of adhesive dispensed is determined by the applied pressure and opening time of the valve.

76 Metallurgy of welding

The product may be applied in various forms: as an individual drop or dot, as a series of dots, or as a continuous bead. The complete surface may be coated by means of a spray or it may be transferred by a roller. There are systems where the component is moved relative to the nozzle, and vice versa. Screen printing is used where adhesives are employed as gaskets. The screen consists of a plastic sheet in which holes of the desired form have been punched. This is placed on the metal surface to be gasketed, and the polymer is applied using a roller.

The range of products for which structural adhesives may be used is very wide, and the manner of their application is correspondingly varied.

3.5.4 Quality control

The non-destructive testing of adhesive joints between metal adherends is of little or no practical value. In radiography, the contrast between voids and solid adhesive is completely obscured by the metal. Much effort has been applied experimentally to the use of ultrasonics, but with no significant result. It is necessary therefore to rely on other control methods. The first is the establishment of a written procedure for each operation, analogous to a welding procedure specification. The second, which is applied in the case of bonding sheet metal, is to make lap-shear tests at specified intervals during production; at the beginning and end of each shift, for example. A similar problem arises in the case of resistance spot welding, for which there is no effective non-destructive testing method; in this case a similar control technique is used.

Exceptionally, a non-destructive test may be applied during production. In one critical application, where a gear ring was applied to a drive shaft using a combination of shrink fit with an anaerobic adhesive, an essential step was the application of a uniform layer or adhesive to the shaft. A fluorescent dye was added to the adhesive, and the coated shaft was viewed by an automatic camera using ultraviolet light. Any defect in the coating could be detected and signalled in this way.

The lap-shear test, together with other tests applicable to adhesive joints between metals, is described below.

3.5.5 Tests for adhesively bonded joints

The lap-shear test is illustrated in Fig. 3.13. It was originally made from two pieces of 1.6 mm metal strip one inch in width, bonded together over a length of half an inch. These dimensions have now (in Europe) been replaced by the metric equivalents shown in the figure. This testpiece is loaded to failure in a normal tensile testing machine. In spite of the alignment tabs shown, the central part of the sample is eccentrically loaded, and it deforms in the manner indicated. At failure the adhesive is subject to a combination of shear forces with forces acting at right angles to the metal surface; these are known as **peel** forces.

The use of adhesives for making structural joints 77

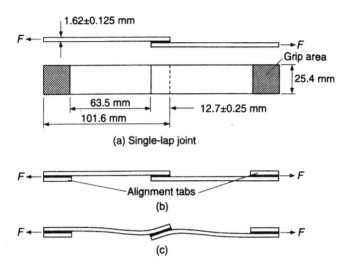

3.13 (a) Lap-shear adhesively bonded joint with dimensions in accordance with ASTM D1002. (b) Optional use of bonded tabs to improve alignment. (c) Typical distortion due to tensile loading. Such distortion occurs also when alignment tabs are used (Adams et al., 1997).

Various tests may be employed to measure the peel strength of an adhesive. The simplest and most commonly used type is that illustrated in Fig. 3.14. Peel strength is the load at fracture divided by the area.

Adhesives that give satisfactory results in the lap joint test (unmodified epoxides for example) may have a low peel strength. The addition of elastomers

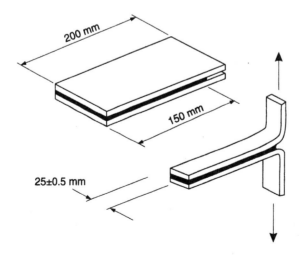

3.14 Peel test in accordance with ASTM D1876.

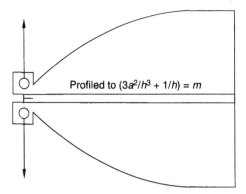

3.15 Double cantilever specimen for the determination of specific fracture energy of an adhesively bonded joint (Adams *et al.*, 1997).

to such types of adhesive usually improves their performance in the peel test. Adhesives treated in this way are said to be **modified**.

It might well be supposed that the addition of rubber to a normally glassy polymer would also improve its fracture toughness. The fracture toughness of an adhesive polymer is measured using a double cantilever beam arrangement, one version of which is shown in Fig. 3.15. The cantilever arms are profiled to give a constant strain energy release rate when fracture is initiated, regardless of the length of the initial crack. This makes it possible to measure the specific fracture energy G_{IC} directly, using the formula

$$G_{IC} = \left(\frac{4F_c^2}{b^2 E}\right) m \qquad (3.11)$$

and

$$m = \frac{3a^2}{h^3} + \frac{1}{h} \qquad (3.12)$$

Here F_c is the applied force at fracture, b is beam width, E is the elastic modulus of the adhesive, a is crack length and h is the beam height at the tip of the crack.

Figure 3.16 shows values of G_{IC} obtained in this way for a bismalemide (high-temperature resistant) adhesive with increasing additions of rubber. Naturally, the higher fracture toughness will be accompanied by a reduction in temperature resistance and lap-shear strength.

To return to the peel test; it will be evident that elastomer additions affect fracture toughness and peel test results in the same way. It seems likely that a low peel test result indicates low fracture toughness.

Another opening-type test that may be utilized for control purposes is the **Boeing wedge test**. The principle of this test, which is standardized in ASTM D3762, is illustrated in Fig. 3.17. A wedge is driven into the bond-line of a lap

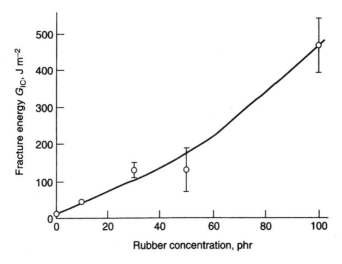

3.16 Specific fracture energy of a modified bismalemide adhesive as a function of rubber content (Shaw and Kinlock, 1985).

3.17 Principle of the Boeing wedge test (Adams *et al.*, 1997).

joint and the rate at which the crack grows is noted visually. Finally the joint is broken open and the proportion of interface failure is measured. This test shows up deficiencies in surface preparation, and may be used to assess the durability of a joint under wet conditions.

3.5.6 Other tests

In routine testing it is customary, notably in the aircraft industry, to use relatively thin adherends. The results obtained may thus be affected not only by flexure of the joint, but also by elastic deformation of the metal to which the bond has been made. Therefore, when it is required to measure the properties of the adhesive a different type of test is required. An apparently simple solution is to make testpieces from bulk samples of the polymer. However, there are difficulties in obtaining a satisfactory cure of such samples, and in practice most tests are carried out on joints between adherends that are thick enough to avoid distortion problems. One type of testpiece is sketched in Fig. 3.18; this is the **napkin ring**

80 Metallurgy of welding

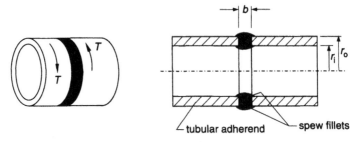

3.18 Napkin ring test (Adams et al., 1997).

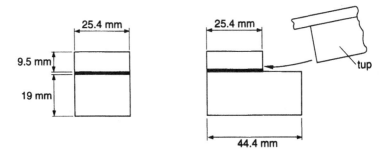

3.19 Impact test specimen for adhesively bonded joint, in accordance with ASTM D950 (Adams et al., 1997).

test, which may be used to measure shear strength and shear modulus. A similar type of test may be made using solid bar instead of rings. Alternatively, a shear-lap joint is made between thick plates and tested using a normal tensile machine, the joint being so disposed that distortion is minimized.

Various dynamic tests have been developed for adhesive joints. A method of measuring the specific fracture energy is described in the previous section. The impact test specimen specified by ASTM is illustrated in Fig. 3.19. This is broken in a pendulum-type impact machine, and the energy absorbed by the fracture is measured. As in fracture energy testing, good impact values are obtained with rubber-modified formulations. Such properties are desirable for adhesive joints used in motorcar bodies. Creep tests are also standardized, in ASTM D1780, for example.

3.6 Joint design and applications

3.6.1 Joint design

It will be evident from the preceding discussion that adhesively bonded joints should be designed to give a sufficient area of overlap to withstand shear forces, combined with a degree of rigidity such that peel forces are minimized. Rigidity

The use of adhesives for making structural joints 81

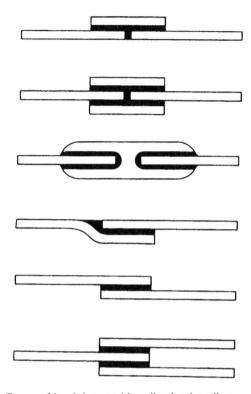

3.20 Types of lap joint used in adhesive bonding.

may be provided by reinforcement; the double lap joint shown in Fig. 3.20 combines reinforcement with maximum bond area. For aluminium construction extrusions may be a simple means of providing such configurations, while at the same time facilitating assembly.

Figure 3.21 shows three examples of design that incorporate these principles. In the first (a) extrusions are used to make butt and corner joints. The second (b) uses box sections with the corners being joined by an internally fitted elbow. The third (c) is a reinforced plate-type corner joint, and the last sketch (d) represents a honeycomb structure, first developed for aircraft use (as floor panels for example) and subsequently used in other applications. Although individual joints in this type of construction are not particularly strong, they are numerous, and rigidity is an inherent characteristic of this configuration.

3.6.2 Applications

The use of structural adhesives is by no means confined to the bonding of metals in the engineering sector of industry. The use of cyanoacrylates to fasten

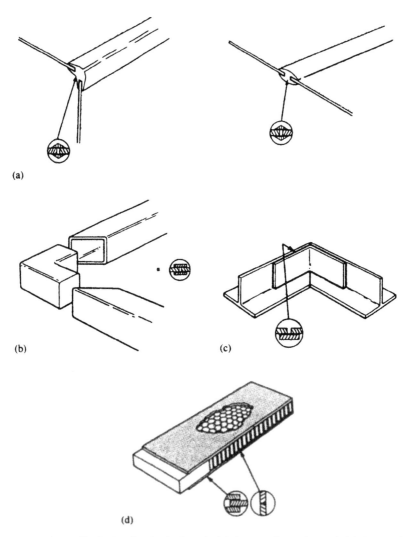

3.21 Typical adhesively bonded joint configurations: (a) joints using profiles; (b) bonded framework, using plug to increase bond area; (c) bonded angle used to increase joint area; (d) bonding of multilayer structure.

components to printed circuit boards has already been mentioned; the same adhesive is used in the manufacture of caps for toiletry bottles, where its low viscosity and instant bonding properties make high production rates possible. This section, however, is concerned with those applications, mainly in the aircraft and automotive industries, where the joints carry a significant load. Such joints may be divided into two main categories; those of cylindrical geometry, on the one hand, and lap joints between sheet metal parts, or between sheet metal and extruded sections, on the other.

Cylindrical joints are mainly required for internal combustion engines, drive shafts, and for bearings. Adhesive bonding makes it possible to dispense with mechanical devices such as keys and cotter pins. In the absence of dynamic loading, it may permit the use of a clearance rather than an interference fit, or it can reduce the amount of shrinkage required, so that the locked-in stresses are lower. One such case was mentioned in the section concerned with quality control; in this instance electron beam welding was considered as an alternative, but was rejected on the grounds of high capital cost, together with possible degradation due to the welding thermal cycle. Such considerations will often favour adhesive bonding as a joining method.

The bonding of flat metal is a more familiar use for adhesives. It is employed for attaching stiffening panels to the bonnets and tail-gates of motor vehicles, for internal panels in road and railway coaches, for the bonding of internal stiffening rings to aircraft fuselages, and for the bonding of honeycomb structures. Some other more specialized applications have been mentioned earlier; the attachment of clutch pads, for example. Adhesive bonding has a number of advantages over riveting or spot welding for the attachment of stiffeners. The joint is continuous and local stress concentrations are eliminated. There is a complete seal, so the risk of crevice corrosion is minimized. There are no surface indentations or discontinuities, and in most cases production costs are lower.

For secondary joints of the type described here adhesives have great advantages, and with continued technological development there is no doubt that increasing numbers of applications will be found. There remains one major limitation: adhesives are not suitable for butt joints, and their strength is one or two orders of magnitude lower than that of metals. Thus they are very unlikely ever to be used for making primary joints in metallic structures. For such purposes alternative methods, such as those described elsewhere in this book, must be sought.

References

Adams, R.D., Comyn, J. and Wake, W.C. (1997) *Structural Adhesive Joints in Engineering*, 2nd edn, Chapman and Hall, London.
Brewis, J.D., Comyn, J. and Trednell, S.T. (1987) *International Journal of Adhesion and Adhesives*, **7**, 30.
Carslaw, H.S. and Jaeger, J.C. (1959) *The Construction of Heat in Solids*, Clarendon Press, Oxford.
Comyn, J., Brewis, D.M., Shalash, R.J.A. and Tegg, J.L. (1979) *Adhesion-3*, p. 13, Chapman and Hall, London.
Comyn, J., Blickley, D.C. and Harding, L.M. (1993) *International Journal of Adhesion and Adhesives*, **13**, 163.
Hondros, E.D. (1985) in *Proceedings of Hard Metal Conference*, Adam Hilger, Bristol.
Kosma, L. and Olefjord, I. (1987) *Material Science and Technology*, **3**, 860–873.
Shaw, S.J. and Kinloch, A.J. (1985) Ibid. **1**, 123.
Simon, H. and Lehmann, H. (1991) *Joining Sciences*, **1**, 23–27.

Further reading

Adams, R.D., Comyn, J. and Wake, W.C. (1997) *Structural Adhesive Joints in Engineering*, 2nd edn, Chapman and Hall, London.

Kinloch, A.J. (1987) *Adhesion and Adhesives—Science and Technology*, Chapman and Hall, London.

4
Soldering and brazing

4.1 Physical aspects

4.1.1 General

In principle, one of the simplest ways of joining two metal surfaces is to flow liquid metal between them and allow it to solidify. There are significant advantages to be gained from such a procedure. There is no bulk melting, distortion is minimized, and the thermal cycle is relatively benign. Two essential requirements must be met if a satisfactory joint is to be made in this way. Firstly, as noted in Chapter 1, the temperature of the solid at the solid–liquid interface must be at least equal to the melting point of the liquid. Secondly, there must be some means of removing the oxide film on the surface of the solid so that the liquid can bond to a clean metal surface.

There are three variants of this method: flow welding, brazing and soldering. In flow welding the composition and melting point of the liquid phase are similar to those of the solid, and the interface is heated to the required temperature by conduction and convection from the liquid. Flow welding processes are described in Chapter 6. The remainder of this chapter is concerned with the brazing and soldering of metals. The use of such processes for joining non-metallic solids to themselves and to metals is covered in the next chapter.

Brazing uses a liquid metal that has a melting point substantially lower than that of the parts to be joined. In the case of steel, for example, copper, copper-based alloys and nickel-based alloys with melting temperatures in the range of 450–1200 °C are employed. The solid is heated by means of a flame or induction coil, or in a furnace or salt bath. Surfaces are cleaned by means of a flux or by a reducing atmosphere. The brazing metal is usually drawn into the joint space by surface tension forces (by **capillary action**). The joints made by brazing are strong, and may be incorporated into structures; for example, brazing is the traditional way to join the parts of a bicycle frame.

Solders are typically lead–tin alloys having a melting point of less than 450 °C. Heating is usually done with an electric resistance element or gas torch, while the solid parts to be joined are, as in flow welding, heated by the liquid metal. Solders have low strength, and are mainly used for electrical connections and for other applications where high strength is not necessary.

4.1.2 Capillary action

The vertical height H to which a liquid rises between two parallel plates separated by a distance d is given by

$$H = \frac{2\gamma_{L/V} \cos \theta}{\rho d g} \tag{4.1}$$

where θ is the contact angle and ρ is the density of the liquid.

Also, the velocity of flow, v, after the liquid has risen to height h into the space between two parallel plates separated by distance d is

$$v = \frac{\gamma_{L/V} d \cos \theta}{4\eta h}$$

where η is the viscosity of the liquid. When $h = 0.5H$, so that the joint is filled to half the theoretical maximum,

$$v = \frac{d^2 \rho g}{4\eta} \tag{4.2}$$

and if d is 0.15 mm, and (for liquid copper) $\rho = 8 \times 10^3 \text{ kg m}^{-3}$ and $\eta = 3.5 \times 10^{-3} \text{ kg m}^{-1} \text{ s}^{-1}$ (Lancaster, 1986) then

$$v = 126 \text{ mm s}^{-1} \tag{4.3}$$

Also for liquid copper at the melting point $\gamma_{L/V} = 1.33 \text{ N m}^{-1}$ (Keene, 1991) and assuming $\cos \theta = 1$

$$H = 220 \text{ mm} \tag{4.4}$$

In practice it is difficult to achieve a penetration of more than 50 mm. In part this is due to interalloying, which decreases the velocity and eventually stops the flow. Also, from equation 3.2

$$\frac{dv}{dT} = \frac{d}{4h\eta^2}\left(\eta \frac{d\gamma}{dT} - \gamma \frac{d\eta}{dT}\right) \tag{4.5}$$

The relevant quantities for liquid copper are $\gamma = 1.33 \text{ N m}^{-1}$, $d\gamma/dT = -2.3 \times 10^{-4} \text{ N m}^{-1} \text{ K}^{-1}$ (Keene, 1991) and $\eta = 3.5 \times 10^{-3} \text{ kg m}^{-1} \text{ s}^{-1}$, $d\eta/dT = -6 \times 10^{-6} \text{ kg m}^{-1} \text{ s}^{-1} \text{ K}^{-1}$ (Lancaster, 1986). Then for $d = 0.15 \text{ mm}$ and $h = 5 \text{ mm}$

$$\frac{dv}{dT} = 4.4 \text{ m s}^{-1} \text{ K}^{-1} \tag{4.6}$$

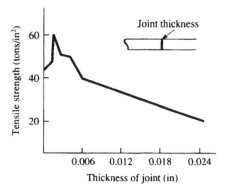

4.1 Relationship of tensile strength to joint clearance of silver–copper–zinc–cadmium joints in stainless steel.

Thus the flow velocity is predicted to rise with temperature. For the stated conditions a rise in temperature of 50 K is predicted to increase the viscosity by 220 mm s^{-1}. In practice there is such a tendency: brazing metal flows to the hottest part of the joint, and if the temperature gradient is excessive and in the wrong direction the brazing metal will flow out of the joint against the natural capillary flow (Connell and Sajik, 1992). This is a potential difficulty for flame brazing, but not for furnace brazing where such temperature gradients are improbable. In flame heating it is easier to obtain a uniform temperature with circular joints than it is with a straight flat seam. Where a flux is used, non-uniform heating tends to result in flux inclusions.

There is an optimum **clearance** (separation of the joint faces) for which the highest strength is obtained. Figure 4.1 shows the tensile strength of a butt joint in austenitic stainless steel made with an Ag–Cu–Zn–Cd (silver solder) brazing filler. The maximum strength is obtained with a clearance of 0.0017 in. (0.04 mm), but with such a gap joint filling would be poor. For steel and copper the optimum clearance lies in the range 0.05–0.40 mm, and for aluminium and magnesium alloys it is 0.125–0.625 mm, depending on the process used. The wider spacing for light alloys is consistent with Equation 4.1.

4.2 Soldering

4.2.1 General

Soldering is an ancient art, and excellent results are obtained by skilled operators using a soldering iron. However, the requirement for increased productivity has led to the introduction of machine soldering, which in turn requires the use of sophisticated techniques in order to maintain quality. This is very much the case in the machine soldering of components mounted on a printed circuit board

(PCB). Such assemblies are used in military and aerospace applications, and the failure of a single joint (typically one of a thousand or more on a single unit) could have serious consequences. Therefore the more responsible manufacturers aim to eliminate defects completely. This objective can only be achieved if all parts are free from contamination by surface films of grease and dirt of any kind, and if oxide layers are thin enough to be completely removed by the flux. Ideally all parts should be freshly tinned prior to the soldering operation.

In practice it is usually sufficient to coat one of the components (in the case of the printed circuit board, the board itself) and if the other surface is clean copper the solder will, in combination with a suitable flux, wet it and make a satisfactory joint. The flux removes residual oxide on the copper surface and enables a metallic bond to be formed. Moreover, tin in the solder alloys with the copper and forms an interlayer of Cu_6Sn_5 (or Cu_3Sn, where the tin has been consumed). Such intermetallic compounds are brittle but do not affect the integrity of the joint unless they become excessively thick, through age or overheating, for example.

In the soldering of ferrous materials one of the functions of the flux is to act as an electrolyte: tin is dissolved from the solder and deposited on the metal surface, where it forms an alloy and allows the solder to spread on a pre-tinned surface.

4.2.2 Joint design

Given proper conditions, the solder will spread with a zero or small contact angle. However, in most soldered joints there is a gap in which the capillary forces act to draw in the solder. Figure 4.2 illustrates some typical soldered joints in sheet metal and for electrical connections. Solder has low strength, and where the joint is subject to tensile forces it is necessary to provide additional support by riveting, spot welding or a mechanical interlock. But such joints are rarely used in soldering.

4.2.3 Solders

Solder is primarily an alloy of lead and tin, and although other elements may be added, it is the tin/lead ratio that governs so far as properties are concerned. Figure 4.3 shows the binary lead–tin constitution diagram. Most solders have a composition close to that of the eutectic (61.9% by mass of tin for pure components: in solder the presence of impurities makes it difficult to define the eutectic composition). There is little advantage gained in increasing the tin content above the eutectic level, and since tin is the costly element, compositions range from 63% Sn downwards. The 63/37 alloy is used for machine soldering, where maximum flowability and minimum risk of solidification cracking due to vibration is required. For manual soldering the

Soldering and brazing 89

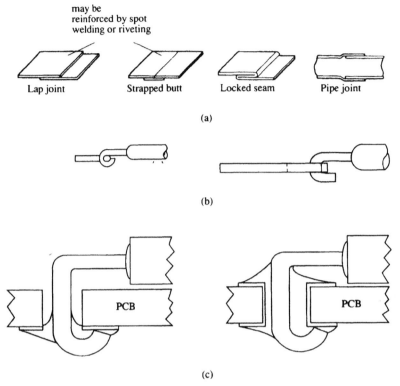

4.2 Soldered joints: (a) sheet metal and tubing; (b) mechanical and pierced electronic connections; (c) printed circuit board connections.

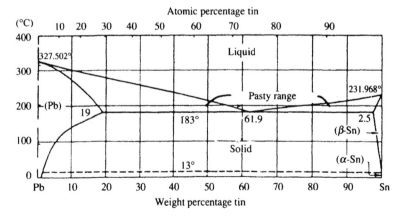

4.3 Binary lead–tin constitution diagram (from Bilotta, 1985).

composition is not too critical and the 50/50 alloy is a good choice. As the tin content is reduced the solidification temperature range increases; there is partial solidification and the effective viscosity is higher. Thus, in times past when lead pipes were used for domestic water, the 40/60 or 35/65 alloy, with a pasty range of about 100 °C, could be employed for making wiped joints.

Some specifications permit a small antimony content, which allows manufacturers to make use of secondary metal derived from bearing metals. Such solders have slightly inferior flowability and higher strength. Silver may be added in order to reduce the solution of silver from silver-plated items. Silver also improves the tensile and creep properties.

In machine soldering it is necessary to control the amount of all impurities in order to achieve the desired 100% reliability level. Table 4.1 shows the limits imposed by the US MIL PIC 815 A together with proposals by Woodgate (1988). At impurity levels approaching the limits shown in the table, the appearance of the joints could be affected but not their integrity. Future developments are likely to be in the direction of still higher purity levels.

Solder is available in bulk form as pig, ingots or bars for use in machine work. Manual soldering is done with either solid or flux-cored wire, the flux being either rosin or one of the organic types. There is a range of preferred shapes for machine work, and it is also supplied as paste, which is a mixture of powdered solder and flux.

4.2.4 Fluxes

Soldering fluxes must dissolve or displace oxide and, if possible, have a positive effect in promoting rapid wetting and spreading. The least active type of flux is composed of a solution of rosin in petroleum spirit. Rosin consists largely of abietic acid, which becomes active at the soldering temperature but reverts to an inert, non-corrosive form on cooling. Various organic compounds are used as fluxes, either alone or mixed with rosin; examples are the hydrochloride of glutamic acid and hydrazine hydrobromide. Such materials decompose at soldering temperatures, and the residues are easily washed off with water. More recently 'low-solid' fluxes have been developed for machine soldering that, it is claimed, need not be removed after soldering, or require minimum cleaning. The soldering of steel requires a zinc chloride flux, which leaves highly corrosive residues and must, therefore, be completely removed by post-soldering cleaning. A solution of zinc chloride in hydrochloric acid, which is used for soldering stainless steel, is even more corrosive.

The removal of flux residues is an essential part of the soldering operation. Even non-activated rosin, which is non-corrosive, may collect dust and dirt in service leading to a risk of deterioration. Active fluxes are corrosive and must be

Soldering and brazing 91

Table 4.1. Permissible impurity levels in machine soldering

Specification	Maximum impurity content (mass %)									
	Al	Sb	As	Bi	Cd	Cu	Au	Ag	Ni	Fe	Zn	Others
MIL Spec PIC 815 A	0.006	0.2 to 0.5	0.03	0.25	0.005	0.30	0.20	0.10	0.10	0.20	0.005	–
Woodgate (1988) proposed	0.005	0.5	–	0.25	0.005	0.20	0.10	0.10	0.10	0.20	0.003	0.05

removed. The method of removal depends on the application; usually washing in hot water is adequate.

Printed circuit boards require special consideration. If the electronic components are fully sealed then water washing may be possible; otherwise the treatment must be confined to the bottom of the board.

4.2.5 Soldering methods

Soldering may be applied by using a soldering iron, by flame heating, resistance heating, induction heating, hot plate heating or oven heating, by dipping or by means of a spray gun. The first three methods are used primarily for manual soldering. The size of heating torch and the composition of the fuel gas is determined by the mass and configuration of the assembly and should be such as to give rapid heating without too much danger of reaching excessive temperatures. Resistance tools employ carbon electrodes and require the solder and flux to be pre-placed.

In dip soldering a complete assembly is degreased, cleaned and fluxed, and is then dipped in a pot of molten solder, thus simultaneously making numerous joints. Induction heating is applied to quantity production of small parts and is capable of complete mechanization. In any mechanical soldering operation it is essential that the joint clearance be accurately maintained until the joint has solidified. Ultrasonic transducers may be used for the fluxless soldering of aluminium and other non-ferrous metals. The solder is melted on the surface to be joined, and ultrasonic vibrations are applied by means of a probe to the surface of the solder. The oxide film on the metal surface is thereby broken up and the metal is tinned. This technique cannot be applied directly to the soldering of lapped or crimped joints.

In electronics, soldering is used to join components to printed circuit boards, and for large-scale production a technique known as **wave soldering** is employed for that purpose. A standing wave is created on the surface of a pool of solder by means of a mechanical or electromagnetic pump, and the board, with components already mounted, is moved so that its bottom surface (on which the circuit is printed) encounters the wave progressively. The principle of the technique is illustrated in Fig. 4.4. Because the solder is continuously recirculated it is subject to contamination by dissolved metals, particularly copper and iron, and the composition must be maintained below specified limits such as those indicated in Table 4.1. Initially the uptake of contaminants is high, but this levels off with time and as additions of fresh solder are made to the bath.

A variant of wave soldering (initially introduced to bypass patents, but now a process in its own right) is **drag soldering**. The board is fluxed, preheated and then traversed across the surface of a solder bath (Fig. 4.5). Currently available soldering machines for either wave or drag soldering incorporate computer

Soldering and brazing 93

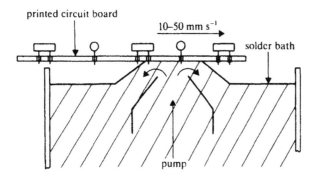

4.4 The principle of wave soldering.

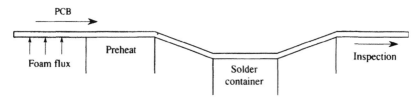

4.5 The principle of drag soldering.

controls. Yet another technique is vapour condensation. Originally intended for printed circuit boards, this method employs an organic liquid vapour which heats the component to soldering temperature by condensing on its surface. Solder is pre-placed above the joints. Air is excluded from the vapour zone so that good quality joints are made with the minimum of flux (Fig. 4.6). Vapour soldering is used for specialized applications.

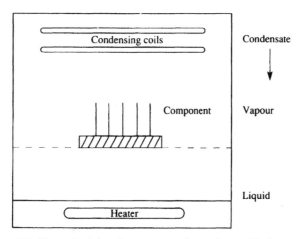

4.6 The principle of vapour condensation soldering.

94   Metallurgy of welding

An alternative to wave or drag soldering for printed circuit boards is **reflow soldering**. Reflow is a term used for the process of heating the board, with or without components, so as to melt the solder and cause it to flow, for example into the drilled holes where component leads are soldered. In reflow soldering the parts are mounted (using adhesive if necessary) in contact with pre-placed solder on the board, and then heated by infrared radiators to melt the solder and make the joints. In this, as in other mass soldering operations, manual soldering with an iron is required to make final assembly joints and also to rework joints that are rejected at the various inspection stages.

Quality control is primarily by visual inspection, which can detect such defects as blowholes and dry joints, but is never completely reliable. Inspection machines are available. One type measures the thermal response from a joint heated by a laser pulse; another compares visual images with a model stored in memory. No system is perfect and much reliance is placed on strict control of the complete sequence of operations. In particular it is necessary to take account of the fact that presoldered parts (for example printed circuit boards that have been subject to reflow) deteriorate on exposure to atmosphere owing to an increasing thickness of oxide film and also to an increase in the thickness of the intermetallic layer, which occurs even at room temperature. Exposure of the intermetallic layer during a soldering operation can result in dewetting and poor joints.

4.2.6 Application to various metals

Soldering temperatures are such that there are no significant metallurgical changes in the parent materials, so that, apart from the danger of subsequent corrosion due to residual flux or to galvanic action, soldering does not have damaging side-effects on metals.

The oxides that form on stainless steel, aluminium–bronze and beryllium–copper are resistant to chemical attack and require the use of acid fluxes. Graphitic cast irons are also difficult to wet because of the graphite exposed at the surface, and electrolytic treatment in a salt bath may be necessary. Aluminium and magnesium both present difficulties. Aluminium may be soldered with a tin–zinc or zinc–aluminium solder but the resulting joint has poor corrosion resistance due to galvanic action between solder and aluminium. The same problem arises with magnesium and, in both cases, soldered joints must be protected against moisture.

4.3   Brazing

4.3.1   Joint design

Typical designs of brazed joint are shown in Fig. 4.7. The strength of sound brazed joints is of the same order of magnitude as that of normal engineering

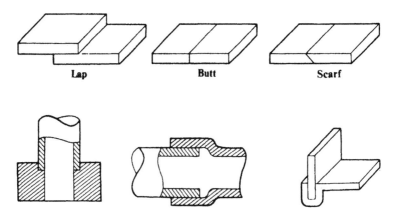

4.7 Typical brazed joints.

materials, so that the additional means of strengthening employed for soldered joints are not required. Butt joints in sheet or plate, however, are rarely used, because it is difficult to ensure freedom from defects. Optimum joint clearance is in the range 0.05–0.125 mm for most brazing materials; exceptions are copper, which gives strongest joints with an interference fit (maximum 0.075 mm), and light metals, for which gaps up to 0.25 mm are satisfactory.

So far as practicable, brazed joints are designed to be self-jigging. Where jigs are used there must be allowance for thermal expansion, notably in flame brazing and furnace brazing.

4.3.2 Brazing metals

Standard specifications for brazing alloys are so numerous as to be confusing, and the same is even more true of proprietary brands. Table 4.2 lists a selection of the more commonly used types listed by the American Welding Society. The common brazing alloy is 50/50 or 60/40 brass, which is widely used for brazing steel where corrosion resistance is not important. The bicycle frame is a classic application (however, the currently popular mountain bicycle requires a heavier frame and is fusion welded). Pure copper is used for the furnace brazing of steel with small clearances: often an interference fit. Copper–phosphorus alloys are used for brazing copper and its alloys; they cannot be used on nickel or ferrous alloys because of the formation of phosphides. Silver-base solders are low-melting alloys (down to 620 °C) that have good corrosion resistance and are particularly useful for austenitic stainless steel. Most contain cadmium, and may generate toxic fumes if ventilation is not good; BAg-7, however, is cadmium-free. The nickel-base brazing alloys depend on additions of boron, phosphorus

Table 4.2. Brazing alloys

Type	AWS designation	Typical composition (%)													Use
		Ag	Al	Au	B	Cd	Co	Cu	Fe	Ni	P	Si	Sn	Zn	Other	
Aluminium–silicon	BAl Si-2		Rem					0.25				7.5				Cladding for 3003 or 6951 sheet. Dip brazing.
	BAl Si-4		Rem					0.30				12				General-purpose filler. All processes.
Copper	BCu-1							99.9 min			0.075					Furnace brazing of steel and ferrous alloys.
Brass (spelter)	RBCuZn-A							60						40		General-purpose brazing.
Nickel silver	RBCuZn-D							48		10				42		Brazing tungsten carbide and ferrous alloys.
Copper–phosphorus	BCuP-2							Rem		7.25						High fluidity suitable for small clearances of 0.03–0.08 mm.
	BCuP-4	6						Rem		7.25						
Silver solder	BAg-1	45				24		15						16		High fluidity. Ferrous and non-ferrous except Al, Mg.
	BAg-3	50				16		15.5		3				15.5		Stainless steel and tungsten carbide tips.
	BAg-7	56						22					5	17		Cadmium-free for low toxicity: food processing. Vacuum tube assemblies.
Gold-base	BAu-1			37.5				Rem								Furnace brazing of superalloys.
	BAu-6			70					22						Pd 8	
Nickel-base	BNi-1			3			14		4.5	Rem		4.5				High-temperature, cryogenic and corrosion-resistant work.
	BNi-2			3			7		3	Rem		4.5				As above. Most commonly used.
	BNi-8							4.5				7			Mn 23	Honeycomb structures. Vacuum braze with low O_2.

or manganese to reduce the melting point, and are typically used for high-temperature applications such as gas turbines. Aluminum and its alloys are brazed using metal based on the aluminium–silicon binary system. The Al–12Si alloy is employed for manual work, but in many applications sheet clad with Al–7.5Si (**brazing sheet**) is preferable. Filler metal may be supplied in various forms. Manual brazing uses a rod. In furnace brazing the metal is pre-placed as rings, washers, discs or other forms to suit the geometry of the parts to be joined. A coating may be used: copper parts may, for example, be electroplated with silver: when heated to the correct temperature the Cu–Ag eutectic provides the brazing alloy. Aluminium alloys may be clad with the Al–Si brazing metal BAl Si–2 (see Table 4.2). Brazing pastes consisting of powdered metal mixed with flux also find a use.

4.3.3 Fluxes and protective atmospheres

Borax has long been used as a brazing flux for ferrous alloys. Borax and boric acid are reduced by chemically active metals such as chromium to form low-melting borides. At the same time, they have the power to dissolve non-refractory oxide films at the brazing temperature. Addition of fluorides and fluoroborates to a borax–boric acid mix lowers the melting point to a level where the flux can be used with silver solders, and increases the activity against refractory films. For aluminium, magnesium, titanium and zirconium the most effective fluxes are mixtures of chlorides and fluorides. Borax flux residues after brazing are often glass-like, and can be removed only by thermal shock (quenching), or abrasive or chemical action. The chloride–fluoride flux residues are water-soluble and must be removed to avoid subsequent corrosion.

Brazing is frequently conducted in a furnace with a protective atmosphere, either with or (preferably) without flux (Table 4.3). The type of atmosphere required may be qualitatively assessed by considering the dissociation of the metal oxide concerned:

$$2MO = 2M + O_2 \qquad (4.7)$$

for which the equilibrium constant is

$$K = \frac{[a_M]^2 [O_2]}{[a_{MO}]^2} = p_{O_2} \qquad (4.8)$$

where p_{O_2} is the **dissociation pressure** of the oxide in question, and a_M and a_{MO} (both equal to 1) are the activities of metal and oxide respectively. Hence

$$\Delta G = -RT \ln p_{O_2} \qquad (4.9)$$

Dissociation pressures for various oxides are obtained (as $RT \ln p_{O_2}$) from Fig. 4.8.

Table 4.3. Furnace atmospheres for brazing

Type	Dewpoint of incoming gas	Approximate composition range (%)				Typical use
		H_2	N_2	CO	CO_2	
Burnt fuel gas	Ambient	10–12	70–75	8–10	6–8	Copper brazing of low-carbon steel.
Burnt fuel gas, low hydrogen	Ambient	2–4	80	2–4	12–14	Brazing copper and copper-base alloys.
Cracked fuel gas	Down to –15°C	43–50	20–28	25–28		Brazing medium- and high-carbon steels and alloy steels.
Cracked ammonia	–50°C	75	25			Alloys containing chromium, including stainless steels; tungsten and platinum contacts.
Dry hydrogen	–60°C	100				Cobalt-base alloys; carbides.
Inert gas, argon						Titanium, zirconium, hafnium.
Vacuum						All non-volatile metals.

Soldering and brazing 99

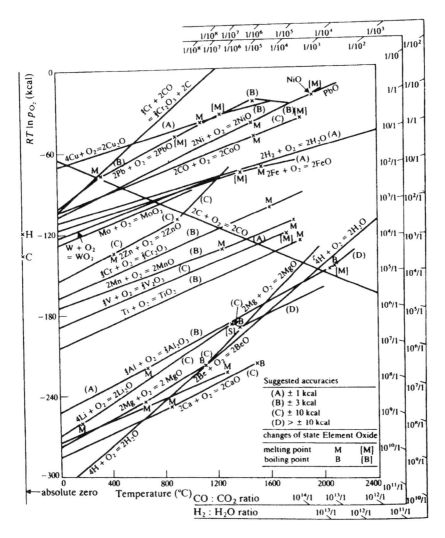

4.8 The free energy of formation/dissociation pressure of oxides and gas mixtures (from Milner, 1958, p. 97, after the original diagram by Ellingham).

The more stable oxides have lower dissociation pressures, and therefore require a lower **oxygen potential** in the protective atmosphere if they are to be effectively reduced. The oxygen potential of the atmosphere results from dissociation of gaseous oxides:

$$2CO_2 = 2CO + O_2 \qquad (4.10)$$

$$2H_2O = 2H_2 + O_2 \qquad (4.11)$$

each of which has an equilibrium constant such as

$$K_{H_2O} = \frac{[H_2]^2[O_2]}{[H_2O]^2} = \frac{p_{H_2}^2 p_{O_2}}{p_{H_2O}^2} = \exp\left(-\frac{\Delta G_{H_2O}}{RT}\right)$$

so

$$RT \ln p_{O_2} = -\left[\Delta G_{H_2O} + 2RT \ln(p_{H_2}/p_{H_2O})\right] \quad (4.12)$$

Values of $RT \ln p_{O_2}$ for the dissociation of steam and carbon dioxide are obtained from Fig. 4.8 by joining the origin (marked H for H_2–H_2O mixtures and C for CO–CO_2 mixtures) to the ratio of partial pressure of the mixtures concerned. Thus the oxygen potential of the atmosphere is reduced, other things being equal, by increasing the hydrogen content and reducing the water vapour content. Similarly from equation 4.10 increasing the CO/CO_2 ratio will reduce the oxygen potential. The dissociation pressure of metal oxides is so small that, if the oxide is in contact with a gas mixture, the value of p_{O_2} will be virtually equal to the oxygen potential of the gas. Therefore, if p_{O_2} for the gas is lower than the dissociation pressure of the oxide, reduction is thermodynamically possible, and vice versa. Consequently, the more stable oxides will only be reduced by hydrogen if the partial pressure of water vapour in the mixture is extremely low. In brazing technology, the water-vapour content of a hydrogen atmosphere is usually measured by the **dewpoint**. The relationship between

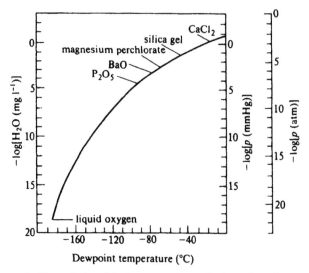

4.9 The relationship between the various scales of measurement of the saturated vapour pressure of ice and the dewpoint (from Milner, 1958, p. 99).

dewpoint and the equilibrium vapour pressure of water or ice is shown in Fig. 4.9.

Burnt fuel gas provides a cheap reducing atmosphere for the flux-free copper brazing of steel and for brazing copper alloys. Cracked fuel gas, which is obtained by passing hydrocarbon gas over a heated catalyst, is a suitable atmosphere for low-alloy steel, but more refractory materials require progressively lower dewpoints, as obtained from dissociated (cracked) ammonia, nitrogen–hydrogen mixtures or dry hydrogen. Increasing use is now made of purified argon or vacuum furnaces for brazing high-temperature alloys and metals such as titanium and zirconium. Flux-free brazing is possible in pressures substantially higher than the dissociation pressure of the oxide concerned; for example, type 304 austenitic stainless steel is wetted by a silver-base brazing alloy at a vacuum of 2.6×10^{-3} N m^{-2} and temperature of 860 °C, whereas the dissociation pressure of chromic oxide at this temperature is, from Fig. 3.8, 1.8×10^{-21} N m^{-2}. The reason for this is that the oxide is reduced by carbon dissolved in the steel:

$$2C + \tfrac{2}{3}Cr_2O_3 = \tfrac{4}{3}Cr + 2CO \tag{4.13}$$

The free-energy change for this reaction is plotted in Fig. 4.8, and if the carbon activity is of the order 10^{-4}, the equilibrium CO pressure is still greater than 2.6×10^{-3} N m^{-2}.

A similar effect is observed in the vacuum brazing of carbon steel, and for both materials it is accompanied by surface decarburization.

4.3.4 Brazing methods

Brazing may be achieved by means of a torch, by resistance or induction heating, by dipping or in a muffle-type furnace. Torch brazing is the common manual technique, an oxygen–coal gas or oxyacetylene torch being used. Induction heating is useful for repetition work suitable for brazing in the open air, the induction heating coil being designed and shaped for each specific application. Dip brazing is applied to aluminium: sheet coated with a 7.5 % Si brazing alloy is formed to the required shape, and then dipped into a flux bath at the correct temperature. In manual brazing, the filler alloy is added in the form of rod, which may be dipped in flux before application. Alternatively, it may be in the form of granules (50Cu–50Zn spelter), or it is pre-placed around or in the joint in the form of wire or metal foil.

4.3.5 Braze welding

Braze welding or **bronze welding** is a fusion-welding process that utilizes 60/40 brass as a filler metal with brazing-type fluxes to aid wetting. It was originally developed for the repair of cast iron parts and is still useful for this purpose because the thermal cycle is less severe than in normal fusion welding, so that the risk of hardening and cracking the cast iron is reduced.

To follow the traditional procedure, the joint is prepared with a vee as for fusion welding. A small area of the parent metal is heated with an oxyacetylene torch. A globule of filler metal is then deposited on this area together with flux, and the torch is removed momentarily to allow the globule to solidify. A second adjacent area is then heated and filler metal added so that it unites with the first globule. The procedure is repeated along the joint. A slightly oxidizing oxyacetylene flame is used in order to burn off any surface graphite.

In recent years welding with a non-ferrous filler has been used in making various types of sheet metal assemblies. The welding processes employed include carbon arc, tungsten inert gas and plasma arc welding. Metal inert gas welding has also been used in the short-circuiting mode for welding galvanized steel sheet. Filler metals include brass, cupro-nickel, aluminium bronze and the nickel–boron brazing alloy BNi-3. The heat source is used to raise the temperature of the base metal to a level sufficient to allow the filler to wet and bond to the surface. Multipass welds may be made in thicker material.

4.3.6 Brazing applications

Copper that is furnace or torch brazed must be oxygen-free or deoxidized, since tough pitch copper is embrittled by hydrogen. Steel, copper alloys, nickel and nickel alloys are particularly susceptible to intergranular penetration and cracking by brazing material if residual or applied stresses are present during the brazing operation, or if the brazing temperature is too high, so they should be stress-relieved before brazing and supported so as to minimize stress. If stress relief is impracticable, a low-melting silver brazing alloy is used. Austenitic stainless steels that may be exposed to a corrosive environment after brazing must be of the extra-low-carbon, titanium-stabilized or niobium-stabilized type, since during brazing the metal is held in the carbide precipitation range. As in welding, nickel and nickel alloys must be free from surface contaminants – sulphur, lead, bismuth and other low-melting metals – otherwise cracking may occur.

Graphite on the surface of grey, malleable and nodular cast iron may prevent effective wetting, and the cleaning techniques recommended for preparation before soldering must be used.

The strength of a defect-free brazed joint in carbon steel increases as the joint clearance is reduced, and may be substantially higher than the strength of the brazing metal in bulk form. However, when the gap is reduced below a certain level, the joint will contain defects and therefore there is an optimum range of joint clearance for maximum soundness and strength. Typical figures are 0.075–0.4 mm for brazing copper with spelter (60/40 brass) and 0–0.075 mm for the copper brazing of steel.

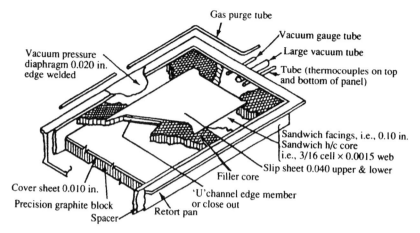

4.10 Retort assembly for brazing precision honeycomb sandwich. 1 in = 25.4 mm (from AWS Brazing Handbook, 1991).

The effect of the brazing heat on the mechanical properties of hardened and tempered steels or precipitation-hardened alloys must be carefully considered. For certain tool steels, the tempering temperature is close to the brazing temperature for silver solder, so that it is possible to combine the tempering and brazing operations. Likewise, for precipitation-hardening materials, it may be possible to combine brazing with solution treatment. Careful selection of the brazing material and heating cycle is necessary if it is required to combine the brazing operation with heat treatment.

The use of adhesive bonding for honeycomb structures in aircraft is mentioned in Section 3.6.1. Where such structures are exposed to elevated temperatures, however, adhesives are not suitable and it is necessary to make brazed joints. Typical applications are engine nacelle and wing panelling for supersonic military aircraft. Outer sheets are titanium or precipitation-hardening stainless steel, while the honeycomb structure itself is metal foil between 0.1 and 0.5 mm thick. Brazing metal for stainless steel is normally one of the nickel-base types (BNi-8 for example, as listed in Table 4.2), and preferably in the form of rapidly solidified amorphous metal foil, which flows into gaps more readily than other types. Brazing must be carried out in a low-oxygen atmosphere, and Fig. 4.10 illustrates a retort assembly in which the required conditions may be achieved.

The improved brazing properties of amorphous metal are considered to be due to uniformity of composition and low oxygen content. It is available in a variety of nickel-base, gold-base and cobalt-base alloys, all of which have applications in this type of work. For titanium, however, the preferred brazing metal is type 3003 (Al–1.25 Mn) alloy.

References

American Welding Society (1991) *Welding Handbook*, 8th edn, *Vol. 2: Welding Processes*, AWS, Miami; Macmillan, London.
Bilotta, A.J. (1985) *Connections in Electronic Assemblies*, Marcel Dekker, New York.
Connell, L.D. and Sajik, W.M. (1992) *Welding and Metal Fabrication*, **60**, 73–76.
Keene, B.J. (1991) *Surface Tension of Pure Metals*, NPL Report DMM(A)39, National Physical Laboratory, Teddington, UK.
Lancaster, J.F. (ed.) (1986) *The Physics of Welding*, 2nd edn, Pergamon Press, Oxford.
Milner, D.R. (1958) *British Welding Journal*, **5**, 90–105.
Woodgate, R.W. (1988) *Handbook of Machine Soldering*, John Wiley, New York.

Further reading

General

American Welding Society (1991) *Welding Handbook*, 8th edn, *Vol. 2: Welding Processes*, AWS, Miami; Macmillan, London.

Soldering

Bilotta, A.J. (1985) *Connections in Electronic Assemblies*, Marcel Dekker, New York.
Hwang, J.S. (1996) *Modern Solder Technology for Competitive Electronic Manufacturing*, McGraw-Hill, New York.
Klein Wassink, R.J. (1984) *Soldering in Electronics*, Electrochemical Publications, Ayr.
Thwaites, C.J. (1977) *Soft-soldering Handbook*. Int. Tin Res. Inst. Publ., no. 533, International Tin Research Institute, London.
Woodgate, R.W. (1988) *Handbook of Machine Soldering*, John Wiley, New York.

Brazing

American Welding Society (1991) *Brazing Handbook*, 3rd edn, AWS, Miami.
Brooker, H.R. and Beatson, E.V. (1953) *Industrial Brazing*, Associated Iliffe Press, London.

5
The joining of ceramics: microjoining

5.1 Scope

Microjoining has been appended to this chapter because it employs a number of techniques already described under the headings of solid-phase welding, adhesive bonding, soldering, brazing and ceramic bonding. It also has applications for fusion welding. Indeed the electronic and miniature fabrication industry probably employs a broader spectrum of joining techniques than any other, and at the same time overcomes what at first sight would appear to be impossibly difficult problems in handling minute parts.

The term 'ceramics' will be interpreted to include those non-metallic, inorganic materials that are used for their resistance to wear or temperature, or for their electrical resistance, and which from time to time must be attached to metals. These include glasses, refractory oxides, carbides and nitrides.

5.2 The properties of ceramics

Whereas metals are characteristically ductile, have high thermal and electrical conductivity, and relatively high coefficients of thermal expansion, ceramics are brittle, have low thermal and electrical conductivity, and their thermal expansion coefficients tend to be somewhat lower than those of metals. There are of course exceptions. Superconducting ceramics have been developed with a critical temperature of up to $-150\,°C$. Dispersions of SiC whiskers in various ceramic matrices (Al_2O_3 for example) have given fracture toughness values of up to $15\,MN\,m^{-3/2}$, which is to be compared with a value of about $20\,MN\,m^{-3/2}$ for flake graphite cast iron (Fig. 5.1). And thermal expansion characteristics are not necessarily unmatchable, as shown in Fig. 5.2.

Nevertheless, the chief obstacle to direct metal–ceramic bonding is the combination of brittleness and thermal expansion mismatch. In consequence, fusion welding is generally regarded as inapplicable to such joints, and although the electron beam process has been used to make joints between, for example,

106 Metallurgy of welding

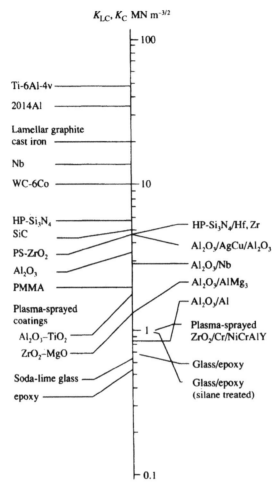

5.1 Fracture toughness K_{IC} for some engineering materials and K_C for joints (after Elssner, 1990).

tantalum and alumina, there is little confidence in the possibility of further progress along such lines.

Much metal–ceramic joining is mechanical in character. Press-fitting or shrink-fitting is extensively used in electrical equipment, for example in spark-plugs.

The attachment of fireproofing and other insulating material to steel is by means of studs or framework welded to the metal surface, and is primarily mechanical. However, the main technical interest lies in those applications where a bond is formed at a metal–non-metal interface. One of the oldest of such applications is glass–metal bonding, which will be considered first.

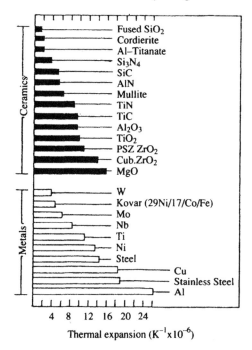

5.2 Comparison of thermal expansion coefficients for metals and ceramics (from Fernie *et al.*, 1991).

5.3 Glass–metal seals

5.3.1 General

The sealing of glass to metal has been in use for many years in the production of incandescent lamps and electron tubes and more recently for vapour lamps and housings for semiconductors. Most early incandescent lamps were sealed with lead zinc borate types of glass, which soften and flow at relatively low temperatures. Currently, there are numbers of properietary glasses which have been developed to match individual metals. These range from soft glass (soda-lime silica or lead oxide/mixed alkali silica) with working temperatures of 800–1000 °C and thermal expansion coefficients above $5 \times 10^{-6} \, K^{-1}$ to hard glasses (borosilicate, aluminosilicate and vitreous silica) with working temperatures of 1000–1300 °C and thermal expansion coefficients below $5 \times 10^{-6} \, K^{-1}$.

5.3.2 Joint design

Glass–metal joints may be divided into two types: matched and unmatched seals. In matched seals the thermal expansion coefficient of the glass is matched as

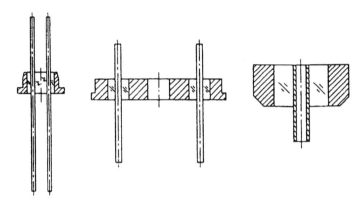

5.3 Typical designs of compression glass–metal seals (from Tomsia and Pask, 1990).

closely as possible to that of the metal, and there is chemical bonding at the glass–metal interface.

Unmatched seals fall into two categories. The first is the compression seal, in which the glass surrounds the metal conductor or duct, and is itself surrounded by a metal ring which, on cooling to room temperature, contains residual tensile stress, so putting the glass in a state of compression. A chemical bond between metal and glass is not necessary but is desirable (Fig. 5.3).

The other joint is the Housekeeper joint, named for its inventor, W.G. Housekeeper, and illustrated in Fig. 5.4. A glass ring is sealed on to a copper tube, the end of which has been tapered to a point. Differential expansion is accommodated by flexure of the tube without placing any significant stress on the

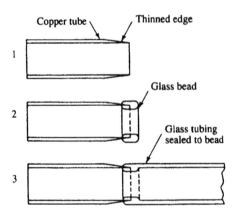

5.4 Steps in making a Housekeeper seal between copper and glass tubing: 1, the copper tube is thinned; 2, glass bead is applied to the edge of the thinned copper tube; 3, the glass tubing is then sealed to the bead (from Tomsia and Pask, 1990).

glass. A glass tube or other shape is then fused to the ring. There is good chemical bonding between copper and glass. Housekeeper seals may be made with other metals but are particularly useful for copper, whose thermal expansion is higher than that of any commercial glass.

5.3.3 Residual strain in a matching seal

On cooling from the liquid state, glass, being an amorphous substance, does not crystallize and solidify at a fixed point but becomes progressively more viscous. Thus there is a **working temperature** at which the viscosity is $10 \, \text{kg m}^{-1} \, \text{s}^{-1}$ and below which the glass can no longer be manipulated. At the **softening temperature** the viscosity is $10^{4.5} \, \text{kg m}^{-1} \, \text{s}^{-1}$ and below this the glass behaves generally as an elastic solid. At the **annealing temperature** it takes longer than 15 min to relieve residual stress and the viscosity is $10^{10} \, \text{kg m}^{-1} \, \text{s}^{-1}$ and finally at the **strain temperature** it takes longer than 4 h to relieve stress and the viscosity is $10^{11.6} \, \text{kg m}^{-1} \, \text{s}^{-1}$. Between the strain point and the annealing point is the **set point**, and in glass–metal systems this point is taken as the temperature below which elastic strain can be generated in the glass if there is a mismatch in the contraction.

Figure 5.5 shows the expansion curves for tungsten and a compatible glass, Corning 772. To obtain the residual strain at room temperature the glass curve is displaced until the set point lies on the tungsten curve; the residual strain is the gap between the curves at 20 °C, in this instance 7×10^{-5}, which is acceptable. The curve also indicates a maximum strain of 3.2×10^{-4} during the initial fall in temperature but this is dissipated on further cooling. In general a final residual

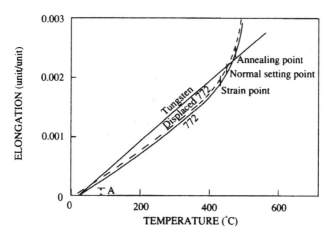

5.5 Thermal expansion curves for tungsten and Corning glass 772. The relative strain at room temperature is A (from Tomsia and Pask, 1990).

strain of less than 1×10^{-4} is safe and up to 5×10^{-4} may be acceptable. Higher differential strains require smaller and thinner assemblies.

5.3.4 Bonding mechanisms

In principle, the chemistry of a glass–metal bond is very simple. The oxide layer on the metal surface is dissolved by the liquid glass, which wets and is directly bonded to the metal surface. In detail, however, the reactions may be complex.

So far as oxide formation is concerned, metals may be divided into three categories: those of low reactivity, a medium-activity group and a high-activity group. The first category includes gold, silver and platinum. Oxidation in air does not produce bulk oxide layers but oxides may be adsorbed; however, these dissociate on heating. In the case of silver, Ag_2O dissolves in the glass and a silver silicate forms at the interface, resulting in a chemical bond. Gold and platinum do not react in this way.

Iron, nickel and cobalt fall into the medium-activity group. Iron is not much used for electrical or electronic applications, but in porcelain and vitreous enamel coatings a powdered glass is applied to an iron or steel substrate and subsequently fired to form a glass overlayer which is bonded to the metal. The oxide layer formed before and during the firing operation dissolves in the glass, and provided that the glass is saturated with FeO at the interface the bond strength is high. When the oxide concentration falls owing to diffusion into the glass, the bond becomes progressively weaker. However, if cobalt is added to the glass, this is reduced by iron at the interface to form FeO:

$$Fe + CoO_{glass} = FeO_{glass} + Co \tag{5.1}$$

The cobalt alloys with iron to form dendrites that grow into the glass and improve bonding. The beneficial effects of cobalt were discovered empirically by porcelain enamellers many years ago.

Kovar (29 Ni–17 Cr–54 Fe) is a low-expansion alloy much used for glass–metal seals. This falls into the medium-activity group but the bond strength is largely governed by the thickness of a preformed oxide layer, as will be seen later.

The high-activity group includes chromium, titanium and zirconium. Chromium in particular is present in many ferrous alloys. Chromium may reduce silicates in glass to form $CrSi_x$ crystals which grow into glass from the interface. In alloys reactions may be complex owing to the formation of multilayer oxides on the metal surface.

5.3.5 Forming a glass–metal seal

The metals used for glass–metal seals include platinum, tungsten, molybdenum, copper and various alloys of iron, nickel, cobalt and chromium. Tungsten and Kovar both have low expansion coefficients, and a number of proprietary

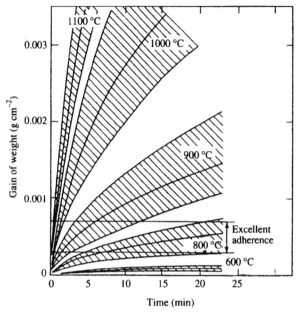

5.6 The oxidation rate of Kovar as a function of time and temperature (from Tomsia and Pask, 1990).

matching glass compositions such as the one illustrated for tungsten in Fig. 5.5 are available for these metals.

The first step in forming a seal is to clean the metal surface by heating in a wet hydrogen atmosphere at 1100 °C; this removes hydrocarbons and other contaminants and results in some etching and roughening. Secondly an oxide layer is formed by heating in air; Fig. 5.6 shows the optimum degree of oxidation for Kovar.

The glass is applied as a powder coating on the metal surface, and the assembly heated to the **sealing temperature**, typically 1000 °C. The bulk glass is then sealed to the coating. Alternatively metal pins and powdered glass may be put together in a metal frame and then heated to the sealing temperature. The assembly is then cooled at a controlled rate to room temperature. The critical temperature is around the set point, and the glass must be cooled slowly through this temperature range, or annealed, in order to avoid damaging thermal stresses.

5.4 Glass-ceramics

5.4.1 General

Glass has the useful property that, when molten, it will wet and bond to an oxidized metal surface. However, being an amorphous substance, it softens at elevated temperature and lacks the rigidity and strength which most ceramics exhibit under such conditions. With glass-ceramics it is possible to combine these desirable characteristics by forming and bonding the substance when it is in a glassy state, and then causing it to crystallize and transform into an elastic, non-viscous solid.

Crystallization (devitrification) of a common sodium silicate glass is well known as a form of deterioration; the process starts at the surface and results in an opaque, coarse-grained substance that has little coherence. The crystallization of glass-ceramics, however, is nucleated internally so as to produce a uniform, fine-grained polycrystalline material.

The majority of the substances that form glass-ceramics are silicates, and the range of compositions now available is very wide. They are made by melting the raw materials together in the temperature range 1000–1700 °C to form a glass. Suitable nucleating agents such as P_2O, TiO_2 and ZrO_2 are incorporated. The molten glass is then worked to the required form by normal glass-making techniques and cooled to room temperature. Conversion to the polycrystalline

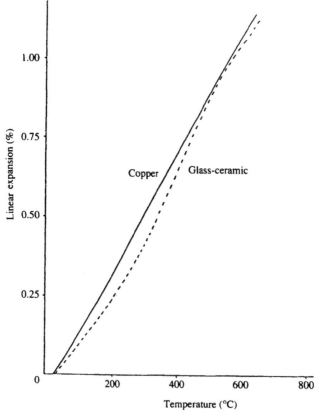

5.7 Thermal expansion of copper and matching glass-ceramic (from Partridge, 1990).

form is accomplished by first heating to a temperature at which the nuclei precipitate, and then to a higher temperature at which the crystallization takes place. In some cases both nucleation and crystallization occur at the same temperature. For details of the composition and properties of glass-ceramics the reader is referred to the textbooks listed at the end of this chapter (McMillan, 1979; Strnad, 1986). So far as metal–ceramic joints are concerned one advantage of glass-ceramics is the wide range of thermal expansion properties that are available, making it possible, for example, to match those of copper (Fig. 5.7). Another example is lithium aluminosilicate, which is compatible with Inconel 718, and is suitable for use in glass turbine parts up to 700 °C.

5.4.2 Making glass-ceramic seals

One bonding technique is illustrated in Fig. 5.8. The metal parts are cleaned, roughened and pre-oxidized and then placed in the mould. Molten glass is then introduced and formed by means of a plunger. When set, the assembly is transferred to a furnace where the nucleation and precipitation heat treatments are carried out. In a second method, the metal parts are placed together with preformed glass in a graphite mould. The whole assembly is heated in an inert atmosphere to a temperature at which the glass flows and bonds to the metal. In this technique the glass-ceramic transformation usually takes place on cooling in the mould.

Glass-ceramic seals are used for high-voltage and high-vacuum devices, and for thermocouple terminations.

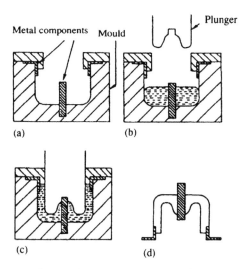

5.8 Stages in the preparation of glass-ceramic to metal seal (from Partridge, 1990).

5.4.3 Coatings

Glass-ceramics may be applied as a coating to metals in much the same way as vitreous enamelling. The metal is prepared and glass is applied by spraying a powder suspension either as a uniform coat or through a wire mesh screen where it is required to develop a pattern. Alternatively dry powder is deposited electrostatically. The coatings are dried when necessary and then the assembly is fired so as to flow the glass and make a bond. Crystallization normally takes place during the firing cycle.

Glass-ceramic coatings have the advantage over vitreous enamel that they are refractory. Thus they can withstand the firing temperature of 850–900 °C which is required for the application of thick-film circuitry.

5.5 Brazing

5.5.1 General

Many of the technically important ceramics that have been developed in recent years, including alumina, zirconia, beryllia, silicon carbide, titanium carbide, silicon nitride, boron nitride and sialons, are not wetted by the commonly used brazing metals. An exception to this rule is tungsten carbide, which is used for machining, and is attached to a steel shank by silver brazing.

There are two ways of overcoming the non-wetting problem. The first is to **metallize** the surface of the ceramic and then make a brazed joint between the metallized layer and the bulk metal. The second method is to employ a brazing alloy incorporating an ingredient that interacts chemically with the ceramic surface. This procedure is known as **active metal brazing**.

5.5.2 Metallization

The metallization technique most commonly used, particularly for alumina, is the manganese–molybdenum process. The primary metallizing layer is applied initially as a paint consisting of a mixture of molybdenum, manganese oxide, a glass frit, a carrier vehicle and a solvent. This coating is fired at 1500 °C in an atmosphere of hydrogen or cracked ammonia containing water vapour. During this operation the glass bonds with the ceramic (which itself must contain a glassy intergranular phase) and the metallic particles sinter together. The second step is to nickel-plate the surface either by electroplating or using an electroless process, after which the component may be once again sintered in a hydrogen atmosphere or go directly to brazing. The brazing filler metal is usually the 72 Ag–28 Cu eutectic, which has a low melting point (780 °C) and good ductility to accommodate any thermal contraction mismatch. For higher-temperature applications a nickel-based brazing metal may be used.

The manganese–molybdenum process requires the presence of a glassy phase in the ceramic. Where this is not acceptable or not practicable the metal coating must be applied directly, for example by vapour deposition or by ion sputtering.

5.5.3 Active metal brazing

The active element in all commercially available ceramic–metal brazing alloys is titanium. The relevant reaction is, with alumina for example

$$3[Ti] + Al_2O_3 \leftrightarrow 2[Al] + 3TiO \tag{5.2}$$

where square brackets indicate solutes. This reaction goes to the right for quite low concentrations of titanium: theoretically 1 atomic % in copper at 1400 K. TiO is metallic in character and is wetted by brazing metal. Similar reactions would appear to occur with carbides and nitrides; at any rate these types of ceramics are wetted by Ti-bearing alloys. Figure 5.9 shows the effect of Ti on the contact angle of Ag–Cu brazes with alumina, silicon nitride and aluminium nitride. Three filler metal compositions are indicated in the figure; in addition there is the eutectic silver–copper alloy to which 5–10% Ti is added mechanically; for example as cored wire. Other alloys contain indium, which is thought to improve wettability by reducing the value of γ_{sv}.

Prior to brazing, both ceramic and metal must be clean and free from organic contaminants. Rough ceramic surfaces are less easily wetted, and for strong joints it is desirable to finish the surface by lapping or by refiring at high temperature.

Brazing is carried out in a vacuum furnace at a pressure of less than $0.1\,\mathrm{N\,m^{-2}}$, preferably $10^{-2} – 10^{-3}\,\mathrm{N\,m^{-2}}$. Alternatively an atmosphere of purified argon or helium, with an oxygen content of less than 1 ppm, or a hydrogen atmosphere with a dewpoint lower than $-51\,°C$ may be used. Assemblies are heated at a rate of $10–20\,°C\,\mathrm{min}^{-1}$ and the temperature is held for at least 15 min below the solidus temperature of the solder to ensure uniformity. Then it is raised slowly to

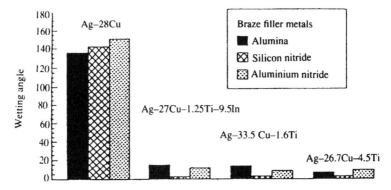

5.9 Wetting angle of Ag–Cu brazing metal with and without titanium additions on various ceramic substrates (from Fernie and Sturgeon, 1992).

116   Metallurgy of welding

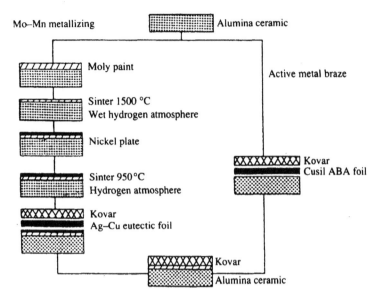

5.10 Alternative routes for brazing alumina to Kovar (from Fernie and Sturgeon, 1992).

50 °C above the liquidus and held for 30 min or longer. The wetting reaction is sluggish and the fluidity of the brazing metal is relatively low.

The two alternative routes for making a metal–alumina brazed joint are shown diagrammatically in Fig. 5.10.

5.6 Other techniques

5.6.1 Diffusion bonding

As compared with the diffusion bonding of metal–metal joint, the ceramic–metal joint presents several additional problems. Firstly, only one component, the metal, is capable of significant plastic deformation. Secondly, there is no interdiffusion. Thirdly, there must be some means of accommodating thermal mismatch. The last condition usually requires the use of interlayers, which are thin wafers of metal having a thermal expansion coefficient intermediate between those of the metal and the ceramic. Thus alumina has been bonded to copper for electronic applications using a gold interlayer. However, the use of diffusion bonding for metal–ceramic joints has been somewhat limited. One of the largest applications to date is the accelerator tube for the 20 MV van de Graaff generator at the Danesbury Laboratory, UK. The tube consists of a sandwich structure of insulating rings which form an evacuated envelope for the ion beam (Fig. 5.11). It is built up of 144 stacks each containing 16 alumina-titanium sandwiches, 196 mm long. Details of materials, dimensions and diffusion bonding conditions are shown in Table 5.1.

The joining of ceramics: microjoining 117

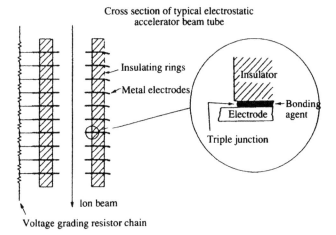

5.11 Arrangement for accelerator tube in van de Graaff generator (from Joy, 1990).

Table 5.1. Materials and procedures for ceramic/metal diffusion bonding in the fabrication of a van de Graaf accelerator tube

Submodule length	196 mm
Insulator	16 per submodule, two of these for thermal balancing
	Material 96.7% isostatically pressed alumina, type UL300 (Wade, Ireland)
	Diamond ground on all faces
	Surface finish 1 µm R_a
	Thickness, 10.00 mm
	Outside diameter, 190.00 mm
	Inside diameter, 150.00 mm
Electrodes	Flat commercially pure titanium
	Thickness, 1.00 mm
	Outside diameter, 203 mm
	Inside diameter, 112 mm
	Internal edge radiused
Inserts	Pure titanium, polished and vacuum baked
Construction	High-temperature diffusion bonded using 0.1 mm Al interlayer; pressed at 550°C for 1 h at 56 MPa
Vacuum integrity of submodule	$<10^{-10}$ at 10 bar He pressure
Production rejection rate at this leak rate	5% of submodules
Connecting flanges	Pure titanium
Seals	Pure aluminium (Daresbury design)

Source: Joy (1990).

118   Metallurgy of welding

The thermal match between titanium and alumina is relatively good, but it was decided nevertheless to use an aluminium interlayer. Early bonding trials were made in high-purity argon but it proved impossible to maintain the oxygen level required to avoid damage to the titanium electrodes, and eventually a press was designed which operated in a vacuum below 10^{-3} N m^{-2} at 650 °C, equivalent to an inert gas purity of 0.01 ppm by volume. The individual subassemblies were subject to a final test capable of detecting leaks of 10^{-10} mbar 1 s^{-1} for helium, and the reject rate was maintained below 10%.

This tube has operated successfully and no failures of the metal–ceramic joints have been experienced either in commissioning or in service.

5.6.2 Electrostatic bonding

This process is also known as **field-assisted bonding**.

If a smooth metal plate and a smooth ceramic surface are heated in contact in a vacuum, and if a voltage is applied to the joint, metal ions will migrate from the ceramic material into the metal, leaving an ion-depleted zone close to the interface. This, in turn, gives rise to an electrostatic field which generates a bonding force. The tension P is

$$P = \tfrac{1}{2}\varepsilon\varepsilon_0 E^2 \tag{5.3}$$

where ε is the relative dielectric constant of the ceramic material, ε_0 is the permittivity of free space (8.854×10^{-12} F m^{-1}) and E is the electric field intensity. If the depth of the ion-depleted zone is δ, then $E = V/\delta$, where V is the voltage applied during bonding. Hence

$$P = \tfrac{1}{2}\varepsilon\varepsilon_0 V^2/\delta^2 \tag{5.4}$$

Applied voltage is typically in the range 100–500 V, ε is typically about 10, and the depth of the ion-depleted zone is of the order of micrometres. For $\delta = 1$ μm and $V = 100$ V, $P = 0.44$ N m^{-2}. This is a very small tension and it may be that the applied voltage improves bonding in some other way, for example by promoting interdiffusion across the interface.

Figure 5.12 shows an electron-probe microanalysis survey across aluminium–glass and silver–glass joints made by the field-assisted process. There is significant interdiffusion, and silver appears to be particularly mobile, having replaced sodium in the ion-depleted zone. In the reaction bonding of Al–2.5 Mg to zirconia, the interdiffusion is similar to that in the aluminium–glass joint; there is a gradient of both aluminium and magnesium contents on the zirconia side of the joint. The bonding conditions are given in Table 5.2.

The current that flows during this operation is very small, of the order of microamperes in the case of an aluminium/glass bond. For such a joint, a

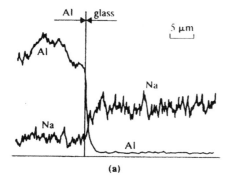

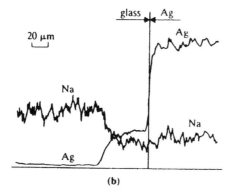

5.12 An electron probe microanalysis survey across the interface of: (a) aluminium–glass and (b) silver–glass bonds (from Arata *et al.*, 1984).

Table 5.2. Bonding conditions for field-assisted metal/ceramic joints

Anode metal	Bonding conditions				Current (μA)	Amount of electricity (mC)	Bonding observed
	Pressure (MN m^{-2})	Temp. (°C)	Voltage (V)	Time (min)			
Ag	0.687	500	200	7	1850–8550	3480	Yes
Al	0.687	400	200	40	3–27	10	Yes

minimum quantity of electricity must pass before bonding will occur. The amount of electricity required is reduced if the axial pressure during the bonding operation is increased.

Field-assisted bonding has been applied to the joining of metals to glass, β-alumina and other electrical insulators.

5.6.3 Friction welding

The technique used is similar to that for metal–metal joints; usually the ceramic is held stationary and the metal is rotated. Rotation continues for a predetermined period, then it stops and the joint is upset, giving the characteristic flash. The nature of the bond is uncertain and may in part be mechanical. Various metal–ceramic joints have been made experimentally and this technique may find a use where aluminium heat sinks need to be attached to AlN substrates.

5.6.4 Ceramic–ceramic bonds

Glasses and glass ceramics may be used as bonding media for oxide ceramics and for those ceramics that have an intergranular glassy phase. They cannot accommodate a thermal mismatch and are therefore most suitable for joining a material to itself. Alumina has been joined using glasses of Al_2O_3–MnO–SiO_2 and Al_2O_3–CaO–MgO–SiO_2 compositions. Silicon nitride has been joined using an Al_2O_3–MgO–SiO_2 glass, which simulates the intergranular phase. Bonding was carried out in a nitrogen atmosphere at a temperature between 1550 and 1650 °C. Alternatively, bonds can be made using sintering additives (MgO–Al_2O_3 or Y_2O_3–Al_2O_3), again in a nitrogen atmosphere and at temperatures above 1500 °C, with or without pressure.

Ceramics may also be joined by metallizing and brazing or diffusion bonding: however, such joints are not refractory and may not be suitable for elevated-temperature applications.

5.7 Microjoining

Microjoining is employed in the assembly and fabrication of electronic devices. The soldering of components to printed circuit boards has been discussed in Section 4.2.5. There are many other joining requirements, however. In most instances active electronic devices must be protected by means of a **package**, which is a sealed metal or ceramic box. The silicon chips must be bonded to a substrate, and interconnecting wires bonded to the chips and to other circuitry. Joints must be made between devices and heat sinks. Some of the joining methods and their applications are listed in Table 5.3.

Johnson (1985) gives an example of the complexity of the joints required in microelectronics (Fig. 5.13). In this case a high-power silicon chip is soldered to a BeO heat spreader. BeO has high thermal conductivity combined with electrical insulating properties: hence its use for this purpose. The heat spreader has been metallized on the underside by the manganese–molybdenum process (Section 5.5.2) and then gold-plated. This is then diffusion bonded to the copper heat sink at 300 °C with a pressure of less than $20\,N\,mm^{-2}$ and with a bonding time of

Table 5.3 Joining processes used in microjoining and their applications

Method	Application	Limitations
Gas tungsten arc	Joining sheet and foil for the production of bellows, diaphragms and meshes. Not for circuitry.	Arc unstable at currents below 10 A.
Microplasma welding	As above.	Range of use extended by pulse arc operation.
Electron beam	Bellows, diaphragms, electrode assemblies, sealing packages.	Capital cost. Need for high vacuum.
Laser	Microsoldering, brazing, seam welding packages, butt welding wires.	Capital cost. Reflectivity: e.g. of solders. Need for safety screening.
Resistance welding (spot, seam, projection and butt)	Sealing of packages, wire attachment, component assembly, sheet fabrication.	Electrode size may limit scope of application.
Hot pressure welding	Bonding wire to metallized pad on silicon chip.	
Diffusion bonding	Bonding silicon chips to substrates by gold–silicon eutectic heat sinks.	
Electrostatic bonding	Joining silicon to glass for pressure transducers.	Good surface finish required. Experience limited.
Ultrasonic welding	Bonding wires, joining sheet and foil, packing. Widely used in microjoining.	May be applied hot (thermosonic welding).
Brazing and soldering	Mainly soldering: for attaching components to printed circuit boards. Wire connections.	Need to use heat. Possible flux residues.
Glass fusion	Joining ceramic lids to substrate.	Possible thermal mismatch.

5–10 min. An intermediary foil of silver may be desirable to reduce contraction strains.

The lid of metal packages may be bonded to the base by various methods. Figure 5.14 shows the example of nickel-plated mild steel which is brazed to the base by projection (resistance) welding. The nickel plating acts as a brazing

122 Metallurgy of welding

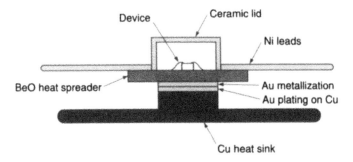

5.13 Diagram of a power device incorporating a BeO heat spreader and a copper heat sink (from Johnson, 1985).

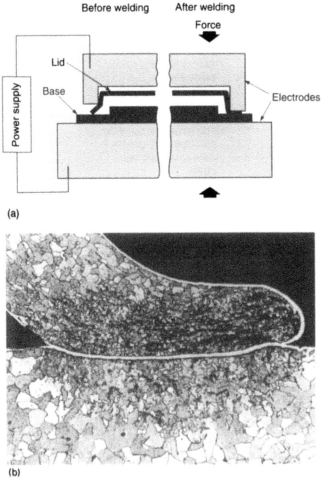

5.14 Projection welding of a nickel-plated steel lid for packaging: (a) diagrammatic section; (b) microsection (from Johnson, 1985).

metal. Gold plating is used for the same purpose. Figure 5.15 illustrates cold pressure welding of copper.

Attachment of silicon chips to the substrate is by **silicon die bonding**. Three techniques are illustrated in Fig. 5.16. The first is diffusion bonding using the gold–silicon eutectic. Adhesive bonding is made with an epoxy or polyamide adhesive loaded with metal powder where electrical or thermal conductivity is required. In such cases the chip may be metallized with an Al–Cr–Ni–Au alloy.

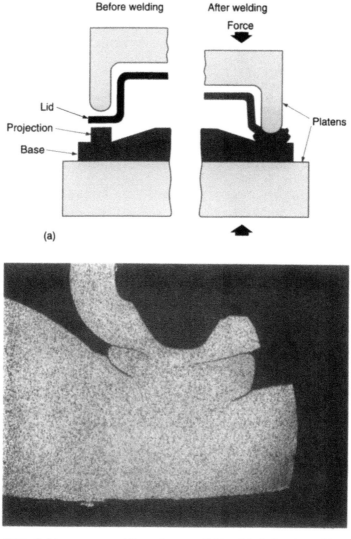

5.15 Cold pressure welding of copper lid to nickel-plated steel base for packaging: (a) diagrammatic section; (b) microsection (from Johnson, 1985).

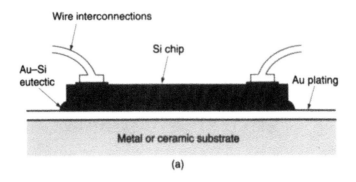

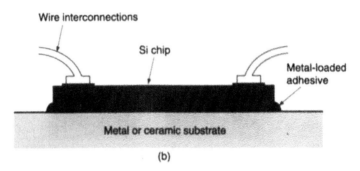

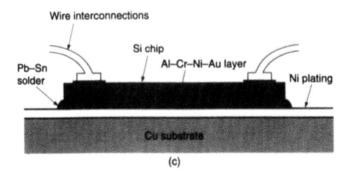

5.16 Arrangements for die bonding of silicon chip to substrate: (a) diffusion bonding by gold–silicon eutectic; (b) metal-loaded epoxy adhesive; (c) solder (from Johnson, 1985).

The same type of metallization is used for soldering. A 95Pb5Sn solder preform is used and may be fused manually with flux or furnace (reflow) soldered in a hydrogen atmosphere.

The other important type of joint is that between wire and, on the one hand, the aluminium pad on the surface of the chip and, on the other hand, the circuit track on the substrate. Aluminium or gold wires are used, and they are attached by

The joining of ceramics: microjoining 125

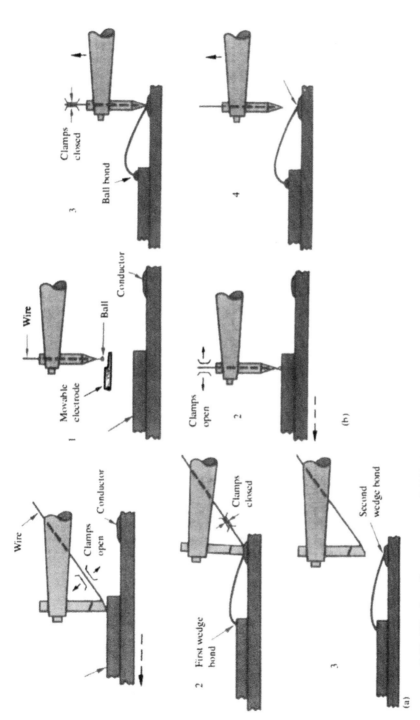

5.17 Wire bonding sequences. (a) Wedge–wedge: 1, first bond, clamps open, circuit is moved; 2 second bond, clamps closed; 3, wire breaks off after second bond. (b) Ball-wedge: 1, ball is formed; 2, first (ball) bond, circuit is moved; 3, second (wedge) bond; 4, wire breaks off after second bond (from Johnson, 1985).

ultrasonic welding. The joint is made either with a flattened length of wire (**wedge bonding**) or with a ball formed at the wire tip by melting (a **ball bond**) (Fig. 5.17).

High-power semiconductor devices are used for switching and regulating current in welding power sources, in heavy-duty motor drives and other large electrical systems. Such devices are capable of switching over 1000 A in microseconds, and groups of them may be used to handle a power of several megawatts. They consist essentially of a circular disc of silicon containing the required diffused-in junctions, normally 0.5 mm thick and up to 125 mm in diameter. During operation heat is generated at a rate of 1 kW in a 50 mm diameter device, and this is dissipated through a copper heat sink. The thermal mismatch between copper and silicon is too great to permit direct bonding, however, so normal practice is to braze the silicon to an intermediate plate of molybdenum using an aluminium–silicon eutectic brazing alloy. A mechanical pressure joint is then made beween the copper and the molybdenum, to produce a copper–molybdenum–silicon–molybdenum–copper sandwich.

This technique is satisfactory for diameters up to about 75 mm but for larger sizes and for higher efficiency it has disadvantages. In order to wet the molybdenum it is necessary to use a brazing temperature of 680–750 °C and at this level some of the silicon is dissolved and the wafer may be partly eroded. Therefore new joining methods are being developed. One technique is to use vapour deposition to apply a layer of the aluminium–silicon eutectic to the silicon wafer, and to coat the molybdenum with a layer of $MoSi_2$. This makes it possible to braze at a lower temperature. Alternatively the two elements are joined by **diffusion soldering**. This method employs a silver or silver–indium solder which is initially liquid and wets both surfaces. The joint is then held at about 200 °C to diffuse out the solute element, leaving an intermediate layer which is mainly silver and has a high remelt temperature, as required for this type of joint. These and other possible techniques are discussed by Jacobson and Humpston (1991).

Microjoining is characterized by a great diversity of techniques and of metal–metal and metal–non-metal combinations, of which only a few examples can be given here. For more detail the reader is referred to Johnson (1985).

References

Arata, Y., Ohmori, A., Sano, S. and Okamoto, I. (1984) *Transactions of the Japanese Welding Research Institute*, **13**, 35–40.

Elssner, G. (1990) in *Joining of Ceramics*, ed. M.G. Nicholas, Chapman and Hall, London.

Fernie, J.A., Threadgill, P.L. and Watson, M.N. (1991) *Welding and Metal Fabrication*, **59**, 179–184.

Fernie, J.A. and Sturgeon, A.J. (1992) *Metals and Materials*, **8**, 212–217.

Jacobson, D.M. and Humpston, G. (1991) *Metals and Materials*, **7**, 733–739.
Johnson, K.I. (1985) *Introduction to Microjoining*, TWI, Cambridge, UK.
Joy, T. (1990) in *Joining of Ceramics*, ed. M.G. Nicholas, Chapman and Hall, London.
Partridge, G. (1990) in *Joining of Ceramics*, ed. M.G. Nicholas, Chapman and Hall, London.
Tomsia, A.P. and Pask, J.A. (1990) in *Joining of Ceramics*, ed. M.G. Nicholas, Chapman and Hall, London.

Further reading

Johnson, K.I. (1985) *Introduction to Microjoining*, TWI, Cambridge, UK.
McMillan, P.W. (1979) *Glass-ceramics*, 2nd edn, Academic Press, London.
Nicholas, M.G. (1990) *Joining of Ceramics*, Chapman and Hall, London.
Strnad, Z. (1986) *Glass-ceramics Materials: Glass Science and Technology 8*, Elsevier, Amsterdam.

6
Fusion welding processes and their thermal effects

6.1 The development of fusion welding

The earliest fusion welding technique, mentioned briefly in the previous chapter, known as **flow welding** or **burning together**, has been practised since the early Bronze Age. A mould is made around the two bronze parts to be joined, and molten bronze is poured between and sometimes around them. This method was used for repairing swords and it was also employed for the fabrication of a bronze chariot discovered in 1980 near the tomb of the Chinese Emperor Qin Shi Huang. As practised in earlier times, this was an unreliable process; nevertheless, some of the welds in Emperor Qin's chariot remained intact when it was dug up after two millennia.

Flow welding is used to this day, in the form of **thermit welding** for joining rails *in situ*. A mould is set up around the rail ends with a crucible on top containing a mixture of iron oxide and aluminium powders. The rails are preheated to about 500 °C, then the thermit mixture is ignited, producing highly superheated steel. This melts through a plug in the base of the crucible, flows into the cavity between the rails and makes a weld. A similar technique is employed for joining copper power cables in the field. Sound, reliable joints are obtained using the thermit process.

Other fusion welding techniques rely on the use of a heat source that is intense enough to melt the edges as it traverses along a joint. Such heat sources first became available on an industrial scale at the end of the nineteenth century, when **gas welding**, **arc welding** and **resistance welding** all made their appearance. Gas welding was made possible by the supply of oxygen, hydrogen and acetylene at an economic price, and by the invention of suitable torches and gas storage techniques. By 1916, **oxyacetylene welding** was a fully developed process capable of producing good-quality fusion welds in thin steel plates, aluminium and deoxidized copper, and differed only in detail from the process as it is known today. Arc welding with a fusible electrode – the most important of the fusion processes – is more complex in character and developed more slowly. Initially, bare wire electrodes were used, but the resultant weld metal was high in nitrogen content and therefore brittle. Wrapping the wire with asbestos or paper improved the properties of weld deposits, and led eventually to the modern type of arc welding electrode, which is coated with a mixture of minerals, ferro-alloys and in some cases organic materials, bonded with sodium or potassium silicate.

In the early stages of this development, fusion welding was used primarily as a means of repairing worn or damaged metal parts. But during World War I, government departments and inspection agencies such as Lloyd's Register of Shipping initiated research into the acceptability of the technique as a primary means of joining metals, and prototype welded structures were made. Nevertheless, riveting remained the predominant means of joining steel plates and sections until the outbreak of World War II.

In other fields fusion welding made more progress. The girth welds in oil pipelines laid in the USA during the 1920s were welded with paper-coated electrodes. The welds were porous and leaky, but this was overcome by wrapping or burying in concrete. Aluminium and copper vessels used in the brewing and food industries were fabricated by oxyacetylene welding. In the USA, the fuselage of aircraft consisted of a framework of high tensile steel tubing which was joined by welding, and over which linen cloth was stretched.

During and after World War II welding for shipbuilding, for chemical, petroleum and steam power plant, and for structural steelwork became widespread and replaced riveting for the majority of fabrications.

Largely because the metallurgical problems that beset fusion welding do not arise to the same degree in resistance welding (at least so far as carbon steel is concerned), this latter group of processes was established in production long before arc welding was generally accepted. **Spot** and **seam welding**, which are used for making lap joints in thin sheet, and **butt welding**, used for chain making and for joining bars and sections, were well established by 1920.

The introduction of argon-shielded tungsten arc welding during the 1940s has led to a remarkable multiplication of fusion welding processes. First there came metal inert gas welding and then a number of techniques, including CO_2-shielded dip transfer welding, flux-cored, metal-cored and open-arc welding, all of which employ a continuously fed fine wire as the electrode. Recently the electronic control of power sources has increased the versatility and range of gas-shielded welding processes. Developments in steelmaking have also been influenced by welding requirements and, in their turn, are influencing welding technology. One of the most important of these changes has been the radically improved control over the level of impurities in steel, which makes possible a combination of better properties (particularly notch-ductility) and good weldability.

6.2 The nature of fusion welding

6.2.1 The classification of fusion welding processes

The way in which a fusion weld is made may have a profound effect on the properties of the joint. The three most important characteristics of a fusion welding process in this respect are the **intensity** of the heat source, the **heat input rate** per unit length of weld and the type or effectiveness of the method used to **shield** the weld from the atmosphere. In Table 6.1, which sets out the characteristics of the more important fusion welding processes, the method of shielding has been used (arbitrarily) as a means of grouping the different methods. The first group is that employing slag, or a combination of slag plus gas generated by decomposition of the slag, or a combination of slag and gas supplied from an external source as the protective medium. In the second group, the shield is gas alone, from an external source. Finally, there is a group in which it is not necessary to shield the weld from the atmosphere, e.g. electron-beam welding carried out in a vacuum.

6.2.2 Shielding methods

The drop at the tip of a fusible electrode may reach boiling point, and the centre of the weld pool, although much cooler, is substantially above the melting point. At such temperatures, reaction with oxygen and nitrogen from the atmosphere is rapid and, to avoid embrittlement by these elements, some form of **shielding** is required. This protection may be obtained by means of flux, gas or a combination of the two, or it may result from physical shielding (as in spot welding) or from evacuation of the atmosphere (as in electron-beam welding). The only process in which flux alone provides the shield is electroslag welding. In manual metal arc and submerged arc welding, flux is provided, but the gases generated by vaporization and chemical reaction at the electrode tip are also protective.

Electrode coatings are composed of a mixture of minerals, organic material, ferro-alloys and iron powder bonded with sodium or potassium silicate. **Cellulosic electrodes** (BS Class C or AWS 6010) contain cellulose, rutile and magnesium silicate; **rutile electrodes** (BS Class R or AWS 6013) contain a small amount of cellulose with rutile and calcium carbonate. Because of the organic content, these types of coating are baked after application at 100–150 °C and a substantial amount of moisture is retained. Consequently, the gas that is generated during welding by decomposition of the coating contains a high proportion (about 40%) of hydrogen. **Basic electrodes** (BS Class B or AWS 7015, 7016 and 7018), on the other hand, contain no organic matter, and therefore are baked at 400–450 °C, which drives off the bulk of the moisture content. The mineral content of such a coating is largely calcium carbonate,

Table 6.1. Fusion welding and cutting processes

Process	Heat source	Power source and polarity	Mechanics	Shielding or cutting agent	Typical applications		Industrial use
					Metals	Thickness range	
Electroslag welding	Resistance heating of liquid slag	Alternating or direct current	Automatic; joint set up vertically; weld pool and slag contained by water-cooled shoes; filler wire fed into slag pool and melted by resistance heating; no arc	Slag	Carbon, low-alloy and high-alloy steel	50 mm upwards	Welding thick sections for press frames, pressure vessels, shafts, etc.; foundry and steelworks applications; general engineering
Submerged arc welding	Arc	Alternating or direct current	Automatic or semi-automatic; arc maintained in cavity of molten flux formed from granular material	Slag and self-generated gas	Carbon, low-alloy and high-alloy steels; copper alloys	1 mm upwards (but generally over 10 mm)	Downhand or horizontal vertical joints suitable for automatic welding; boilers, pressure vessels, structural steel; horizontal joints in storage tanks

(*Continued*)

Table 6.1 (continued)

Process	Heat source	Power source and polarity	Mechanics	Shielding or cutting agent	Typical applications		Industrial use
					Metals	Thickness range	
Manual metal arc welding (coated electrodes)	Arc	Alternating or direct current; electrode positive or negative	Short lengths of wire coated with flux; manual operation	Slag and self-generated gas	All engineering metals and alloys except pure Cu, precious metals, low-melting and reactive metals	1 mm upwards	All fields of engineering
Gas metal arc welding (flux-cored wire)	Arc	Direct current; electrode positive	Flux is enclosed in tubular electrode of small diameter; automatic or semi-automatic; wire fed continuously through a gun with or without a gas shield	Slag and gas, either self-generated or from external source (normally CO_2)	Carbon steel	1 mm upwards	Sheet metal welding; general engineering

Gas metal arc welding (solid wire)	Arc	Direct current; electrode positive	As above, but using solid wire; free flight metal transfer	Argon or helium, argon–O_2 or argon–CO_2	Non-ferrous metals; carbon, low-alloy or high-alloy steel	2 mm upwards	Welding of high-alloy and non-ferrous metals; pipe welding; general engineering
	Arc	Direct current; electrode positive	As above, but in short-circuiting metal transfer mode	Argon–O_2, argon–CO_2 or CO_2	Carbon and low-alloy steel	1 mm upwards	Sheet metal; root pass in pipe welding; positional welding
	Pulsed arc	Direct current; electrode positive; 50–100 Hz pulse superposed on low background current	Pulse detaches drop at electrode tip and permits free flight transfer at low current	Argon, argon–O_2 or argon–CO_2	Non-ferrous metals; carbon, low-alloy and high-alloy steels	1 mm upwards	Positional welding of relatively thin carbon or alloy steel
Gas welding	Oxyacetylene flame		Manual; metal melted by flame and filler wire fed in separately	Gas (CO, CO_2, H_2, H_2O)	Carbon steel, copper, aluminium, zinc and lead; bronze welding	Sheet metal and pipe up to about 6 mm	Sheet metal welding, small diameter pipe
Gas cutting	Oxyacetylene/ oxygen flame		Oxygen jet injected through flame oxidizes and ejects metal along the cutting line	Oxygen	Carbon and low-alloy steel		Cutting and bevelling plate for welding; general engineering appications

(*Continued*)

Table 6.1 (continued)

Process	Heat source	Power source and polarity	Mechanics	Shielding or cutting agent	Typical applications		Industrial use
					Metals	Thickness range	
Gas tungsten arc welding	Arc	Alternating current with stabilization for aluminium, magnesium and alloys; direct current; electrode negative for other metals	Manual or automatic arc maintained between non-consumable tungsten electrode and work; filler wire fed in separately	Argon, helium or argon–helium mixtures	All engineering metals except Zn and Be and their alloys	1 mm to about 6 mm	Non-ferrous and alloy steel welding in all engineering fields; root pass in pipe welds
Pulsed gas tungsten arc welding	Arc	Direct current; electrode negative with low-frequency (1 Hz) or high-frequency (1 kHz) current modulation	Low-frequency pulse allows better control over weld pool behaviour; high-frequency pulse improves arc stiffness	Argon	As above	1 mm to about 6 mm	Automatic gas temperature welding of tubes or of tubes to tubesheets to improve consistency of penetration or (high frequency) prevent arc wander

Plasma welding	Arc	Direct current; electrode negative	As for gas tungsten arc, except that arc forms in a chamber from which plasma is ejected through a nozzle; improved stiffness and less power variation than gas tungsten arc welding	Argon, helium or argon–hydrogen mixtures	As above	Usually up to about 1.5 mm	Normally low-current application where gas tungsten arc lacks stiffness; also used at higher currents in keyholing mode for root runs
Plasma cutting	Arc	Direct current; electrode negative	As for welding, but higher current and gas flow rates	Argon–H$_2$	All engineering metals	1 mm upwards	Used particularly for stainless and non-ferrous metal but also carbon and low-alloy steel
Stud welding	Arc	Direct current; electrode negative for steel, positive for non-ferrous	Semi-automatic or automatic; arc drawn between tip of stud and work until melting occurs and stud then pressed on to surface; weld-cycle controlled by timer	Self-generated gas plus ceramic ferrule around weld zone	Carbon, low-alloy and high-alloy steel; aluminium; nickel and copper alloys require individual study	Stud diameters up to about 25 mm	Shipbuilding, railway and automotive industries; pressure vessels (for attaching insulation); furnace tubes and general engineering

(*Continued*)

Table 6.1 (continued)

Process	Heat source	Power source and polarity	Mechanics	Shielding or cutting agent	Typical applications		Industrial use
					Metals	Thickness range	
Spot, seam and projection welding	Resistance heating at interface of lapped joint	Alternating current; transformer with low-voltage, high-current output	Lapped sheet clamped between two copper electrodes and welded by means of high-current pulses; weld may be continuous (seam) or intermittent (spot and projection)	Self-shielded plus water for resistance welding Mo, Ta and W	All engineering metals except Cu and Ag; Al requires special treatment	Sheet metal up to about 6 mm	Automobile and aircraft industries; sheet metal fabrication in general engineering
Electron-beam welding	Electron beam	Direct current; 10–200 kV; power generally in range 0.5–10 kW; workpiece positive	Automatic welding carried out in vacuum; beam of electrons emitted by cathode focused on joint; no metal transfer	Vacuum (~10^{-4} mm Hg)	All metals except where excessive gas evolution and/or vaporization occurs	Up to about 25 mm normally but may go to 100 mm	Nuclear and aerospace industries; welding and repair of machinery components such as gears

| Laser welding | Light beam | None | As for electron beam except different energy source | Helium | As for electron beam | Up to 10 mm | Potentially as for electron beam; cutting non-metallic materials |
| Thermit welding | Chemical reaction | None | A mixture of metal oxide and aluminium is ignited, forming a pool of superheated liquid metal, which then flows into and fuses with the joint faces | None | Steel, austenitic CrNi steel, copper, copper alloys, steel–copper joints | Normally up to 100 mm | Welding rails and copper conductors to each other and to steel |

which breaks down during welding to provide a $CO-CO_2$ shield. The volume of gas so generated is lower than with cellulosic or rutile coatings and this is one reason why it is necessary to use a shorter arc with basic electrodes than with other types.

In welding with coated electrodes the drops that transfer to the weld pool are coated with slag and the weld pool itself has a slag coating. The self-generated gas is the main protective agent during metal transfer, whereas the slag protects the solidifying and cooling weld metal.

In the American Welding Society (AWS) designations for coated electrodes cited above, the first two numbers represent the minimum ultimate tensile strength of deposited weld metal in thousands of pounds per square inch (ksi). The second pair of numbers indicates the type of coating; for example, 15, 16 and 18 are for basic coatings. This system, which has the great merit of simplicity, is being accepted to an increasing extent in other national and in international standards, with ksi being replaced by metric or SI units. To accommodate the need for improved quality requirements, the AWS has recently introduced 'optional supplemental designations'. E7018-R is the new designation for basic-coated rods with improved resistance to moisture pick-up; E7018-1 has increased ductility and notch-toughness, and so forth. Details will be found in the relevant standards.

Submerged arc flux is a coarse granular powder that is dispensed on to the workpiece immediately ahead of the arc. This flux melts around the arc to form a bubble, which bursts periodically and re-forms. Metal may be transferred directly across the cavity so formed or it may transfer through the wall of molten flux. The shield in this case consists of gas generated from the electrode plus molten flux, plus the unmelted granular layer around the molten flux. The granular flux may be made by agglomeration, sintering or fusing and may be chemically basic, neutral or acid. It may also contain ferro-alloys as deoxidants or additive elements.

Shielding by means of an external gas supply is used in gas metal arc, gas tungsten arc and plasma welding. The gas is dispensed through a nozzle surrounding the electrode, although in the welding of reactive metals an extended shield, or a gas-filled box, may be required. For non-ferrous metals and gas tungsten arc welding, the inert gases argon or helium are used. Argon–hydrogen mixtures may be used for the gas tungsten arc welding of nickel and for plasma torches, where the hydrogen increases the arc power. In the gas metal arc welding of steel, the pure inert gases are not suitable because cathode spots wander over the surface of the workpiece, resulting in irregular weld pool formation. Additions of oxygen or carbon dioxide inhibit cathode spot wander and promote regularity of operation. Typical mixtures are 98%Ar–2%O_2 and 80%Ar–20%CO_2. Spatter increases with increasing CO_2 addition, and above 20%CO_2 it is necessary to use short-circuiting metal transfer to avoid excessive spatter. Pure CO_2 may be used for such **short-circuiting** or **dip-transfer welding** techniques.

In gas metal arc welding, increased use is made of flux-cored wires. Such wires consist of a metal tube or envelope containing flux and drawn down to the same diameter as solid wire: typically 1.2 or 1.6 mm. Deoxidants may be included in the flux, and in this way improved mechanical properties and resistance to weld defects such as porosity may be obtained. Flux-cored arc welding may employ $Ar-O_2$, $Ar-CO_2$ or CO_2 shielding, as with solid wires, the advantage of argon-rich shielding gas being that spatter is less. A special type of flux-cored wire containing aluminium and sometimes more volatile metals is used for **self-shielded** or **no-gas welding**, where there is no externally provided gas shield. The advantage of this process is cost and portability; its disadvantage is that the notch ductility of the weld metal is lower than with a gas shield, and that the arc generates a heavy fume, which impairs visibility. Self-shielded welding may be used in the site fabrication of steel structures.

In resistance welding, the weld pool is generated internally and is protected by the surrounding solid metal. Stud welding relies partly on physical shielding, but the stud tip may be coated with deoxidant. The vacuum in which electron-beam welding is carried out provides complete protection against the external atmosphere, but if the metals being welded contain dissolved gas, this may be evolved, causing porosity. These three processes, which in other respects are very different, all rely to some degree on a mechanical rather than a chemical shield.

6.2.3 Surface and penetrating heat sources

A minimum **heat source intensity** is required before it is possible for a fusion weld to be made. It will be evident that if the total heat flux across the **fusion boundary** is q, then as a minimum there must be heat liberation at an equal rate within the area of the molten weld pool. As the heat source intensity increases, a point is reached (at about $10^9 \, W \, m^{-2}$) where the source is capable of vaporizing the metal as well as melting it. The pressure generated by the emergent metal vapour then depresses the weld pool until this pressure is balanced by hydrostatic pressure and surface tension forces. This type of **penetrating heat source** is characteristic of electron beam and laser welding. The arc force may also exert a pressure on the weld pool surface, and where this is high (as in high-current gas metal arc welding) it may produce a deep finger-like penetration. In plasma welding the arc force is augmented by a gas jet which can produce a condition halfway between cutting and welding. There is a hole in the weld pool but the molten metal flows together at the rear: a technique known as **keyhole** welding.

In principle the use of a penetrating heat source in fusion welding is attractive since the fused and heat-affected zones are narrow and this minimizes both metallurgical damage and distortion. However, the joint fit-up prior to welding must be accurate and the motion of the torch relative to the joint must be equally accurate. Even so, such deep-penetration welds may be subject to defects such as **cold shuts** or **lack of fusion**, **porosity** and **hot cracking** if correct procedures are not followed. Because of the requirement for high precision and because of the high cost of equipment, this type of welding is used mainly for joining pre-machined components and in capital-intensive industries such as automobiles and aerospace. Likewise, keyhole welding, although more tolerant as regards fit-up, is not an easy process to control, and is encountered more frequently in the research laboratory than in production.

Most welding is carried out using processes that generate a surface heat source and produce a weld pool of roughly semicircular cross-section. It is possible to make a weld pool of any size, simply by increasing the heat input rate. However, the weld pool may become difficult to control as it becomes larger, and the grain size may become undesirably large. Therefore, it is customary to limit the size or penetration of a weld to, say, 20 mm or less. When it is required to fill a joint of greater thickness, it is necessary to make a number of successive passes or runs. This technique is known as multipass or multi-run welding (Fig. 6.1). Note that a multipass weld contains a multiplicity of heat-affected zones which run through both parent metal and weld metal.

The electroslag and electrogas processes are capable of welding very thick plate in a single pass. In these processes the electrode or electrodes are fed downwards into a gap between the two plates, and a weld pool is formed that occupies the entire thickness. This is held in place by water-cooled dams that move upwards with the weld. Welds made in steel by such processes are coarse-grained unless given a normalizing heat treatment, and may not be acceptable for applications where there is a notch-ductility requirement.

6.2.4 Selection of welding process

For the most part, the process to be used for a particular application will be dictated by a combination of economics and feasibility. For example, in the shop welding of pressure vessels, it is practicable to make the main seam welds in the flat position. At the same time, since the plate to be welded is usually fairly thick, it is necessary to maximize the rate of deposition of weld metal. For such a purpose, submerged arc welding is both feasible and economic, and this is the process most commonly selected.

Welding with a parallel-sided or slightly tapering edge preparation (**narrow gap welding**) is used in the fabrication of heavy-walled pressure vessels, usually

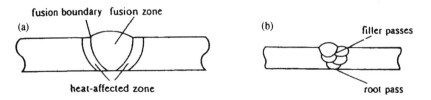

6.1 (a) Single-pass and (b) multipass fusion welds.

with submerged arc but sometimes employing gas metal arc welding. Electroslag welding is possible, but the joints (in steel) require normalizing, and normalizing a complete pressure vessel shell is not usually feasible. Electron-beam welding would be economic, but not feasible because of the large space to be evacuated and because of fit-up problems. Such restraints apply in many cases, and restrict the applicability of individual processes as indicated by Table 6.1.

Welding with coated electrodes remains the most widely used welding process. It is infinitely versatile and can be used in any position, for deep sub-ocean work and for welding in outer space. However, gas metal arc welding is more economic (largely because it uses a continuous electrode) and is gaining ground wherever it can be used.

Automatic systems with feedback control to obtain optimum welding conditions (e.g. to achieve consistent penetration) are increasingly used, as are robots. Gas-shielded welding processes may be employed in such applications: gas tungsten arc for thin material, and gas metal arc for laying down heavier welds and particularly for robot-controlled fillet welds. Robots are used for spot welding in the automobile industry.

6.2.5 Welding power sources

The electronic revolution has had a profound effect on the versatility and controllability of the type of machine used to supply current in electric arc welding and cutting. A brief outline of these developments, which have a considerable potential for improvement in weld quality, is given below (Lancaster, 1997).

Simple machines consist of a transformer with, typically, a three-phase AC output with an open-circuit voltage of 65–100 V: a level usually considered low enough to be safe. Current control is obtained either by taking different tappings on the secondary coil or by moving the transformer core. Alternatively, for site work, a rotating power source may be driven by an internal combustion engine. In the past, AC was more commonly used in Europe and Japan but US fabricators preferred DC for quality work. Almost all welding was carried out using stick electrodes and the machines were traditionally designed to have a drooping characteristic. In this way, the change in current produced by an inadvertent change in arc length (with corresponding change in arc voltage) is minimized so helping to maintain steady welding conditions (Fig. 6.2). The availability of solid-state rectifiers has led to a more general use of DC, and simple transformer–rectifier sets are often employed for welding with coated electrodes.

More recent developments owe much to a need to improve the operating characteristics of pulsed arc welding. The first pulsed arc machines were only capable of pulsed frequencies of either mains frequency or multiples thereof, and were dynamically slow, such that the process was difficult to use (Fig. 6.3).

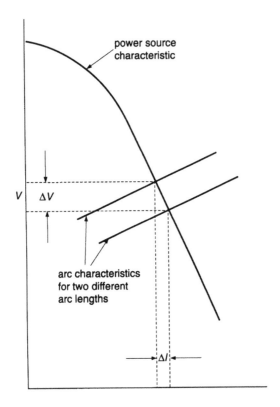

6.2 Effect of changing arc length on welding current.

Variable frequency sources were required, and these became available as a result of the development of a transistor chopper-controlled regulator in the early 1970s. Figure 6.4(a) illustrates a typical circuit. Such machines were, however, large and expensive, and largely used for research and development. An improved version, which became commercially available, utilized solid-state power devices on the output side of the transformer to generate pulses.

The third generation of pulsed arc machines (Fig. 6.4c) is a radical departure. The three-phase power supply is first rectified, and then inverted to high-frequency AC. The AC is fed to a transformer, the output of which is rectified and provides either a pulsed or steady DC source. For pulsed current, the switching is carried out using power transistors or thyristors on the primary side of the transformer, and this makes possible a nearly square wave shape for the pulse. When the switch is on the secondary side, however, the waveform is trapezoidal (Fig. 6.3). The high-frequency transformer requires a smaller core than a normal mains frequency type, such that the weight and volume of an inverter power source is about half that of a conventional type, and the price, although higher, may well be justified by improved performance.

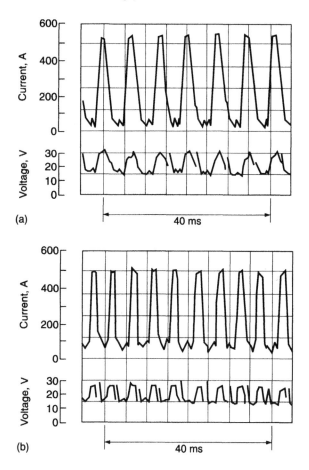

6.3 Oscillograms of voltage and current for: (a) conventional pulsed power source (trapezoidal wave); (b) inverter type (rectangular square wave) (Daihen Corporation).

Apart from pulsed arc welding, current pulsing may be useful in gas tungsten arc or plasma welding, in short circuiting and flux-cored gas metal arc welding, and in high-speed plasma cutting. Inverter power sources may also be used primarily for portability, for example in low-current (below 50 A) plasma-cutting machines and for site welding.

Thus the number of types of power source has multiplied remarkably, as has the number of welding processes. Figure 6.5 shows a classified list of power sources used in Japan, in which the listing relates to the control system. Within such general categories, there are various types adapted to particular processes or to a range of processes; for example, in 1989 one company listed seven types of

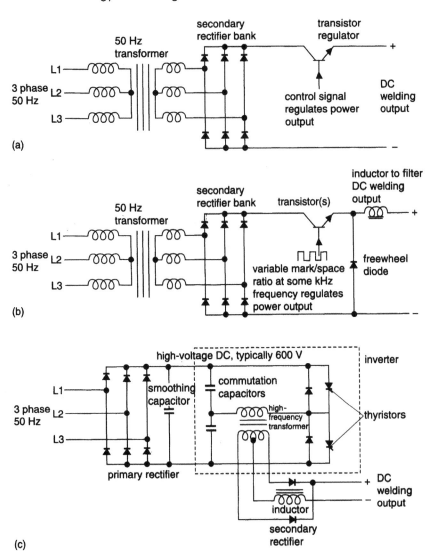

6.4 Various types of pulsed welding power sources: (a) 50 Hz secondary transistor regulator; (b) secondary transistor switched mode regulator; (c) primary switched inverter.

inverter source, while the old moving core or stepped voltage AC welding set was at that time still available. The general tendency with all power sources, however, is towards electronic control both on the input side (to compensate for fluctuations in mains voltage) and on the output side to provide a minimum stepless control of both current and voltage.

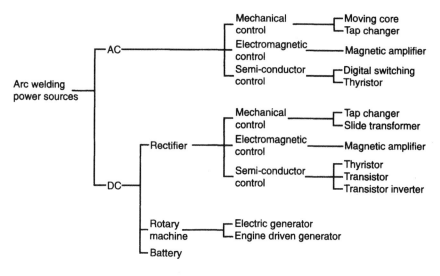

6.5 Classification of arc welding power sources.

6.3 Types of fusion-welded joint

Fusion welds may be made in a variety of forms. The joint illustrated in Fig. 6.1 is a **butt weld** and in this instance the metal is fused to the full thickness of the plates being joined; it is therefore a **full penetration weld**. For some engineering purposes, **partial penetration welds** are also used. These and other types of joint that are made by fusion welding are illustrated in Fig. 6.6.

Note that a **spot-welded lap joint** may be made by resistance welding, in which case the fused zone is formed at the joint interface, or by arc welding, when the fused zone penetrates from the surface of the top sheet to the underlying sheet. Resistance welding may also be used to make a continuous weld (known as a **seam weld**) by forming a series of overlapping spot welds.

In order to achieve the required degree of penetration, it is usually necessary to shape the edges of plates that are to be butt welded, and sometimes to leave a gap between them (Fig. 6.7). In principle, it is possible, by using a heat source of sufficient power, to fuse through the complete section even of very thick plate. However, the large weld pool so produced is difficult to control unless, as in electroslag welding, special equipment is available for this purpose. Also, the weld metal and heat-affected zone of such welds have a relatively coarse grain and their mechanical properties may be adversely affected. Therefore it is normal practice to make multi-run welds in all but the thinnest material. Edge preparation is required in multi-run welds so as to give access to the root and all other parts of the joint, thereby ensuring that each run may be properly fused to the parent metal and to the run below. The metal required to fill the groove so formed is supplied

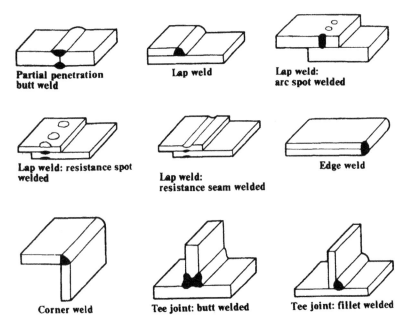

6.6 Types of fusion-welded joint.

either by **filler rod**, which is added independently of the heat source, as in oxyacetylene and gas tungsten arc welding, or by the melting of an electrode in other fusion welding processes. A fusion weld is most readily made when the plate lies in a horizontal plane and the welding is carried out from the top side (**flat welding**). Quite frequently, however, it is necessary to make the joint when the two members are set up vertically or at an angle to the horizontal, or it may be required to weld horizontal plates from the underside. The operation is then

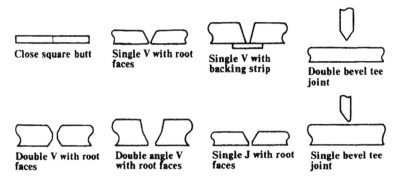

6.7 Typical edge preparations for welding.

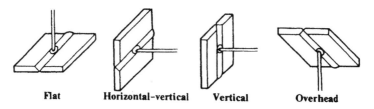

6.8 Welding positions.

known as positional welding and the nomenclature of the various positions is indicated in Fig. 6.8.

Position is an important variable in the qualification of welding operators and procedures. The ASME (American Society of Mechanical Engineers) boiler and pressure vessel code and the structural welding code AWS D1.1 give detailed recommendations.

6.4 Heat flow in fusion welding

6.4.1 General

The equations that govern the flow of heat in an isotropic solid are linear in character, and it is therefore possible to find analytic solutions that give the temperature distributions for a number of different configurations and types of heat source. The models used in such derivations are usually of simplified form; for example, heat is assumed to be emitted at a point, or along a line, or across a plane. Real heat sources are finite in extent, and for this and other reasons the actual temperature distributions obtained in welding deviate somewhat from those calculated from theoretical expressions. This is not important, however, because calculations of heat flow in welding are not required for any practical purpose, but rather to promote understanding of the nature of the process, and to predict trends in general terms. For the most part, therefore, it is sufficient to consider solutions obtained using an idealized model of the process in question.

6.4.2 Stationary heat sources

In this and the following section Cartesian coordinates are used, with orthogonal axes x, y and z. In this system the equation for the conduction of heat in an isotropic uniform solid is

$$\nabla^2 T - \frac{1}{\alpha}\frac{\partial T}{\partial t} = 0 \qquad (6.1)$$

or

$$\frac{\partial^2 T}{\partial x^2} + \frac{\partial^2 T}{\partial y^2} + \frac{\partial^2 T}{\partial z^2} - \frac{1}{\alpha}\frac{\partial T}{\partial t} = 0 \tag{6.2}$$

where T is temperature, t is time and α is the *diffusivity of heat* for the solid in question. Diffusivity is equal to thermal conductivity divided by the specific heat per unit volume:

$$\alpha = \frac{K}{\rho c} \tag{6.3}$$

where K is thermal conductivity, ρ is density and c is specific heat. In SI units the dimensions of α are $m^2 s^{-1}$.

Consider firstly the case where a quantity of heat Q is released instantaneously at a point in an infinite solid. It will be convenient to define the strength of this **heat source** as Q_s, where

$$Q_s = \frac{Q}{\rho c} \tag{6.4}$$

Then if the heat source is located at the origin, the temperature distribution is given by

$$T = \frac{Q_s}{8(\pi \alpha t)^{3/2}} \exp(-r^2/4\alpha t) \tag{6.5}$$

where $r^2 = x^2 + y^2 + z^2$.

At any point distance r from the source the temperature rises from zero to a maximum and then falls to zero again. The time to reach the maximum is $t = r^2/6\alpha$.

In the case where heat is emitted instantaneously along a line at Q' heat units per unit length, the corresponding source strength is $Q'_s = Q'/\rho c$. The temperature distribution is then

$$T = \frac{Q'_s}{4\pi \alpha t} \exp(-r^2/4\alpha t) \tag{6.6}$$

where $r^2 = x^2 + y^2$. It is assumed here that the source lies along the z axis.

The corresponding expression for a plane heat source of strength Q''_s per unit area and lying in the y–z plane is

$$T = \frac{Q''_s}{2(\pi \alpha t)^{1/2}} \exp(-x^2/4\alpha t) \tag{6.7}$$

Note that heat flow associated with point, line and plane heat sources is, respectively, three-, two- and one-dimensional. Denoting the number of such dimensions as D, the temperature distribution for all three cases may be expressed as

$$T = \frac{[Q_s]}{2^D(\pi\alpha t)^{D/2}} \exp(-r^2/4\alpha t) \tag{6.8}$$

where $[Q_s]$ is Q_s, Q'_s or Q''_s as appropriate. Similarly the time to reach peak temperature at any radius r (obtained by putting $\partial T/\partial t = 0$) is

$$t = \frac{r^2}{2D\alpha} \tag{6.9}$$

For two-dimensional flow the formula is valid for a plate or sheet, provided that heat flow across the surfaces is small enough to be ignored. The temperature distribution for a plate of thickness w is then

$$T = \frac{Q}{4\pi w K t} \exp(-r^2/4\alpha t) \tag{6.10}$$

For the purpose of calculation, it is convenient to express temperature in terms of a reference radius r_0, which represents the boundary of the fused zone. At this radius the peak temperature is the melting point T_m, and the elapsed time is

$$t_0 = \frac{r_0^2}{2D\alpha} \tag{6.11}$$

and

$$T_m = \frac{[Q_s]}{2^D(\pi\alpha t)^{D/2}} \exp\left(-\frac{D}{2}\right) \tag{6.12}$$

In the case of an instantaneous line source, for example

$$T = \frac{Q'}{\pi r_0^2} \exp(-1) \tag{6.13}$$

at the time t_0 the radial temperature distribution is

$$T = \frac{Q'_s}{4\pi\alpha t_0} \exp(-r^2/4\alpha t_0)$$
$$= T_m \exp(1 - r^2/r_0^2) \tag{6.14}$$

Likewise, at radius r, the temperature as a function of time is

$$T = T_m \frac{t_0}{t} \exp(1 - t_0/t) \tag{6.15}$$

Equations 6.14 and 6.15 have been evaluated for steel with $T_m = 1500\,°C$ and the results are plotted in Fig. 6.9 and 6.10 respectively. The physical properties of some metals that may be used in heat flow calculations are listed in Table 6.2.

In electric resistance spot welding, the joint is made by passing a high current for a period of less than one second, so that the instantaneous heat source could

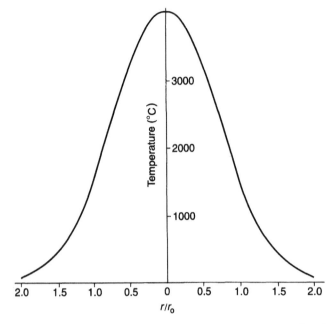

6.9 Temperature distribution due to an instantaneous line heat source at time t_0 when the peak temperature at radius τ_0 is equal to 1500 °C.

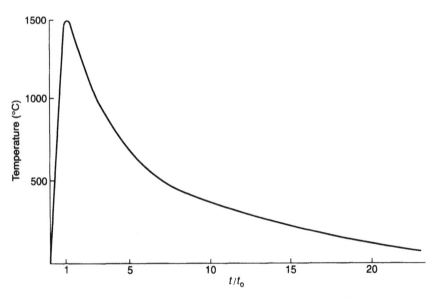

6.10 Time variation of temperature at radius r_0 due to a line heat source when the peak temperature at r_0 is 1500 °C.

Table 6.2. Physical properties of metals

Metal	Mass density (kg m^{-3})	Specific heat c (J kg^{-1} K^{-1})	Thermal conductivity K (J m^{-1} K^{-1} s^{-1})	Thermal diffusivity α (m^2 s^{-1})	Melting point T_m (°C)	Boiling point T_b (°C)
Aluminium	2.7×10^3	880	205	8.6×10^{-5}	660	2400
Brass (60/40)	8.55×10^3	370	121	3.8×10^{-5}	965	–
Copper	8.96×10^3	380	390	1.14×10^{-4}	1083	2580
Nickel	8.9×10^3	444	91	2.3×10^{-5}	1453	2820
Silver	10.5×10^3	232	418	1.7×10^{-4}	981	2180
Steel, mild	7.85×10^3	450	50	1.4×10^{-5}	1500	3000
Steel, austenitic	7.9×10^3	500	16	4.05×10^{-6}	1500	3000
Titanium 6Al4V	4.4×10^3	610	5.8	2.2×10^{-6}	1700	3300

152   Metallurgy of welding

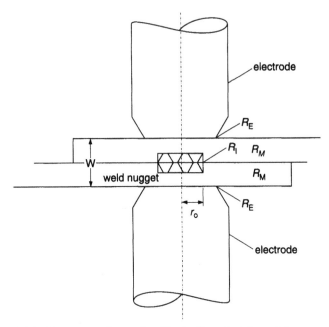

6.11 Diagrammatic section illustrating electric resistance spot welding of sheet metal.

provide a reasonable model for this process. Figure 6.11 indicates some essential features of spot welding. Overlapping sheets of metal are clamped between two water-cooled copper alloy electrodes. This arrangement forms part of the secondary loop of a low-voltage transformer. When contact is made, heat is generated by resistance at the locations indicated, R_E being the surface resistance between electrode and metal, R_M the resistance of the bulk metal, and R_I that of the interface of the surfaces between the two sheets. Of these, the highest is R_I and it is here that the metal melts to form the **weld nugget**. The heat generated at the interface is

$$Q = I^2 R_I t \tag{6.16}$$

where I is electric current and t is the time for which it passes.

Some typical figures for the spot welding of mild steel sheet are given in Table 6.3. These include the radius of the fused zone, which is the radius at which the peak temperature just reaches the melting point.

Although the heat source is disc-shaped, and lies in a plane parallel to the surface, heat flow outside the fused zone must, because of the geometry of the arrangement, be largely radial and two-dimensional. Therefore the line heat source is used as a suitable model. In terms of previous nomenclature, the radius

Table 6.3. Conditions for resistance spot welding of mild steel

Sheet thickness (mm)	Electrode face diameter (mm)	Force (kN)	Welding current (A)	Weld time (s)	Weld radius (mm)	Pressure over weld area (MN m^{-2})
0.5	4.7	1.76	8 800	0.117	1.25	360
1.5	7.8	4.95	15 000	0.235	2.92	166
3.0	9.4	9.3	21 000	0.43	3.81	204

Source: recommendations from AWS *Welding Handbook*, Vol. 2, converted to SI units.

of the fused zone is r_0 and the time required to form it is t_0. The quantity of heat required to make the weld is then, from equation 6.6

$$Q = 4\pi \rho c T_m w \alpha t_0 \exp(r_0^2/4\alpha t_0) = \pi \rho c T_m w r_0^2 \exp(1) \qquad (6.17)$$

since for a peak temperature $r_0^2 = 4\alpha t_0$. From this same expression the weld time t_0 is $r_0^2/4\alpha$. The surface resistance R_1 is then equal to $Q/I^2 t_0$.

From these various expressions it is possible, using the weld radii listed in Table 6.3, to estimate the quantity of heat required to make the weld, the electrical resistance, and the weld time for the three sheet thicknesses. The results are given in Table 6.4. The calculated figures are low, as will be seen from the comparison of actual and calculated weld times. The same applies to the figures for surface resistance. Measured values for this quantity, made under good welding conditions, lie typically in the range 50–250 µΩ. The discrepancy may in part be due to heat loss to the electrodes, for which no allowance was made. On the whole, however, it is sufficient, bearing in mind the idealized and somewhat unrealistic model, that the figures obtained are of the correct order of magnitude and show correct trends.

Table 6.4. Calculated variables for spot welding

Sheet thickness	Quantity of heat (J) Calculated	Surface resistance (µΩ) Calculated	Weld time (s) Calculated	Weld time (s) Actual
0.5	71	35	0.028	0.117
1.5	1137	34	0.15	0.235
3.0	3941	34	0.26	0.43

6.4.3 Moving heat sources

Most fusion welding processes employ a heat source of high power density that is traversed at a steady rate along the joint. In analysing the thermal effect of such moving heat sources, it is customary to regard the source as being fixed, and located at the origin of the coordinate system. The solid then moves with velocity v in the positive direction along the x axis. The resulting temperature distribution resembles the bow wave and wake of a ship, as seen from the ship. Relative to the source, this temperature distribution is stationary, and the condition is known as the **quasi stationary state**. Distances upstream of the source are negative and vice versa.

The basic assumptions are that the solid is isotropic and moves at a uniform rate, that thermal properties are constant, and that the effect of latent heat may be ignored. The equation for diffusion of heat is then

$$\nabla^2 T - \frac{v}{\alpha}\frac{\partial T}{\partial x} = 0 \qquad (6.18)$$

Solutions of this equation for the three-, two- and one-dimensional cases respectively are, in terms of rectangular coordinates (x, y, z), as follows:

1. A point heat source of power q on the surface of a semi-infinite body that is moving with velocity v:

$$T = \frac{q}{2\pi K r} e^{-v(r-x)/2\alpha} \qquad (6.19)$$

   where $r^2 = x^2 + y^2 + z^2$.

2. A line source of power q' per unit length penetrating an infinite plate that is moving with velocity v:

$$T = \frac{q'}{2\pi K} e^{vx/2\alpha} K_0\left(\frac{vr}{2\alpha}\right) \qquad (6.20)$$

   where $r^2 = x^2 + y^2$.

3. A plane heat source of power q'' per unit area lying at right angles to a semi-infinite rod moving with velocity v:

$$\begin{aligned} T &= \frac{q''}{\rho c v} && \text{if } x > 0 \\ T &= \frac{q''}{\rho c v} e^{vx/\alpha} && \text{if } x < 0 \end{aligned} \qquad (6.21)$$

Solution 1 represents an approximation to the conditions in multipass welding on thick plate; 2 approximates to single-pass welding of thin sheet or welding with a penetrating heat source such as an electron beam or a laser beam; 3 gives the temperature distribution in the solid portion of a welding electrode where circulation in the liquid portion is such as to make it more or less transparent to heat.

The point and line source solutions are those relevant to the temperature distribution in fusion welding. The **thermal cycle** is obtained by plotting T as a function of t (equal to x/v) for a fixed distance from the centreline $y = 0$.

One of the most important variables in fusion welding is the **cooling rate** in the weld and heat-affected zone. The theoretical cooling rate for any point (x, y, z) relative to the source as origin may be obtained by differentiation of equations 6.19 and 6.20 to give, for the three-dimensional case,

$$\frac{\partial T}{\partial t} = \frac{vT}{r}\left[\frac{x}{r} - \frac{vr}{2\alpha}\left(1 - \frac{x}{r}\right)\right] \tag{6.22}$$

and for the two-dimensional case,

$$\frac{\partial T}{\partial t} = \frac{v^2 T}{2\alpha}\left[\frac{(x/r)K_1(vr/2\alpha)}{K_0(vr/2\alpha)} - 1\right] \tag{6.23}$$

where $K_n(z)$ is a tabulated function.

For simplicity, consider a point along the central axis of the weld at the rear boundary of the weld where $r = x = x_1$ and $T = T_m$ the melting temperature. Then for three-dimensional flow,

$$\frac{\partial T}{\partial t} = -\frac{vT_m}{x_1} \tag{6.24}$$

while for two-dimensional flow,

$$\frac{\partial T}{\partial t} = -\frac{v^2 T_m}{2\alpha}\left[\frac{K_1(vx_1/2\alpha)}{K_0(vx_1/2\alpha)} - 1\right] \tag{6.25}$$

Equation 6.24 predicts that the cooling rate at the downstream edge of the weld pool will increase as the welding speed increases, and for similar welding speeds will be higher for small weld pools than for large. The cooling rates for two-dimensional flow follow a similar trend but are generally about half those for similar conditions in three-dimensional flow (Fig. 6.12 and 6.13). Equation 6.24 may also be expressed as follows:

$$\frac{\partial T}{\partial t} = -\frac{2\pi K T_m^2}{q/v} \tag{6.26}$$

Thus the cooling rate along the axis at the rear of the weld pool is inversely proportional to the parameter q/v. This quantity is known as the **heat input rate** (expressed in joules per unit length) and may be used as a means of standardizing or comparing the heat-flow conditions of welds. Both weld pool size and heat input rate are readily observable indicators of cooling rate, and their significance will be discussed in a later section. Note that in all cases the temperatures referred to are those in excess of the temperature of the metal being

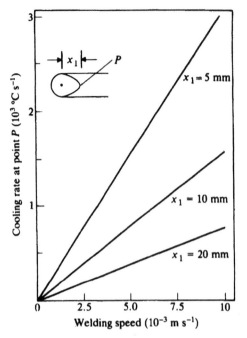

6.12 The theoretical cooling rate at the rear of a weld pool for three-dimensional heat flow in carbon steel, assuming a point heat source.

welded. Designating temperatures as θ so that the melting temperature is θ_m and the plate temperature θ_0, we have

$$T = \theta - \theta_0$$
$$T_m = \theta_m - \theta_0$$
(6.27)

If the metal is **preheated** prior to welding, θ_0 is the **preheat temperature**.

6.4.4 Calculating the thermal cycle

It is instructive to calculate the thermal cycles for different welding processes and, in particular, to compare that of electroslag welding, which has very slow heating and cooling rates, with those for other techniques. Table 6.5 lists values of the welding variables that are typical of submerged arc, electroslag and manual metal arc welding. Since the electroslag process makes the weld in a single pass, the quantities for the other two processes are also those appropriate to single-pass welding, and the relevant heat flow equation is that for a line heat source.

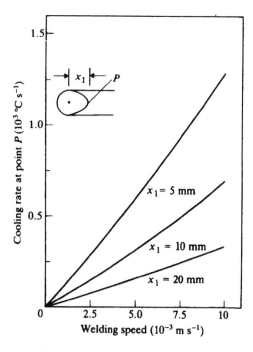

6.13 The theoretical cooling rate at the rear of a weld pool for two-dimensional heat flow in carbon steel, assuming a line heat source.

Table 6.5. Process variables and calculated weld dimensions

Process	Submerged arc		Electroslag		Manual metal arc	
Welding voltage (V)	34		52		20	
Welding current (A)	1100		700		150	
Welding speed v (m s^{-1})	8×10^{-3}		4×10^{-4}		2.5×10^{-3}	
Plate thickness (line source) w (m)	1.27×10^{-2}		7.5×10^{-2}		6×10^{-3}	
Operating parameter, n	22.68		1.10		0.483	
Type of heat source	Point*	Line**	Point	Line	Point*	Line**
$\bar{r}_0 = V r_0 / 2\alpha$	9.2	23.04	na	0.76	0.36	0.60
$\bar{x}_0 = v x_0 / 2\alpha$	8.30	22.55	na	0.48	0.10	0.36
$\bar{y}_0 = v y_0 / 2\alpha$	3.97	4.70	na	0.59	0.36	0.48
Calculated weld width, $B = 2y_0$ (mm)	27.8	32.9	na	82.6	8.1	10.8

*Representing a multipass weld run on thick plate.
**Representing a single-pass weld using a backing plate.
na, not applicable.

158 Metallurgy of welding

The first step is to establish the power of the source. This is

$$q' = \frac{\eta VI}{w} \qquad (6.28)$$

where V is voltage, I is current, w is plate thickness and η is the source efficiency. Figure 6.14 shows values of η for arc welding processes. In the case of electroslag welding there is some loss of heat to the water-cooled shoes that support the weld, but figures for this quantity are lacking, and it will be assumed that $\eta = 1$. The same value will be taken for submerged arc welding, and that of 0.85 for manual metal arc.

The equations for the temperature distribution due to a moving heat source lend themselves to expression is non-dimensional form, and this is helpful both

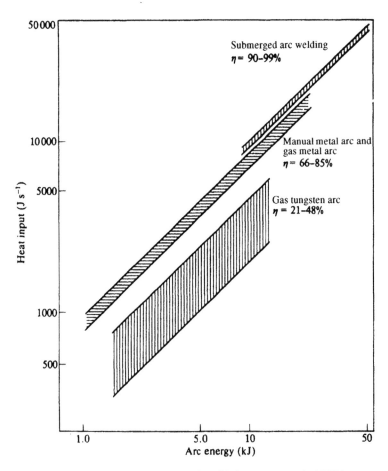

6.14 Measured arc efficiency η. (After Christensen et al., 1965.)

Fusion welding processes and their thermal effects 159

for presentation and for computation. Thus, we put, for the three coordinates

$$\bar{x} = \frac{vx}{2\alpha}, \quad \bar{y} = \frac{vy}{2\alpha}, \quad \bar{z} = \frac{vz}{2\alpha} \tag{6.29}$$

while the non-dimensional plate thickness is

$$\bar{w} = \frac{vw}{2\alpha} \tag{6.30}$$

The temperature is normalized in relation to the melting point T_m:

$$\bar{T} = \frac{T}{T_m} \tag{6.31}$$

The relationship for three-dimensional flow then becomes

$$\bar{T} = \frac{qv}{4\pi K\alpha T_m \bar{r}} \exp[-(\bar{r} - \bar{x})] \tag{6.32}$$

The quantity $qv/(4\pi K\alpha T_m)$ is also non-dimensional, and was designated the **operating parameter** by Christensen et al. (1965), whose work is reported later in this chapter. So we put

$$n = \frac{qv}{4\pi K\alpha T_m} \tag{6.33}$$

and hence

$$\bar{T} = \frac{n}{\bar{r}} \exp[-(\bar{r} - \bar{x})] \tag{6.34}$$

where $\bar{r}^2 = \bar{x}^2 + \bar{y}^2 + \bar{z}^2$.

The corresponding expression for two-dimensional flow is

$$\bar{T} = \frac{n}{\bar{w}} e^{\bar{x}} K_0(\bar{r}) \tag{6.35}$$

where $\bar{r}^2 = \bar{x}^2 + \bar{y}^2$.

The boundary of the fused zone is located along a line or plane where, as in the case of a stationary heat source, the peak temperature just reaches the melting point. The relevant quantities will be given the suffix zero. Thus the weld width is $2y_0$. In two-dimensional flow y_0 is independent of the depth z, but in the three-dimensional case the value at the surface ($z = 0$) will be used. Thus, in both cases the weld boundary is defined by

$$\frac{d\bar{T}}{d\bar{x}} = 0, \quad \bar{y} = \bar{y}_0, \quad \bar{r}^2 = \bar{x}^2 + \bar{y}_0^2 \tag{6.36}$$

whence

$$\frac{d\bar{r}}{d\bar{x}} = \frac{\bar{x}}{\bar{r}} \tag{6.37}$$

Applying these conditions to equation 6.34 leads to

$$\bar{x}_0 = \frac{\bar{r}_0^2}{(1 + \bar{r}_0)}; \quad \bar{y}_0 = \bar{r}_0 \frac{(1 + 2\bar{r}_0)^{1/2}}{(1 + \bar{r}_0)} \tag{6.38}$$

Equation 6.34 then becomes

$$\bar{T} = 1 = n/\bar{r}_0 \exp\left\{-\left[\frac{\bar{r}_0}{(1+\bar{r}_0)}\right]\right\} \quad (6.39)$$

whence

$$\bar{r}_0 \exp\left[\frac{\bar{r}_0}{(1+\bar{r}_0)}\right] = n \quad (6.40)$$

which may be used to calculate first $\bar{r}_0$, then $\bar{x}_0$ and $\bar{y}_0$, for any value of n. For steel

$$n = \eta V I v / 13.195 \quad (6.41)$$

The boundary of the fused zone at the surface is defined by

$$\bar{T} = 1 = \frac{n}{\bar{r}} \exp[-(\bar{r} - \bar{x})] \quad (6.42)$$

which is more convenient in the form

$$\bar{x} = \bar{r} - \ln\left(\frac{n}{\bar{r}}\right) \quad (6.43)$$

Fused zone profiles have been calculated for the submerged arc and the manual metal arc variables listed in Table 6.5, and these are plotted in Fig. 6.15(a). In both cases the variables have been taken to be applicable firstly to a point heat source, and secondly to a line source. Figure 6.15(a) and (b) show the elongated shape typical of a high heat input rate process. The profile for the line source (b) is unrealistic, since this type of weld would need to be supported by a backing bar, which would invalidate the assumption of a line heat source. Figure 6.15(c) and (d) have the rounded shape typical of low heat input rate processes such as manual metal arc or gas tungsten arc.

Having obtained a figure for $\bar{y}_0$, the thermal cycle at the weld boundary is obtained from equation 6.34, with $\bar{x} = (\bar{r}^2 - \bar{y}_0^2)^{1/2}$ and $t = x/v = \bar{x} \times 2\alpha/v^2$. However, in order to include electroslag welding, the line source equation must be used. The non-dimensional form of this equation is given by equation 6.35:

$$\bar{T} = \frac{n}{\bar{w}} e^{\bar{x}} K_0(\bar{r}) \quad (6.35)$$

This is derived by putting $q' = q/w$ in equation 6.20 and then proceeding as for the point source. At the fusion boundary, as before $\bar{r}^2 = \bar{x}^2 + \bar{y}_0^2$, and $d\bar{r}/d\bar{x} = \bar{x}/\bar{r}$. However, the condition $d\bar{T}/d\bar{x} = 0$ leads to

$$\frac{\bar{x}_0}{\bar{r}_0} = \frac{K_0(\bar{r}_0)}{K_1(\bar{r}_0)} \quad (6.44)$$

$K_0(r)$ and $K_1(r)$ are Bessel functions, which in general are solutions of the type of differential equation that is appropriate to systems of cylindrical geometry.

It is now necessary to obtain $\bar{r}_0$ from

$$\bar{T} = 1 = n/\bar{w}\{\exp[K_0(\bar{r}_0)/K_1(\bar{r}_0)]\bar{r}_0\}K_0(\bar{r}_0) \quad (6.45)$$

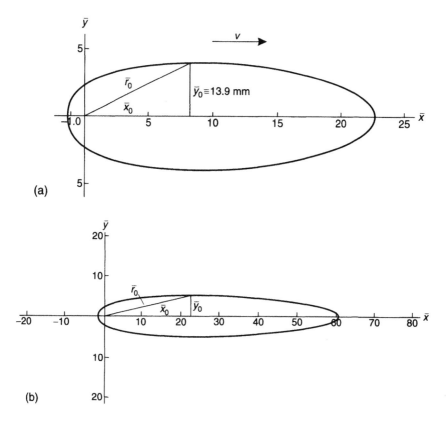

6.15 Calculated surface outline of the fused zone in steel plate due to a moving heat source: (a) submerged arc, point source; (b) submerged arc, line source; (c) manual metal arc, point source; (d) manual metal arc, line source. Variables are those listed in Table 6.5.

and hence

$$\bar{y}_0 = \left[1 - \frac{K_0^2(\bar{r}_0)}{K_1^2(\bar{r}_0)}\right]^{1/2} \bar{r}_0$$

The thermal cycle along the fusion boundary may then be obtained from equation 6.35 with, as before, $\bar{x} = (\bar{r}^2 - \bar{y}_0^2)^{1/2}$ and $t = \bar{x} \times 2\alpha/v^2$.

Plots derived in this manner for the three sets of variables given in Table 6.5 are shown in Fig. 6.16. Note the long dwell time at high temperature in the case of electroslag welding; this produces the coarse as-welded structure illustrated in Fig. 8.24.

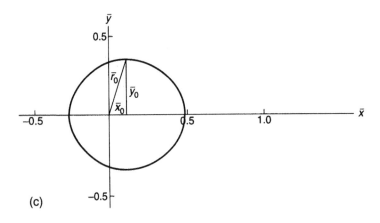

(c)

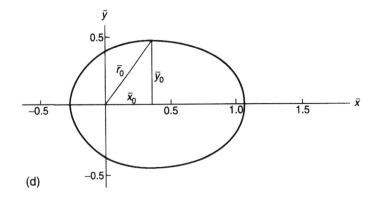

(d)

6.15 *(Continued)*

6.4.5 Calculations for a moving line source of heat

Whereas it is relatively straightforward to obtain accurate results using the equation for a moving point source of heat, this is not so for the line source case, where values of the modified Bessel function of the second kind, $K_0(z)$ and $K_1(z)$ are required. Although tabulated values of these functions are available, it is much more satisfactory to compute the figures directly. These may be obtained using the following expressions:

$$K_0(z) = -\left[\gamma + \ln\left(\frac{z}{2}\right)\right] + \sum_{r=1}^{\infty} \frac{(z/2)^{2r}}{(r!)^2}\left\{-\left[\gamma + \ln\left(\frac{z}{2}\right)\right] + \left(1 + \frac{1}{2} + \frac{1}{3} + \cdots + \frac{1}{r}\right)\right\} \quad (6.46)$$

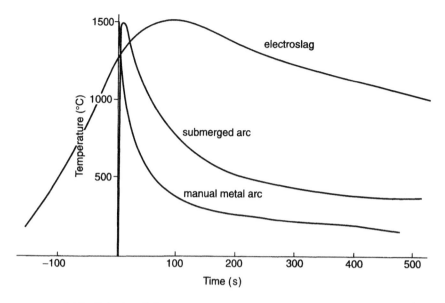

6.16 Calculated time–temperature charts for single-pass welds in steel using the manual metal arc (shielded metal arc), submerged arc and electroslag processes. The welding variables used in calculation are given in Table 6.5.

$$K_1(z) = \frac{1}{2} + \left[\gamma + \ln\left(\frac{z}{2}\right) - \frac{1}{2}\right]\frac{z}{2} + \sum_{r=1}^{\infty} \frac{(z/2)^{2r+1}}{r!(r+1)!}$$
$$\times \left\{\left[\gamma + \ln\left(\frac{z}{2}\right) - \frac{1}{2}\right] - \frac{1}{2}\left[\left(1 + \frac{1}{2} + \frac{1}{3} + \cdots + \frac{1}{r}\right)\right.\right.$$
$$\left.\left. + \left(\frac{1}{2} + \frac{1}{3} + \frac{1}{4} + \cdots + \frac{1}{r+1}\right)\right]\right\} \tag{6.47}$$

In these formulae γ is Euler's constant

$$\gamma = \lim_{m \to \infty}\left[1 + \frac{1}{2} + \frac{1}{3} + \cdots + \frac{1}{m} - \ln(m)\right]$$
$$= 0.577\ 156\ 649\ 015\ 328\ 606\ 1\ldots \tag{6.48}$$

Equations 6.46 and 6.47 have been set out in a form suitable for use in an iterative computer or calculator program. Subject to the limitations discussed below, results correct to four decimal places are obtained if the program branches when the final term has a value less than 1×10^{-5}.

For large values of z the series for $K_0(z)$ and $K_1(z)$ may converge slowly, but in addition, the sum of the first few terms may greatly exceed the final result. Thus, when $z = 9$, $K_1(z) = 5.364 \times 10^{-5}$, whereas the sum of the first three terms is 140.7.

To obtain a final value correct to the third decimal place, it is necessary in such a case to work to at least eight significant figures. Euler's constant must be expressed to the same degree of accuracy.

This problem escalates with higher values of z and it is then preferable to use an approximation known as 'asymptotic expansion':

$$K_n(z) \approx \left(\frac{\pi}{2z}\right)^{1/2} e^{-z} \left\{ 1 + \frac{4n^2 - 1^2}{1!(8z)} + \frac{(4n^2 - 1^2)(4n^2 - 3^2)}{2!(8z)^2} + \cdots + \frac{(4n^2 - 1^2)\cdots[4n^2 - (2r-3)^2]}{(r-1)!(8z)^{r-1}} \right\} \quad (6.49)$$

This series is not convergent; successive terms decrease in value to a minimum, and then increase indefinitely. The smallest term occurs where $r = 2z$ or close thereto, and greatest accuracy is obtained by computing up to this point. This operation does not present any significant difficulties. However, for calculating weld isotherms and the like it is usually sufficient to take the first three terms of the series, such that

$$K_0(z) \approx \left(\frac{\pi}{2z}\right)^{1/2} e^{-z} \left[1 - \frac{1}{8z} + \frac{9}{128z^2} - \frac{75}{1024}z^3 \right] \quad (6.50)$$

and

$$K_1(z) \approx \left(\frac{\pi}{2z}\right)^{1/2} e^{-z} \left[1 + \frac{3}{8z} - \frac{15}{128z^2} + \frac{103}{1024}z^3 \right] \quad (6.51)$$

Computing down to the minimum term, or the use of equations 6.50 and 6.51 will be accurate enough for values of z down to 5.

The length of the weld pool downstream of the origin is given by

$$\bar{T} = 1 = \frac{n}{w} e^{\bar{x}} K_0(\bar{r}) \quad (6.35)$$

with $\bar{r} = \bar{x} = \bar{x}_1$.

For large values of the operating parameter n we may use equation 6.50 to calculate $K_0(\bar{r})$. Ignoring the expression within the square brackets (which is justified in this case)

$$\bar{T} = 1 \approx \frac{n}{\bar{w}} \left(\frac{\pi}{2\bar{x}_1}\right)^{1/2}$$

and

$$\bar{x}_1 \approx \frac{\pi}{2} \left(\frac{n}{w}\right)^2 \quad (6.52)$$

This approximation may be useful in making a first estimate of the theoretical weld pool length.

6.4.6 Comparison with experimental plots

Christensen *et al.* (1965) made an extensive series of experiments to test the validity of the theoretical point source relationship. Surface weld runs were made on aluminium and steel plate using the submerged arc, manual metal arc and gas tungsten arc processes. Measurements included weld width, weld depth and the time required for the temperature to fall from 800 °C to 500 °C at a specified point. This latter quantity is used as a measure of the cooling rate for the austenite–ferrite transition of steel.

The results are expressed in terms of the non-dimensional form of the point source equation (equation 6.34). Thus, results for both aluminium and steel are plotted on the same diagram. Generally speaking, there is good qualitative agreement between calculated and measured results. For example, Fig. 6.17 shows that even over a range of four orders of magnitude for the operating parameter n, the weld bead non-dimensional width follows the same trend as the theoretical curve. In detail, however, the deviations may be quite substantial: up to twice the theoretical value. The scatter in the case of weld pool depth is over a factor of 5, but once again the general trend follows the theoretical curve. Bearing in mind the geometry of the molten pool, as discussed earlier, it is not surprising to find some scatter in the dimensions of the fused zone.

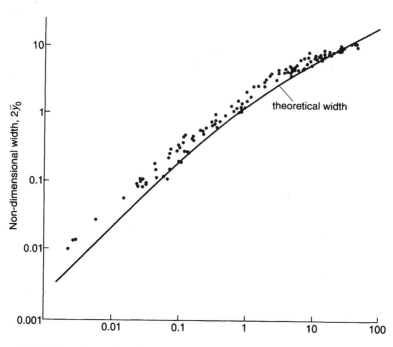

6.17 Non-dimensional weld width as a function of operating parameter *n*.

6.5 Weld defects

6.5.1 General

An unwanted consequence of fusion welding is the presence, on occasions, of defects in the fused zone. These may take the form of discontinuities such as cracks, lack of fusion, porosity and slag inclusions, or irregularity in weld profile, namely undercut and lack of penetration. These defects are illustrated diagrammatically in Fig. 6.18. Any fusion welding process, whether manual or automatic, may be subject to such problems if correct procedures are not followed, or if unforeseen circumstances arise. However, the process that is most subject to weld defects is manual welding with coated electrodes, and their incidence is usually higher in site welding than under factory conditions.

Some defects, porosity for example, result from processes that commonly take place during fusion and re-solidification, and these are considered in the chapters concerned with individual metals. Others, such as lamellar tearing, are due to defective parent metal. For the most part, however, weld defects are caused by faulty manipulation or other operational errors.

6.5.2 The metallurgical background

Cracking is the most serious defect and in steel it is almost invariably caused by hydrogen. Hot cracking, due, for example, to pick-up of sulphur from low-quality steel, is a rare possibility which should not be ignored. Hydrogen cracks resulting from the use of damp electrodes or submerged arc welding flux usually occur in the weld metal and their orientation will depend on the direction of the

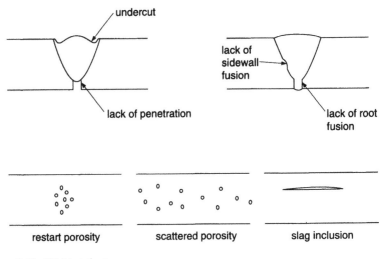

6.18 Weld defects.

predominant shrinkage stress. In fillet welds joining thick plates, for example, these stresses act longitudinally, and cracking is transverse.

Hydrogen cracking does not affect non-ferrous metals. Solidification cracking remains a possibility, notably with some aluminium alloys, but standard filler alloys are crack-resistant, and this problem is rare. Similarly, fully austenitic chromium–nickel steel weld deposits are inherently crack-sensitive, but electrode manufacturers have developed formulations that give sound deposits.

Localized porosity in the manual metal arc welding of steel is most likely to occur at restarts, and be due to hydrogen. Scattered porosity, on the other hand, is most likely to be the result of nitrogen evolution. Porosity in nickel and nickel-base alloys is almost certainly a nitrogen problem. Aluminium welds made using argon-shielded gas metal arc welding often contain fine porosity that results from dissociation of the hydrated oxide film and subsequent hydrogen evolution.

Undercut, lack of penetration and slag inclusions are normally caused by faulty manipulation. Persistent undercut when reworking old steel may, however, be due to excessive oxygen or sulphur contents.

6.5.3 The acceptability of weld defects

Where butt welds carry a high primary stress in service, and particularly for pressure-containing equipment, it is normal practice for the applicable code to set an upper limit to the extent of any given type of defect. In many countries such requirements have the force of law. Cracks and lack of fusion are unacceptable regardless of size, while other defects may be allowable up to a certain size or extent. In a number of welding applications, pipelining and refinery construction, for example, non-destructive examination and the assessment and repair of weld defects forms a major part of the operation. Moreover, the **repair rate** provides a measure of weld quality generally. In process plant construction most site welding consists of circumferential butt welds in piping, each of which is counted as one weld, regardless of diameter. The repair rate is the percentage of such welds that are subject to repair. Up to 5% would be considered acceptable, but above this figure remedial action might be considered necessary. Such action could include retraining of welders, increasing supervision, and possibly reviewing the methods of welder training. Where a particular type of defect predominates, it may be desirable to reconsider welding procedures.

Standards for weld defect acceptability are arbitrary and represent a quality level that can be achieved by good welding practice. Experience demonstrates that welds produced to such acceptance levels will perform satisfactorily in service, provided that there is no metallurgical deterioration.

Experience also indicates that it is safe to operate at the defect level indicated by codes regardless of the method of inspection. For example, ultrasonic testing shows up cracks more readily than radiography, but welds tested to code requirements by either of these techniques will operate satisfactorily. It is normal practice to employ radiography as the inspection method because this gives a permanent record that may be used in the case of later modifications or repairs. If ultrasonics is to be used for inspection, this must be established and agreed before work starts. Once the method of weld examination has been agreed, no arbitrary change should ever be made during the course of the contract.

It may happen that cracks do develop in service due to a process unrelated to the presence of defects in the as-welded condition, or it could be that a more critical examination of existing records discloses the presence of unacceptable defects after the item has been installed or has gone into service. Under such circumstances it may be desirable to obtain a measure of the failure risk that would result from continued operation. Standard procedures exist for this purpose, and are outlined in Chapter 11.

References

Christensen, N., Davies, V. de L. and Gjermundsen, K. (1965) *British Welding Journal*, **12**, 54–74.

Lancaster, J.F. (1997) *Handbook of Structural Welding*, Abington Publishing, Cambridge.

Further reading

American Society for Metals (1983) *Metals Handbook*, 9th edn, Vol. 6, ASM, Metals Park, Ohio.

American Welding Society (1991) *Welding Handbook*, 8th edn, *Vol. 2: Welding Processes*, AWS, Miami; Macmillan, London.

Houldcroft, P.T. (1977) *Welding Process Technology*, Cambridge University Press, Cambridge.

Houldcroft, P.T. (1990) *Which Process?*, Abington Publishing, Cambridge.

Hua, Z. and Fan, P. (1988) *Joining and Materials*, **1**, 280–281.

Singer, C., Holmyard, E.J. and Hall, A.R. (1954) *A History of Technology*, Clarendon Press, Oxford.

The Welding Institute (1978) *Advances in Welding Process*, TWI, Cambridge.

The Welding Institute (1979) *TIG and Plasma Welding*, TWI, Cambridge.

The Welding Institute (1981) *Development in Mechanised, Automated and Robotic Welding*, TWI, Cambridge.

7
Metallurgical effects of the weld thermal cycle

A number of reactions may take place in the liquid weld metal: first in the liquid drop at the electrode tip, secondly during transfer from electrode to weld pool, and thirdly in the weld pool itself. These reactions include:

1   solution of gas, causing gas–metal reactions or reaction with elements dissolved in the liquid metal;
2   evolution of gas;
3   reaction with slag or flux.

Generally, but not invariably, gases other than the inert gases have an unfavourable effect on weld metal properties and it is the object of **shielding** methods (discussed in Chapter 6) to minimize any metallurgical damage from this cause. Slag–metal reactions occur mainly during the welding of steel, and will be discussed under that heading.

7.1   Gas–metal equilibria

The nature of fusion welding is such that thermodynamic equilibrium between the liquid metal of the weld pool and its gaseous atmosphere is rarely possible. This is especially true of arc welding, where the gas at the arc root is itself not in a condition of equilibrium. Nevertheless, the quantity of gas absorbed by a weld pool follows the same trend as equilibrium solubility, so this aspect of the subject will be discussed first. The mechanism of absorption will be considered later.

The gases that are used as a protective shield in arc welding are argon, helium, hydrogen, water vapour, carbon dioxide and carbon monoxide, the last-named being formed by decomposition of CO_2, flux or electrode coating. Oxygen is added to argon in the gas metal arc welding of steel in order to stabilize the arc, and is also formed by decomposition of CO, CO_2 and H_2O in the arc. Nitrogen has been used as a shielding gas in the tungsten-arc welding of copper.

Argon and helium do not dissolve in or react with liquid metals although they may suffer occlusion; an argon content of 10^{-5}–10^{-4} mass% has been measured in gas–metal arc welds in steel. Argon may also be formed in steel exposed to a

7.1.1 Diatomic gases; hydrogen

The equilibrium solubility of a diatomic gas is governed by Sievert's law:

$$s = A p_g^{1/2} \quad (7.1)$$

where s is solubility, p_g is the partial pressure of the gas and A is a constant. The equilibrium reaction is

$$G_2 = 2[G]_D \quad (7.2)$$

where the suffix D indicates that the gas is dissolved. The corresponding equilibrium constant K may be expressed as

$$K = a_g / p_g^{1/2} \quad (7.3)$$

where a_g is the activity of dissolved gas

$$a_g = s f_g \quad (7.4)$$

where f_g is the activity coefficient. For a dilute solution, $f_g = 1$ and hence

$$s = K p^{1/2} \quad (7.5)$$

as in equation 7.1. Also

$$K = e^{-\Delta G / RT} \quad (7.6)$$

where ΔG is the free energy of solution, R is the gas constant and T is temperature on the absolute scale. Generally, ΔG is a function of temperature:

$$\Delta G = B + CT \quad (7.7)$$

so that the solubility can be expressed as

$$\ln(s/p^{1/2}) = -(B/RT + C/R) \quad (7.8)$$

For example, the solubility of hydrogen in liquid aluminium, expressed in ml/100 g metal, is

$$\ln(s/p^{1/2}) = -6355/T + 6.438 \quad (7.9)$$

Figure 7.1 shows the solubility of hydrogen in various metals above and below the melting point for $p = 1$ atm.

As the temperature of the metal approaches the boiling point, its vapour pressure p_m becomes appreciable, and for a total of 1 atm, with p_g etc. in atmospheres, the solubility becomes

$$s = A[p_g(1 - p_m)]^{1/2} \quad (7.10)$$

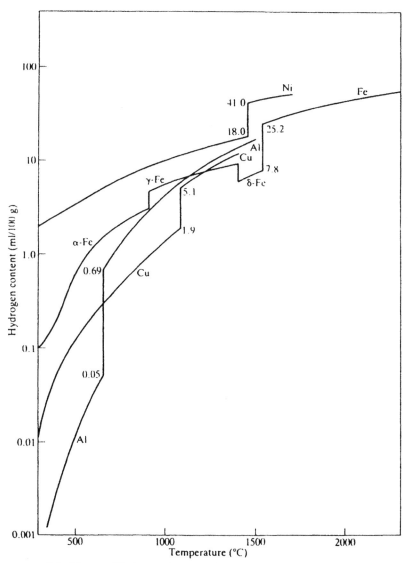

7.1 The equilibrium solubility of hydrogen in aluminium, copper, iron and nickel at 1 atm pressure (data from Brandes, 1983).

Thus the solubility reaches a maximum, and then falls away to zero at the boiling point where $p_m = 1$. A similar effect has been observed with hydrogen, where a spray of metal drops produced by gas evolution causes an effective reduction in the gas partial pressure (Section 7.2.3). Figure 7.2 shows calculated curves for hydrogen solubility in a number of metals as a function of temperature and indicates the **maximum solubility**.

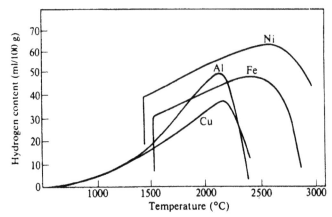

7.2 Curves of hydrogen solubility versus temperature at a hydrogen pressure of 1 atm. The low-temperature part of the curves is derived from known experimental data; the high-temperature end is obtained by extrapolating these data and then applying a correction for metal vapour (from Howden and Milner, 1963).

In fusion welding the liquid metal temperature is normally well below the boiling point, except possibly at the electrode tip in gas metal arc welding with an argon shield.

7.1.2 Oxygen

Oxygen combines exothermically with iron to form FeO and its solubility in liquid iron is governed by the reaction

$$FeO(l) = Fe(l) + [O]_D \tag{7.11}$$

for which the equilibrium constant is

$$K = \frac{[O]_D}{a_{FeO}} = \exp\left(-\frac{1.455 \times 10^4}{T + 2.943}\right) \tag{7.12}$$

a_{FeO} is the activity of the oxide, which may be taken as 1. The solubility of oxygen in liquid iron is shown in Fig. 7.3.

The equilibrium between dissolved oxygen and gaseous oxygen is

$$\tfrac{1}{2}[O_2]_{gas} = [O]_D \tag{7.13}$$

and

$$K = \frac{[O]_D}{(p_{O_2})^{1/2}} = \exp\left(\frac{1.4055 \times 10^4}{T - 3.076}\right) \tag{7.14}$$

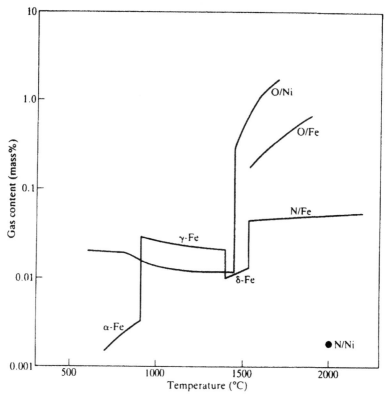

7.3 The equilibrium solubility of nitrogen and oxygen in iron and nickel at 1 atm pressure (sources: nitrogen in iron and oxygen in nickel, Brandes, 1983, Uda and Wada, 1968b; oxygen in iron, Taylor and Chipman, 1943; nitrogen in nickel, Ohno and Uda, 1981).

From equation 7.14 it may be calculated that the pressure of gaseous oxygen in equilibrium with a dissolved oxygen content of, say, 0.20% is of the order of 10^{-9} atm. This is the dissociation pressure of FeO, and may also be calculated from Fig. 4.8.

The reaction 7.11 goes to the left as the temperature falls, and if the oxygen content is equal to the equilibrium value at a given temperature, then any further drop will result in the precipitation of FeO. The metallurgical effect of such precipitates will be discussed in Chapter 8.

7.1.3 Nitrogen

The equilibrium solubility of nitrogen in iron is plotted in Fig. 7.3. Liquid iron at the melting point dissolves 0.044% by mass of nitrogen when the pressure is 1 atm; this is equivalent to about 70 ml/100 g metal measured at normal temperature and pressure (NTP). The solubility of nitrogen in nickel is very small, being equal to 0.0018% by mass in liquid nickel at 1600 °C. Copper likewise appears to have a very small solubility for nitrogen under equilibrium conditions below about 1400 °C. In arc welding, however, nitrogen may dissolve in both these metals (see Chapter 10). Nitrogen dissolves exothermically in aluminium to form aluminium nitride.

7.1.4 The effect of alloying additions

Gas solubility may be changed by the addition of alloying elements to a metal. The change produced by adding a given amount of alloy is expressed in terms of an **interaction parameter**. Generally, we may express the activity a_g of a solute gas in a pure liquid metal as

$$a_g = x_g f_g \tag{7.15}$$

where x_g is the weight per cent of the gas and f_g is the activity coefficient. If other solute elements j, k, l, etc., are now added, the activity coefficient is given by

$$\log f_g = x_j e_g^{(j)} + x_k e_g^{(k)} + \cdots \tag{7.16}$$

where x_j, etc., are the concentrations of the solute elements, usually expressed as weight per cent, and $e_g^{(j)}$, etc., are the interaction parameters. These are defined for solute j as

$$e_g^{(j)} = \left(\frac{\partial \log f_g}{\partial x_j}\right)_{x_j \to 0} \tag{7.17}$$

That is to say, they represent the slope of the curve of activity coefficient for the gas in the binary alloy concerned as a function of alloying element concentration. The effect of alloy additions on the solubility of nitrogen in iron is shown in Fig. 7.4. Interaction parameters for nitrogen and oxygen in steel are tabulated in Table 7.1 together with results obtained for nitrogen in steel with gas metal arc welding in a nitrogen atmosphere. Note that the figures for welding are very similar to the equilibrium values.

From equation 7.3 the activity of the gas in solution is

$$a_g = K p_g^{1/2} \tag{7.18}$$

For any given temperature and gas pressure, the right-hand side of equation 7.16 is constant and therefore the activity of the dissolved gas is also constant. The solubility, therefore, from equation 7.15 is

$$s = x_g = \frac{a_g}{f_g} = \frac{\text{constant}}{f_g} \tag{7.19}$$

Metallurgical effects of the weld thermal cycle 175

Table 7.1. Interaction parameters for the solution of oxygen and nitrogen in binary iron alloys

Solute element	$e_0^{(m)*}$	$e_N^{(m)}$	
		Equilibrium at 1600 °C†	Gas metal arc welding at 150–350 A in a nitrogen atmosphere‡
Al	0.006	0.0025	not measurable
C	−0.44	0.25	0.16
Co	0.007	0.011	0
Cr	−0.041	−0.045	−0.047
Cu	0.0095	0.009	not measurable
Mn	−0.02	−0.02	−0.01
Mo	0.0035	−0.011	−0.05
Nb	−0.14	−0.067	−0.13
Ni	0.006	−0.01	0.011
Si	−0.131	0.047	0.06
Ta		−0.034	−0.04
Ti	−0.187	−0.53	−0.56
V	−0.11	−0.10	−0.09
W	0.009	−0.002	not measurable
Zr		−0.63	−0.40

* Uda and Ohno (1975).
† Pehlke and Elliott (1960), Evans and Pehlke (1965).
‡ Kobayashi et al. (1972).

and hence the solubility increases as f_g decreases. Thus, where the interaction parameter is negative, the solubility increases, as will be evident in the case of nitrogen from a comparison of Table 7.1 with Fig. 7.4.

The numerical value of the interaction parameter decreases with increasing temperature; that is to say, as temperature rises, the effect of alloying elements on solubility (either negative or positive) decreases. A similar effect is observed for the solubility of other gases in iron, and for the solubility of gases in other metals.

7.1.5 Deoxidation

There is an apparent anomaly in that the interaction parameters for oxygen shown in Table 7.1 for known deoxidants such as manganese, silicon and titanium are all negative. It is indicated that, under equilibrium conditions, these elements will increase the oxygen solubility. Such is not the case during deoxidation because the deoxidation reaction is arrested by solidification before it has gone to completion. In steelmaking, deoxidants are added to the melt just before pouring; in welding, a sufficient amount of deoxidant is added to allow for some loss by oxidation. Figure 7.5 shows how the oxygen content of iron weld metal is affected by the presence of various elements. This diagram is quantitatively true only for a particular set of welding variables, but it indicates the relative

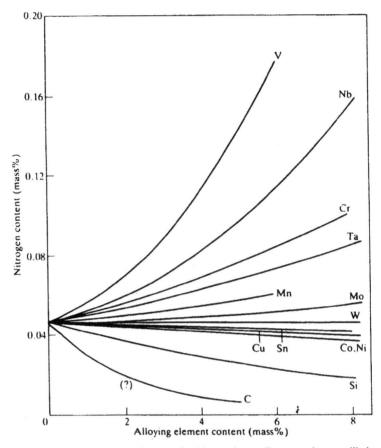

7.4 The solubility of nitrogen in binary iron alloys under equlibrium conditions at 1600 °C and 1 atm nitrogen pressure (from Pehlke and Elliott, 1960).

deoxidizing power of different additives. The oxygen content of steel weld metal is also affected by slag–metal reactions, and this question is discussed further in Chapter 8.

7.2 Gas–metal reactions in arc welding

In non-arc welding processes the amount of gas absorbed by the weld pool is relatively low. For example, the hydrogen content of steel weld metal made using the oxyacetylene torch is in the range 2–3 ml/100 g. With arc welding, it is relatively high. The remainder of this section will be concerned with arc welding.

Investigations into the solution of gases during arc welding have, for the most part, been made using the gas-shielded processes in a closed chamber, with argon–hydrogen, argon–oxygen, argon–nitrogen and argon–CO_2 mixtures. Much

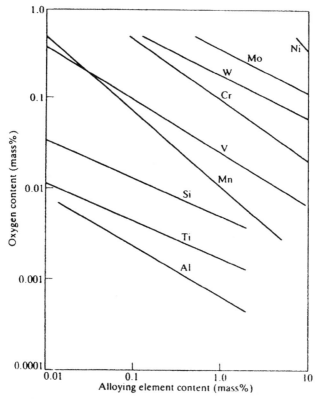

7.5 The influence of alloying elements on the oxygen content of iron weld metal (from Kasamatsu, 1962).

of this work has been done in Japan on the argon–nitrogen system, primarily in connection with the use of self-shielded welding.

7.2.1 Nitrogen

Figure 7.6 shows the relationship between nitrogen partial pressure and the nitrogen content of pure iron for static arc melting, gas-shielded welding and for non-arc (levitation) melting. In the absence of an electric arc, the solution of nitrogen in steel obeys Sievert's law, with a nitrogen solubility of about 0.05% at a liquid metal temperature of 2200 °C and at atmospheric pressure. In arc melting, the solubility is in accordance with Sievert's law up to a nitrogen partial pressure of 0.036 atm, but it is 20 times that for non-arc melting. Above $p = 0.036$ atm, the amount of gas absorbed is constant, independent of pressure, at a level somewhat above the equilibrium value for 1 atm. Under conditions where the nitrogen absorption is constant, gas bubbles out of the liquid metal, generating a spray of fine (10–100 nm diameter) metal particles. The same pattern of

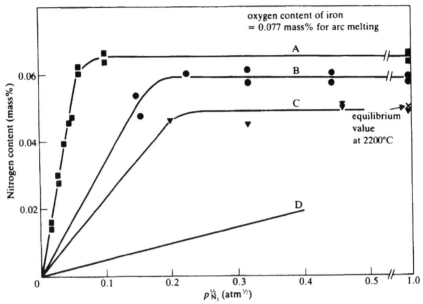

7.6 The nitrogen content of pure iron weld metal as a function of partial pressure of nitrogen in a nitrogen–argon mixture of total pressure 1 atm: curve A, stationary arc melting; B, gas tungsten arc welding; C, gas metal arc welding, electrode positive; D, non-arc levitation melting (after Uda and Wada, 1968a; Kuwana and Kokawa, 1983; Kobayashi et al., 1971).

absorption is observed for welding, but the increase in apparent solubility at low nitrogen partial pressures is somewhat less. Most other welding variables, such as welding speed and arc voltage, do not have much effect on the nitrogen content, but increasing the welding current will in most cases decrease the amount of nitrogen in the weld metal.

Surface-active agents, if present in sufficient quantity, cause an increase in the nitrogen content of the weld. Figure 7.7 shows the effect of increasing the oxygen content of iron in arc melting in an argon–5% nitrogen atmosphere. Above 200 ppm oxygen, the amount of nitrogen in the melt increases sharply. Sulphur and selenium in rather larger amounts (about 0.5 mass%) have a similar effect. The amount of nitrogen absorbed when welding in nitrogen–oxygen mixtures (air, for example) is increased to an even greater extent. Figure 7.8 shows the nitrogen content of gas metal arc welds in carbon steel and austenitic chromium–nickel steel, and demonstrates that in such an atmosphere the maximum nitrogen content occurs at about 50% N_2–50% O_2. Austenitic chromium–nickel steel absorbs more than twice as much nitrogen as carbon steel, as would be expected from the relatively large negative interaction parameter for chromium, shown in Table 7.1. The effect of increasing current in reducing nitrogen absorption is

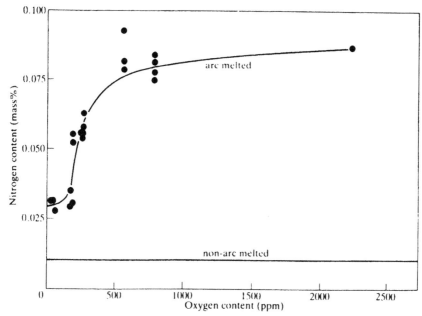

7.7 The nitrogen content of arc-melted iron–oxygen alloy in argon–5% nitrogen at a total pressure of 1 atm as a function of oxygen content (after Uda and Ohno, 1973).

particularly marked for gas metal arc welding in air, as shown for carbon steel and stainless steel in Fig. 7.9.

Oxygen may also modify the effect of alloying additions on the apparent nitrogen solubility. Figure 7.10 shows the nitrogen content of gas metal arc welds made in air with various binary iron alloy wires. The curves for niobium and vanadium are of particular interest. Under equilibrium conditions (Fig. 7.4) these alloying elements increase the nitrogen solubility in a uniform manner; in air welding, however, the nitrogen content first decreases, and only starts to increase when the alloy content is over about 1%. Niobium and vanadium are both nitride-forming and deoxidizing elements; thus, when added in small amounts, they tend to neutralize the effect of oxygen and therefore reduce the nitrogen content, whereas in larger amounts the affinity for nitrogen dominates. With stronger deoxidants such as Ta, Ti, Al and Zr, the deoxidizing effect is the most important, at least up to the alloy content shown.

7.2.2 Oxygen and CO_2

The absorption of oxygen in the gas metal arc welding of pure iron has been studied using argon–oxygen atmospheres in a closed chamber, and the results, as shown in Fig. 7.11, have the same general form as for nitrogen. The amount of

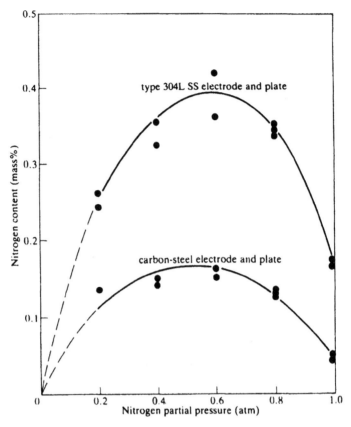

7.8 The effect of nitrogen partial pressure on the nitrogen content of gas metal arc welds made in nitrogen–oxygen mixtures at a total pressure of 1 atm with electrode positive, 25 V, 250 A (after Kobayashi et al., 1971).

oxygen absorbed at lower oxygen partial pressures is approximately proportional to $p_{O_2}^{1/2}$, whereas at higher partial pressures it becomes constant at 0.23%, somewhat lower than the equilibrium value at 1 atm oxygen pressure. The oxygen content at lower partial pressures decreases with increasing current, which again shows the same pattern as for nitrogen.

Similar tests with CO_2 show an increase in both carbon and oxygen contents with CO_2 partial pressure. For both elements, the rate of increase is lower with increasing current. Possible reactions between CO_2 and iron are numerous, but the governing equation is probably

$$CO_2 = CO + \tfrac{1}{2} O_2 \qquad (7.20)$$

$$CO = [C]_D + [O]_D \qquad (7.21)$$

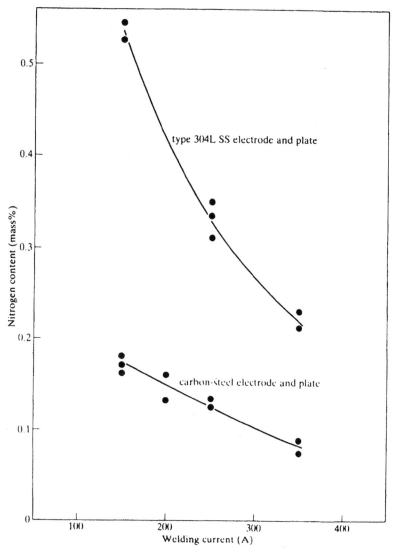

7.9 The effect of welding current on the nitrogen content of gas metal arc welds made in air at a pressure of 1 atm with electrode positive, 25 V (after Kobayashi *et al.*, 1971).

and in the latter case

$$K = p_{co}/a[C]_D a[O]_D \tag{7.22}$$

Assuming the activity coefficient in each case is unity, then for a CO pressure of 1 atm

$$[C]_D[O]_D = 1/K \tag{7.23}$$

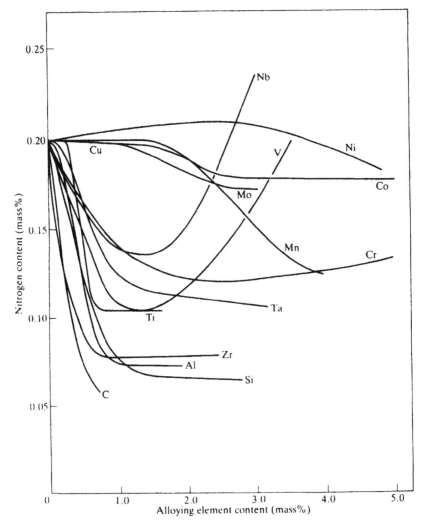

7.10 The effect of alloying elements in electrode wires on the nitrogen content of gas metal arc weld metals in an air welding atmosphere at a pressure of 1 atm, with electrode positive, 25 V, 150 A (after Kobayashi *et al.*, 1968).

and expressing the concentrations as mass per cent

$$\log K = 1160/T + 2.0 \tag{7.24}$$

Figure 7.12 shows the equilibrium relationship between carbon and oxygen concentrations in liquid iron according to equations 7.23 and 7.24, together with the results of the tests described earlier. Note that these results are not representative of practical CO_2-shielded gas metal arc welding, which employs

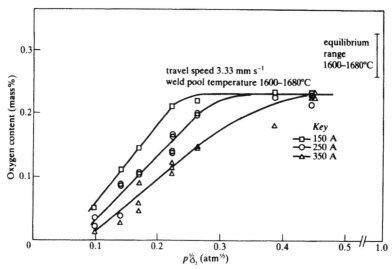

7.11 The oxygen content of pure iron as a function of partial pressure of oxygen in an oxygen–argon mixture of total pressure 1 atm and at various welding currents. Gas metal arc welding with electrode positive (after Kuwana and Sato, 1983a).

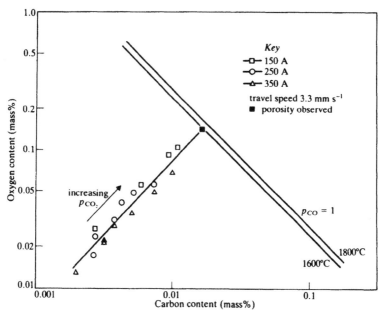

7.12 Carbon content versus oxygen content of gas metal arc welds in pure iron made in an argon–CO_2 atmosphere with up to 0.15 atm CO_2 at a total pressure of 1 atm with electrode positive, 25 V (after Kuwana and Sato, 1983b).

wire containing deoxidants. The oxygen content of the weld metal is therefore reduced and the carbon content increased, typically to about 0.12%.

As will be seen from Fig. 7.12, the weld metal becomes porous when $[C]_D[O]_D$ corresponds to the quantity in equilibrium with CO at 1 atm pressure. The CO partial pressure in contact with the liquid metal surface is that due to dissociation of CO_2 and at the metal temperature is quite low; therefore gas is being evolved when the solubility product is equivalent to the atmospheric pressure.

7.2.3 Hydrogen

Early investigators found that the hydrogen content of gas tungsten arc (GTA) welds made in argon–hydrogen mixtures obeyed Sievert's law, and subsequent work has confirmed this. Figure 7.13 shows the relationship between hydrogen content and partial pressure of hydrogen in arc melting and gas tungsten arc

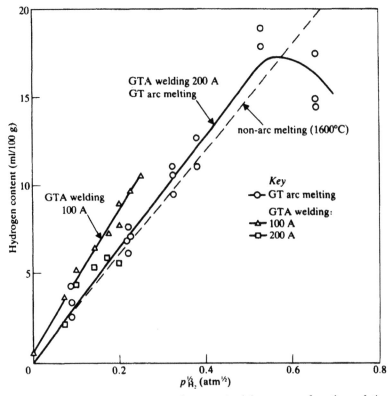

7.13 The hydrogen content of arc-melted iron, as a function of the square root of hydrogen partial pressure in a hydrogen–argon atmosphere at a total pressure of 1 atm compared with the equilibrium solubility at 1600 °C (source: arc melting, Uda, 1982; gas tungsten arc welding, Chew and Willgoss, 1980).

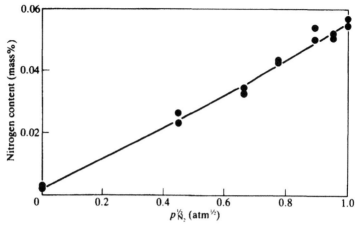

7.14 The nitrogen content of carbon steel weld metal, as a function of partial pressure of nitrogen in a nitrogen–hydrogen mixture, at 1 atm total pressure. Gas metal arc welding with carbon steel electrode, electrode positive, 25 V, 250 A (after Kobayashi et al., 1971).

welding, where again the absorption appears to conform to Sievert's law. However, in arc melting it was found that a metal spray, similar to that observed during arc melting in argon–nitrogen atmosphere, was generated down to a hydrogen partial pressure of 0.1 atm. It is implied that the amount of hydrogen dissolved at the arc root is higher than the equilibrium value, and that this excess hydrogen is evolved from the outer parts of the molten pool. The reduction in the hydrogen content of arc-melted iron above $p_{H_2} = 0.3$ atm is thought to be due to metal spray reducing the effective partial pressure of hydrogen.

The results for gas tungsten arc welding in argon–hydrogen mixtures plotted in Fig. 7.13 show that in this case also the hydrogen content of the weld is in accordance with Sievert's law and is not far from equilibrium values. It is, however, slightly lower at 200 A than at 100 A.

The absorption of nitrogen is also affected by the presence of hydrogen. Figure 7.14 shows the amount of nitrogen absorbed as a function of nitrogen partial pressure for gas metal arc welding in nitrogen–hydrogen atmospheres. Under these circumstances, nitrogen behaves in much the same way as hydrogen in argon–hydrogen mixtures.

7.3 The mechanism of gas absorption in welding

At least some of the phenomena described in Section 7.2.1 and illustrated in Fig. 7.7 may be explicable in terms of surface-tension-induced (Marangoni) flow. At present there is no direct evidence to link gas content of the weld metal with flow, so the following comments must be regarded as speculative. In a pure metal,

186 Metallurgy of welding

Maragoni flow is outwards across the weld pool surface, but the presence of surface-active agents may cause the gradient of surface tension as a function of temperature to reverse and become positive. The surface flow is then inwards, towards the arc root. When this is so, the rich solution will flow downwards. If the nitrogen content of the solution is greater than that corresponding to the ambient pressure, which is 1 atm, nitrogen bubbles will form. These will have a pressure slightly higher than 1 atm, so the surrounding liquid will have a nitrogen content slightly higher than the equilibrium value for pure nitrogen at 1 atm pressure. If, however, surface-active agents are absent the flow is outwards across the surface. The rich solution then comes directly in contact with an atmosphere containing (for the Fig. 7.7 experiment) 5% N_2. It will then degas to a level which corresponds more or less to the equilibrium value for this partial pressure.

The flow reversal is predicted to occur in the iron–oxygen system at an oxygen content of 17 ppm by mass and the gradient of surface tension with temperature becomes positive in the range 60–100 ppm (Keene, 1983). In Uda's experiments (Fig. 7.7) the transition from low-nitrogen to high-nitrogen melts occurred at about 200 ppm oxygen. The low-nitrogen melts had a nitrogen content of 0.03%, as compared with an equilibrium value for 5% nitrogen of 0.01%, while the high-nitrogen melts had a content of 0.085%N as compared with the equilibrium value for pure nitrogen of 0.05%. These figures are consistent with the proposed mechanism.

It must be recognized, of course, that taking fusion welding as a whole, divergent flow in a weld pool is a rare phenomenon. Nevertheless, similar conditions may prevail in other cases; for example, in welding with coated electrodes the rich solution flows as a thin stream along the sides of the crater and may lose some nitrogen by contact with the gas mixture away from the arc root.

The reason for the abnormally high absorption of nitrogen in arc melting and arc welding has not been fully established. However, Uda has suggested that at the arc root (in the anode drop zone in the case of a tungsten arc), the gas may be in an activated condition. Suppose that the free energy of the gas is increased by an amount $\Delta G'$. Then equation 7.6 becomes

$$K' = e^{-(\Delta G - \Delta G')/RT} \tag{7.25}$$

and the solubility is now, from equation 7.5,

$$s' = K' p^{1/2} \tag{7.26}$$

For any given partial pressure of gas, the increase in solubility compared with the equilibrium value is

$$s'/s = K'/K = e^{\Delta G'/RT} \tag{7.27}$$

For arc melting in argon–nitrogen mixtures, $s'/s = 20$, and taking the temperature as 2200 °C (2473 K) and $R = 8.314$ J mol^{-1} K^{-1} gives

$$\Delta G' = RT \ln(s'/s) = 61.6 \text{ kJ mol}^{-1} \quad (7.28)$$

It is possible that this energy results from the diffusion of dissociated nitrogen atoms from the arc column across the anode drop zone to the liquid metal surface. The dissociation energy for nitrogen is 942 kJ mol^{-1}, so the partial pressure of dissociated atoms at the surface would need to be about 6% of the total pressure.

The increment of energy $\Delta G'$ may alternatively be expressed as electron volts per molecule:

$$\Delta G' \text{ (eV)} = 6.16 \times 10^4 / N_o e \quad (7.29)$$

where N_o is Avogadro's number and e is the energy of one electron volt in joules. This leads to $\Delta G' = 0.638$ eV. In arc melting with a tungsten electrode such an increment could be due to elastic collisions between electrons and gas molecules in the anode drop zone, since the energy acquired per electron in falling through the anode drop is in the range 1–2 eV per electron.

The practical effect of an augmented absorption of nitrogen in steel is to increase the susceptibility to strain-age embrittlement and in some cases, to porosity. However, any increased absorption is, under practical welding conditions, normally due to defective shielding or incorrect welding procedures.

7.4 Porosity

There are two main types of porosity observed in weld metal. The first (**globular porosity**) consists of spherical, or near-spherical, pores, which can vary in diameter from 0.05 to 5.0 mm. In the second type, the cavities are elongated. Sometimes these elongated cavities follow the solidification pattern, while in others they nucleate near the centre of the weld and grow lengthwise, sometimes increasing in diameter at the same time. The second type of pore is known as a **blowhole**, **wormhole** or **tunnel pore**.

Porosity results from the desorption of hydrogen or nitrogen. Oxygen is soluble in silver and its evolution may cause porosity in pure silver welds. Gold and platinum do not dissolve oxygen, while other metals react with it to form oxides. In steel the desorption of oxygen takes the form of oxide precipitation.

7.4.1 Globular porosity

Systematic work on porosity has been less extensive than that on gas absorption, but Uda and his colleagues have made some valuable contributions.

Table 7.2 shows the levels of hydrogen and nitrogen at which porosity was observed in test welds. This suggests that nickel welds have a high tolerance for hydrogen. Practical experience shows that steel behaves in the same way; the gas

188 Metallurgy of welding

Table 7.2. Critical nitrogen or hydrogen concentration by volume for porosity in arc welding and levitation melting

System	Arc welding	Non-arc melting (2000 °C)
Fe–H$_2$		10% H$_2$–Ar
Ni–H$_2$	50% H$_2$–Ar	5% H$_2$–Ar
Ni–N$_2$	0.025% N$_2$–Ar	100% N$_2$
Ni–H$_2$–N	1% H$_2$ 0.025% N$_2$–Ar	
	10% H$_2$ 0.05% N$_2$–Ar	

Sources: Ohno and Uda (1981), Uda *et al.* (1976).

from rutile and cellulosic electrodes can be up to 50% H$_2$ without any porosity in the welds. The tolerance for nitrogen is much lower: 0.025% for nickel and in arc melting, 4% for iron. Adding hydrogen to the gas increases the critical level of nitrogen for nickel; this is also in accordance with practical experience.

It will also be evident from Table 7.2 that the tolerance for hydrogen in non-arc (levitation) melting and casting is much lower than in welding. This is associated with the growth of blowholes, the mechanism of which is discussed in the next section.

Taking the evidence as a whole, it seems likely that where significant amounts of hydrogen are present in the arc (say over 10%) there is a vigorous bubbling of gas such that at the rear of the weld pool the hydrogen content is not far from the solubility limit and solidification takes place without further gas evolution. Of course, it is possible to freeze the bubbling zone, and this may occur owing to irregular movement and at stops and starts. Otherwise, in steel the weld pool (possibly because of augmented solution at the arc root) purges itself of hydrogen and to some degree of nitrogen such that porosity is not a serious problem in the fusion welding of steel. Aluminium, however, is very subject to hydrogen porosity owing, it is thought, to the steep gradient of the solubility–temperature curve.

Porosity becomes coarser with an increasing degree of supersaturation. Under marginal conditions, it tends to form at the weld boundary, where the metal is liquid for the shortest time. Porosity may also result from reaction in the melt, owing to the formation of CO, for example. Such cases will be discussed under the heading of the metal concerned.

7.4.2 Tunnel porosity

This type of defect is familiar in ingot casting, where it is known as a blowhole. It takes the form of an elongated cavity, and in welding it is also known as a wormhole. Tunnel porosity is associated with hydrogen, and it may occur at hydrogen contents below the solubility limit. For example, Uda *et al.* (1976)

found that blowholes formed in ingots of pure iron when the hydrogen content was 13 ml/100 g, as compared with an equilibrium solubility of 25.2 ml/100 g at the freezing point.

When iron solidifies the solubility for hydrogen falls abruptly to 7.8 ml/100 g (Fig. 7.1), and the newly solidified metal tends to reject hydrogen into the residual melt. In consequence there is a peak in the curve of hydrogen concentration immediately adjacent to the solidification front. This peak concentration increases as the solidification front advances and eventually reaches a steady value. Assuming that there is no flow the concentration of hydrogen as a function of the distance x from the solid/liquid boundary is

$$C = C_o \left[1 + \frac{(1-k)}{k} e^{-vx/D} \right] \tag{7.30}$$

where C_o is the initial hydrogen concentration in the liquid, v is the velocity of the solidification front (in this case the welding speed) and D is the diffusivity of hydrogen in the liquid metal. The quantity k is the **partition coefficient**, given by

$$k = \frac{\text{solubility in solid at melting point}}{\text{solubility in liquid at melting point}} \tag{7.31}$$

At the interface $x = 0$ and the concentration is C_o/k, so that if C_o exceeds the solid solubility at the melting point the liquid solubility will be exceeded and a hydrogen bubble may form. This bubble persists and moves forwards with the solidification front to form an elongated pore.

In the weld pool the flow velocity is relatively high so that the conditions assumed above are not met. However, there is a boundary layer in which the velocity varies from zero at the interface to that of the bulk liquid. The extent of the enriched zone $\bar{x}$ may be reasonably taken as that over which the concentration falls to one-tenth that at the interface. Putting this value into equation 7.30 leads to

$$e^{-v\bar{x}/D} = \frac{1 - 10k}{1 - k} \tag{7.32}$$

and assuming $v = 3$ mm s^{-1}, $D = 1 \times 10^{-7}$ m^2 s^{-1} and $k = 13/25.2 = 0.516$, a value of 7.2×10^{-5} m is obtained for the thickness of the enriched region. This could therefore be accommodated within the boundary layer, which for a weld pool has a thickness of the order of 10^{-4} m. Thus bubbling at the interface and the formation of a tunnel pore is not impossible even when there is flow in the weld pool.

7.4.3 Oxygen desorption in steel

One of the features of weld metal produced by the submerged arc process or by coated electrodes is the presence of large numbers of small spherical oxide particles. These range in diameter typically from 0.1 to 2 µm and may number 10^6 or more per cubic millimetre. The way in which such inclusions affect the weld metal properties is discussed in Chapter 8.

It is generally assumed in the literature that the inclusions are precipitated in the weld pool and are frozen into the weld metal. The same assumption was made in previous editions of this book. The discussions to date on weld pool flow, however, suggest that this assumption may not be well founded. In the first place, the rapid circulation in the weld pool will tend to accumulate any slag particles at the surface. Secondly, a widespread fine precipitation requires a sharp drop in solubility, such as may be associated with a fall in temperature, whereas the degree of circulation makes sharp temperature gradients in the weld pool itself unlikely. There is, however, a temperature fall in the boundary layer at the rear of the weld pool and a much greater fall in the solidified weld metal. Precipitation of oxides could occur in either of these areas. Information currently available in the literature does not permit any firm conclusion. However, the amount of oxygen precipitated, as judged by the observed volume of oxides, is generally greater than would be predicted from a fall in temperature from the average for the weld pool to the melting point. Also, precipitates sometimes form linear arrays, suggesting that intergranular precipitation has taken place. Therefore the possibility of precipitation in the solidified weld metal must be considered.

7.5 Diffusion

Gas may also escape from the solidified weld metal by **diffusion**. This process is governed by a relationship of the same form as that for the diffusion of heat, except that the constant α for heat diffusivity is replaced by that for diffusivity of gas, D, and temperature is replaced by concentration c, so that

$$\nabla^2 c = \frac{1}{D}\frac{\partial c}{\partial t} \qquad (7.33)$$

There are two solutions to this equation that may be relevant to the escape of gas by diffusion. First, consider a slab of thickness $2w$, the centre of which is the plane $y = 0$ (Fig. 7.15a). Suppose that the concentration of gas in the slab at time $t = 0$ is uniformly c_o and the surface concentration is reduced to zero. The concentration c as a function of time is

$$\frac{c}{c_o} = \frac{4}{\pi}\sum_{n=0}^{\infty}\frac{(-1)^n}{(2n+1)}\left\{\exp\left[-\frac{(2n+1)^2\pi^2 Dt}{4w^2}\right]\cos\left[\frac{(2n+1)\pi y}{2w}\right]\right\}$$

$$(7.34)$$

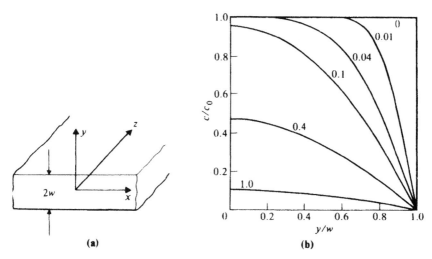

7.15 The diffusion of gas out of a slab of thickness 2w (a); numbers on the curves in (b) are values of Dt/w^2 (part (b) from Carslaw and Jaeger, 1959).

Figure 7.15(b) shows c/c_0 as a function of y/w for different values of Dt/w^2. The concentration along the centreline begins to fall significantly when Dt/w^2 is about 0.1.

For various reasons it may be desirable to carry out a **hydrogen diffusion treatment** after welding, and Fig. 7.15(b) gives some guidance as to how the diffusion time could be estimated. Suppose it is required to reduce the hydrogen content of a weld to half its as-welded level along the centreline. Then the value of Dt/w^2 should be about 0.4 and

$$t = 0.4w^2/D \qquad (7.35)$$

Unfortunately, the value of the diffusivity D is a variable quantity. When the steel is supersaturated with hydrogen, the lattice structure is damaged by the formation of microvoids, which impede normal diffusion. Thus if a specimen is exposed to a high degree of supersaturation and then tested using normal levels of hydrogen pressure, it will have a low or even zero diffusivity. Exposed again to supersaturation hydrogen levels the specimen may show normal diffusivity. Thus the diffusion coefficient shown in Fig. 7.16 varies over a wide range. For elevated temperature and at high hydrogen levels (say above 10 ml/100 g), it may be reasonable to take D as equal to the value for lattice diffusion (shown as a dotted line in Fig. 7.16). However, at room temperature and at lower hydrogen levels the gas diffuses sluggishly and somewhat unpredictably.

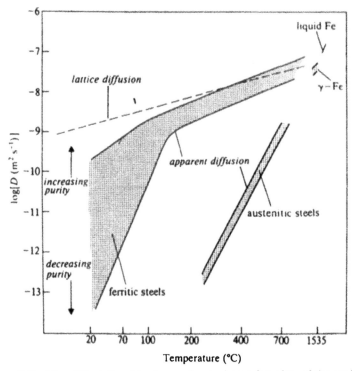

7.16 The diffusivity of hydrogen in steel as a function of the reciprocal of absolute temperature (from Coe, 1976).

7.6 Dilution and uniformity of the weld deposit

In most instances, filler metal is added to fusion-welded joints, and the weld deposit therefore consists of a mixture of parent metal and filler metal. When parent metal and filler metal have the same composition, this is of no consequence; but where they differ, suitable measures must be taken to ensure that the completed weld has the desired composition.

The degree of dilution depends upon the type of joint, the edge preparation and the process used. **Dilution** (expressed as a percentage) may be defined as

$$D = \frac{\text{weight of parent metal melted}}{\text{total weight of fused metal}} \times 100 \qquad (7.36)$$

It is a maximum for single-run welds on thin sections with a square edge preparation, and a minimum in fillet welds or in multipass welds with a normal edge preparation (Fig. 7.17). Dilution is of particular importance for dissimilar metal joints and in the welding of clad material. It may be minimized in such cases by applying a deposit of the composition required to the edges of the joint

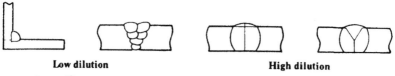

Low dilution **High dilution**

7.17 The dilution of a weld deposit.

before making the weld proper (**buttering**). In some cases, a special alloy is used for this purpose.

Mixing of metal in the weld pool is generally good, and normally the weld bead composition is substantially uniform within the fused zone. If there is a large difference between filler and parent metal composition, some variation may arise, particularly when inert-gas tungsten arc welding is used. Such variation is less likely with processes involving metal transfer. Conversely, alloy steel welds made with coated electrodes occasionally contain partially fused ferro-alloy particles. Some or all of the alloy content of the weld deposit may be introduced in the form of ferro-alloy or pure metal powder mixed with the coating, and the use of too coarse a grade of powder will result in non-fusion or other forms of segregation.

In most metals and alloys, the boundary between the fusion zone and the unmelted part of a fusion-welded joint is quite sharp. At the fusion boundary, the composition changes from that of the parent metal to that of the more or less uniform weld deposit. For a manual weld with coated electrodes, the thickness of this boundary zone lies typically between 50 and 100 µm (between 5×10^{-2} and 10^{-1} mm). In alloys having a long freezing range, partial melting may take place in that portion of the heat-affected zone immediately adjacent to the fusion zone, and some of the liquid pockets so formed may become physically continuous with the weld metal. Alloys that behave in this way are usually difficult to fusion-weld and (except for cast iron) are rarely met with in practice. Hydrogen may diffuse from the fused zone into the heat-affected zone, but other elements do not so diffuse to any significant extent, either from weld metal to unmelted metal or vice versa, during the welding process. Diffusion may occur during postwelding heat treatment or during service at elevated temperature, however.

7.7 Weld pool solidification

The crystals that form during solidification of the weld pool are nucleated by the solid crystals located at the solid–liquid interface. In the welding literature, this type of crystal growth is known as **epitaxial** (the growth modes illustrated in Fig. 7.18 are all epitaxial). Each grain forms initially as a continuation of one of the grains that lie along that part of the fusion boundary where the weld width is greatest. As the fusion boundary moves forwards, grains continue to grow in a columnar fashion. Competition between grains results in some change in relative size, but in general the primary grain size of the weld metal is determined by the

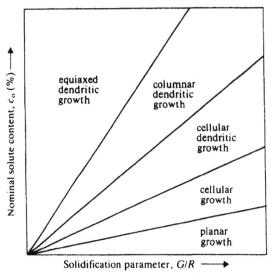

7.18 Factors controlling the growth mode during the solidification of liquid metals.

grain size of the solid metal at the fusion boundary. The factors that, in turn, govern the grain size of the solid will be considered in Section 7.9.1.

A fusion weld has a primary grain structure, and individual grains have a substructure that results from microsegregation. The type of substructure that appears in weld metal depends on the form of the solidification front. This in turn is influenced by the solute content of the liquid weld metal, and by a solidification parameter equal to the temperature gradient G in the direction of solidification divided by the rate of advance R of the solidification front (some authors prefer $\sqrt{R}$ but the simpler relationship will be used here). Figure 7.18 shows how, in general, these factors influence the mode of solidification while Fig. 7.19 illustrates the form of the various microsegregation patterns. For any given solute content, the microstructure tends to become more dendritic as the ratio G/R decreases, while the dendrite spacings tend to increase as the **freezing time** (expressed as $(GR)^{-1/2}$) increases. Eventually, at high values of $(GR)^{-1/2}$, the dendrites nucleate at a point and the structures become equiaxed. For the weld pool, the solidification velocity is equal to the welding speed v multiplied by the sine of the angle ϕ between the tangent to the weld pool boundary and the welding direction:

$$R = v \sin \phi \tag{7.37}$$

The temperature gradient $\partial T/\partial x$ is

$$\frac{\partial T}{\partial x} = \frac{\partial T}{\partial t}\frac{\partial t}{\partial x} = \frac{\partial T}{\partial t} \bigg/ \frac{\partial x}{\partial t} = \frac{\partial T}{\partial t} \bigg/ v = \frac{1}{v}\frac{\partial T}{\partial t} \tag{7.38}$$

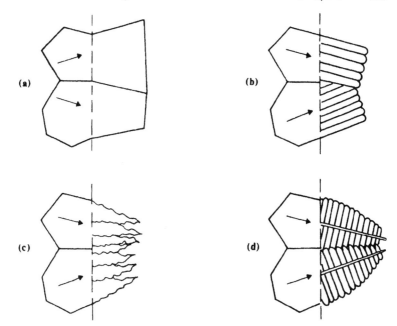

7.19 Solidification modes: (a) planar; (b) cellular; (c) cellular dendritic; (d) columnar dendritic. The arrows indicate the preferred growth direction ($\langle 100 \rangle$ for steel).

while the theoretical value of $\partial T/\partial t$ may be obtained from equations 6.19 and 6.20. For three-dimensional heat flow, the gradient at the rear of the weld pool is, numerically,

$$G = \frac{1}{v}\frac{\partial T}{\partial t} = \frac{T_m}{x_1} \qquad (7.39)$$

where T_m is the melting temperature and x_1 is the distance between the heat source and the rear of the weld pool. At this point $\sin\phi = 1$ and

$$\frac{G}{R} = \frac{T_m}{vx_1} \qquad (7.40)$$

The distance x_1 increases with welding speed and therefore the parameter G/R falls as the welding speed increases. Also, at the boundary of the fused zone, $\sin\phi = 0$ and G/R is theoretically infinite.

Thus the solidification parameter decreases from the fused zone boundary to the central axis of the weld. In practice, the tail of the weld may become sharp at high welding speeds and is generally more elongated than theory predicts. Equation 7.40 may not therefore be quantitatively accurate, but the indicated trend is correct.

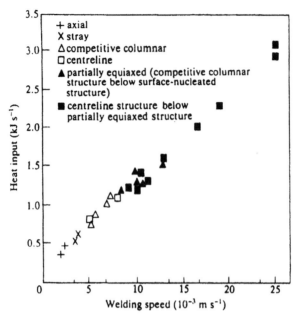

7.20 Weld macrostructure as a function of welding speed and heat input rate for a steel containing 0.038% C (from Ganaha and Kerr, 1978).

The grain structure of the weld appears to depend mainly on three factors: composition (solute content), solidification parameter and shape of the weld pool. Figure 7.20 shows how the grain structure of a gas tungsten arc weld in 1.5 mm thick low-carbon steel sheet varies with welding speed and heat input rate.

The structures and weld pool shapes are shown in Fig. 7.21. At the lowest velocities and high values of the solidification parameter, the central part of the weld is occupied by grains running longitudinally, and this is associated with a nearly circular weld pool. With higher speed and a more elongated pool, the grains established near the fusion boundaries are later blocked by grains growing from the rear of the weld pool, giving a random (stray) grain orientation. At higher speeds still, the weld pool becomes kite-shaped and grains form a herringbone pattern. Equiaxed grains are sometimes observed as shown in Fig. 7.21(e); these are thought to be nucleated by heterogeneous nuclei and are not necessarily related to the solidification parameter. Planar growth structures have been observed at the fusion boundaries, which is consistent with Fig. 7.18 since G/R can have high values in this region. Quite apart from the general effect of solute content, the range of structures obtained may depend on steel composition, i.e. one or more of the structures may not occur in a given steel.

Various techniques for controlling the primary grain size of weld metal have been tested. Ultrasonic vibrations may induce a degree of grain refinement in the case of light metals. Addition of nuclei formers such as titanium may be effective

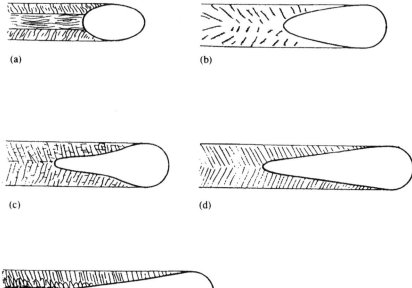

7.21 The macrostructure observed in flat sections of gas tungsten arc welds in low-carbon steel: (a) axial, 2.5 mm s^{-1}; (b) stray, 3.3 mm s^{-1}; (c) competitive, 7.5 mm s^{-1}; centreline, 16.7 mm s^{-1}; (e) partially equiaxed, 8.3 mm s^{-1} (after Ganaha and Kerr, 1978).

for aluminium fusion welds and electroslag welds. Electromagnetic vibration of the arc and pulsed current in gas tungsten arc welding have been successful in reducing grain size in stainless steel, nickel alloys, aluminium alloys and tantalum. In general, however, the grain size is determined by that in the heat-affected zone, which in turn is governed by the weld thermal cycle.

The size of primary grains has little effect on the properties of face-centred cubic weld metal (although finer **dendrite** spacing may increase the strength), but in ferritic steel the ductile–brittle transition temperature increases with grain size, other things being equal, and the yield strength falls.

7.8 Weld cracking

The restrained contraction of a weld during cooling sets up tensile stresses in the joint and may cause one of the most serious of weld defects – cracking. Cracking may occur in the weld deposit, in the heat-affected zone, or in both these regions. It is either of the gross type, which is visible to the naked eye and is termed **macrocracking**, or is visible only under the microscope, in which case it is termed **microcracking** or **microfissuring**. Cracking that occurs during the solidification of the weld metal is known as **solidification cracking** or **hot cracking**. Cracks may form in the heat-affected zone due to liquation of low-melting components: this is known as **liquation cracking**. Embrittlement of the parent metal or heat-affected zone may result in **subsolidus** or **cold cracking**. General features of solidification and liquation cracking are considered in the next section, while subsolidus cracking is dealt with in Section 7.9.2 and, for steel, in Chapter 8.

7.8.1 Solidification cracking

There are two necessary preconditions for the occurrence of cracking during the weld thermal cycle: the metal must lack ductility, and the tensile stress developed as a result of contraction must exceed the corresponding fracture stress. Solidification cracking may take place in two ways. The first, which may affect both castings and fusion welds, relates to the alloy constitution. On cooling a liquid alloy below its liquidus temperature, solid crystals are nucleated and grow until at a certain temperature they join together and form a coherent, although not completely solidified, mass. At this temperature (the **coherence temperature**) the alloy first acquires mechanical strength. At first it is brittle, but on further cooling to the **nil-ductility temperature** ductility appears and rises sharply as the temperature is reduced still further (Fig. 7.22). The interval between the coherence and nil-ductility temperatures is known as the **brittle temperature range**, and in general it is found that alloys possessing a long brittle range are sensitive to weld cracking, whereas those having a short brittle range are not. The second mode of specification cracking, which occurs in the region of the

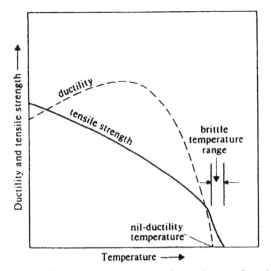

7.22 Mechanical properties of metals as a function of temperature.

solidus, is believed to be due to the presence of continuous intergranular liquid films. The liquid may be a eutectic, as in certain crack-sensitive aluminium alloys, or it may be formed by an impurity such as sulphur in steel. In aluminium alloys, the mechanism of cracking is best understood by reference to the equilibrium phase diagram, as discussed in Section 10.1.2. In those cases where cracking is promoted by impurities, a film will only form if the liquid is capable of wetting the grain boundaries: that is to say, if its surface energy relative to that of the grain boundary is low. Thus manganese which tends to globularize sulphides, helps to inhibit weld cracking due to sulphide films in carbon and low-alloy steel.

It will be evident that the longer the brittle range, the greater the possibility that dangerously high contraction stresses will be set up. In addition, there is in certain alloys a correlation between the tensile strength within the brittle range and the tendency towards weld cracking. In such materials, the slope of the strength–temperature curve is low for crack-sensitive material, but relatively high for a crack-resistant type (Fig. 7.23). An effect of this type is to be expected since the rate of increase of stress with decreasing temperature due to contraction is $xE\alpha$, where E is Young's modulus, α is the coefficient of expansion, and x is a restraint factor. Clearly, if during the brittle range the rate of increase of tensile strength (or fracture stress) is lower than $xE\alpha$, cracking will occur. It also follows that the higher the value of x (i.e. the higher the degree of restraint), the more likely it is that any given alloy will crack. The degree of restraint is a function of the type of joint, the rigidity of the structure, the amount of gap between the abutting edges, the plate thickness, and the relative thickness of plate and weld metal. Maximum restraint is obtained when two rigidly clamped thick plates are joined by a weld of small cross-section, while minimum restraint occurs in a weld of relatively large cross-section between two close-butting thin sheets (Fig. 7.24). Under conditions of extreme restraint, most weld metals will crack, but

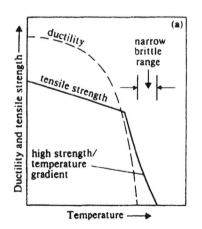
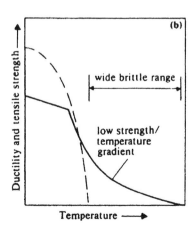

7.23 A comparison of mechanical properties near the solidus: (a) crack-resistant and (b) crack-sensitive alloys.

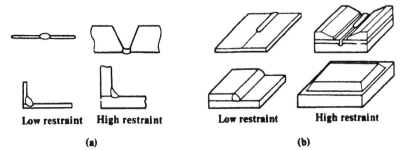

7.24 (a) Restraint dependent on dimensions of weld and plate. (b) Restraint dependent on rigidity of joint or set-up.

with light restraint solidification cracking is only likely in certain alloys, notably fully austenitic chromium–nickel steel, single (α) phase aluminium bronze, and certain aluminium and magnesium alloys. Crack-sensitive materials are usually subject to both micro- and macro-cracking, and microcracks are often present when there are no cracks visible to the naked eye.

Tensile stresses capable of generating solidification cracking may also arise due to bending. In some joint configurations, the weld shrinkage is such as to subject the weld to bending during the solidification period. Cracks may also appear at the fusion boundary, running a short distance into the parent metal. Such **hot tears** are most likely to occur under conditions of overall restraint and when there is a relatively acute angle between the weld reinforcement and the parent plate (an acute **wetting angle**). For the same reason it is very common for a small hot tear, typically about 0.5 mm deep, to form at the toe of fillet welds in carbon steel. Such cracks have a damaging effect on the fatigue strength of the joint, as will be detailed in Chapter 8.

Weld cracking may also result from inadequate **feeding**. The word 'feeding' is used here in the same sense as in foundry technology, i.e. to describe the geometrical arrangement that permits liquid metal to be drawn into and fill any gaps caused by shrinkage during solidification. Normally, fusion welds are self-feeding, but if the penetration is too deep relative to its width, and particularly if the weld is narrower at the top than at its midpoint, **shrinkage cracks** may form (Fig. 7.25). The same type of cracking may occur in electroslag welding if the

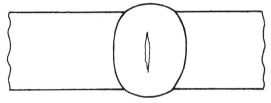

7.25 Solidification cracking due to unfavourable geometry of the fused zone.

welding conditions are set up incorrectly, but in this case the cracks form a herringbone pattern when seen in a radiograph or longitudinal section. Cracks may also form due to localized melting of either the parent metal or (in multi-run welds) the weld metal. This may be due to the fusion of low-melting constituents or phases, or it may be due to segregation of impurity elements to the grain boundaries and melting of the grain boundary regions. Alternatively, the same type of segregation may reduce the cohesion at the grain boundary. Such fusion or near-fusion may result in liquation cracking. Medium-carbon high-tensile steel and certain aluminium alloys may be subject to liquation cracking.

7.8.2 Measuring crack sensitivity

Tests for assessing the susceptibility to solidification cracking usually require that the weld be made under conditions where there is sufficient restraint to ensure cracking. The susceptibility of the weld metal is then assessed by measuring the crack length, the crack area, or the number of cracks. This measurement may be made on the as-welded surface, on longitudinal sections, on transverse sections or by radiography.

Of the many testing techniques that have been used, three are illustrated in Fig. 7.26. The Houldcroft test is useful for gas tungsten arc welding, and was much employed in early investigations of aluminium alloys. The most common method for steel is the Transvarestraint test. The weld is bent over a rooftop-shaped former as it is being made, and the crack sensitivity is expressed by one of the methods given above. The Transvarestraint technique does not discriminate too well between different crack-sensitive alloys such as fully austenitic chromium–nickel steels, and the MISO (measurement by means of *in situ* observation) technique has been developed as a more sensitive method. Strain is applied to the weld by a tensile or bend-test machine, and the progress of cracking is recorded by a high-speed cine camera or video. The surface strain is measured using irregularities on the weld surface as gauge marks, and the minimum strain at which cracking initiates is obtained as a measure of crack sensitivity. This quantity (which may also be obtained from the Transvarestraint test) is probably a more rational means of assessing crack sensitivity than crack length or number. Using the MISO technique, it is also possible to measure the brittle temperature range.

7.9 Metallurgical effects in the parent metal and solidified weld metal

In fusion welding, the heat-affected zone is subject to the full weld thermal cycle and in single-pass welds the solidified weld metal is exposed to the cooling part of the cycle. In multipass welds later weld runs subject the underlying weld metal

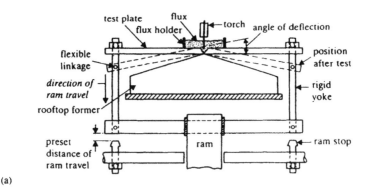

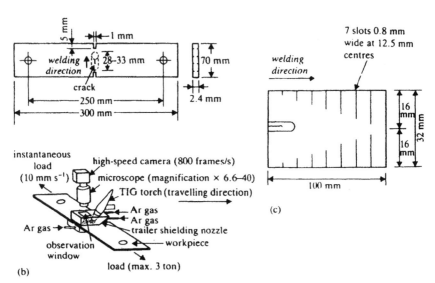

7.26 Solidification cracking tests: (a) Transvarestraint test; (b) MISO test; (c) Houldcroft fishbone test. See Figure 10.11 for modified Houldcroft test (part (a) from Bailey and Jones, 1978; part (b) from Matsuda *et al.*, 1983).

to at least one and, in places, two weld thermal cycles. Heating and cooling rates are usually high, and the heated metal is subject to plastic tensile strain during cooling. It is not surprising therefore that the metallurgical effects of the weld thermal cycle are complex and may in some instances result in an unfavourable change in the properties of the material.

7.9.1 Microstructural changes in the heat-affected zone

In general, the heat-affected zone may be divided into two regions: the high-temperature region, in which major structural changes such as grain growth take

Metallurgical effects of the weld thermal cycle

place, and the lower-temperature region, in which secondary effects such as precipitation (which may nevertheless be important) may occur. In the grain-growth region the final grain size for any given alloy will depend mainly on the peak temperature to which it is exposed and the time of heating and cooling (the **residence** time). For steel, it is found that for isothermal heating the final mean grain size d_t is given by

$$d_t^n = kt + d_0^n \qquad (7.41)$$

where k is a constant, t is time and d_0 is the initial grain size. If the exponent n and the constant k are determined by isothermal heating, the grain growth in the heat-affected zone may be estimated by a stepwise calculation.

However, it will generally be sufficient to note that the final grain size will increase with increasing peak temperature and increasing residence time. In fact, the maximum peak temperature is the melting point and it is the grain size at the fusion boundary in which we are mainly interested, since this determines the grain size in the weld metal. The only significant variable therefore (assuming the same metallurgy) is residence time, which is roughly proportional to the parameter q/v (the heat input rate). Thus welding processes having characteristically high values of q/v, such as electroslag and high-current submerged arc welding, would be expected to generate coarse grain in the heat-affected zone and in the weld metal, and this is indeed the case. For carbon and ferritic alloy steel, we are referring here to the austenitic grain size.

If the metal that is subject to the weld thermal cycle has previously been hardened by cold working, then grain growth may be preceded by recrystallization which, in turn, results in softening. Copper and aluminium are typical of metals that can acquire a useful increase in mechanical properties by cold reduction, and fusion welding will usually destroy the effect of cold working in the heat-affected zone. Similarly, alloys that are hardened by precipitation will usually be softened by fusion welding.

There are two ways in which the residence time can be reduced: first, by using a heat source that generates a deep weld pool having relatively small lateral dimensions and operating at high speed, and secondly by decreasing the heat input rate.

Electron-beam and laser welding have these characteristics and it is one of the advantages of such processes that they minimize the metallurgical disturbance in the heat-affected zone.

7.9.2 Precipitation, embrittlement and cracking in the parent metal

In addition to grain growth, which may in itself cause embrittlement, there are a number of metallurgical changes that result from the weld thermal cycle and may alter the properties of the parent metal. Also, the heat-affected zone may be permeated by hydrogen. Hydrogen does not appear to have any significant effect on the properties of austenitic chromium–nickel steel or non-ferrous metals other than copper. Ferritic steel is embrittled to a greater or lesser degree by hydrogen, and copper, if it is not deoxidized, may be severely embrittled by the reaction of hydrogen with residual oxygen to form steam. The steam precipitates at grain boundaries and generates fissures.

Ferritic alloy steels may be embrittled by the formation of unfavourable transformation products, by carbide precipitation, by grain boundary segregation (temper brittleness) and by strain ageing. Austenitic chromium–nickel steels may embrittle at elevated temperature owing to a strain-ageing mechanism during postweld heat treatment or in service. These processes will be considered in more detail in sections concerned with the particular alloys in question. Their damaging effects can be mitigated by correct choice of materials and by the use of the proper welding procedure and in some cases, by control of the weld thermal cycle.

A number of alloys may be embrittled within a specific temperature range, causing **ductility dip cracking**. These include fully austenitic chromium–nickel steels, some nickel-base alloys and some cupronickels. Figure 7.27 shows a plot of ductility against temperature for a number of cupronickels, which indicates that for nickel contents of 18% and higher the ductility falls sharply to a minimum at about 1000 K. Above and below the ductility trough, fracture is by microvoid coalescence; in the trough itself the fractures are intergranular. Such intergranular failure is normally attributed to segregation of impurity atoms (S or

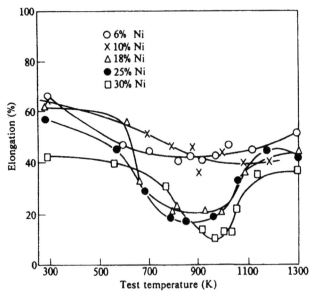

7.27 Ductility test results for binary copper–nickel alloys (from Chubb and Billingham, 1978).

P for austenitic Cr–Ni steel) to the grain boundary. For the cupronickel tests illustrated in Fig. 7.27, high-purity material was used and the segregating element could have been nickel itself.

Ductility dip cracking occurs in multipass welding with a susceptible weld deposit composition. Cracks occur in weld metal below the finishing pass because of tensile straining in the susceptible temperature range (possibly during the heating phase of the weld thermal cycle). Such cracks may not be detected by surface crack detection methods, but would show up in destructive testing, particularly side bend tests. Changes in welding procedure, by modifying the thermal cycle, may eliminate the cracking and in some instances the steps used to avoid solidification cracking may be effective: in austenitic Cr–Ni steel, for example, by reducing S and P, increasing manganese, and adding special deoxidants such as calcium or zirconium.

7.9.3 Contraction and residual stress

The **residual stress** due to a fusion weld in plate arises primarily because the strip of material that has been melted contracts on cooling from melting point to room temperature. If it were unhindered, the longitudinal contraction of the weld would be αT_m where T_m is the difference between the freezing point and room temperature and α is the mean coefficient of thermal expansion over that temperature range. If the plate cross-section is large relative to that of the weld, this contraction is wholly or partly inhibited, so that there is a longitudinal strain of up to αT_m. Assuming only **elastic deformation** occurred, the corresponding residual stress would be approximately $E\alpha T_m$, where E is the mean value of Young's modulus over the relevant temperature range. However, except for metals with very low melting points, the value of $E\alpha T_m$ is greater than the elastic limit, so that some **plastic deformation** of the weld takes place during cooling, and the final residual stress in the weld approximates to that of the elastic limit.

Measurement shows that the residual stress in a thin plate after welding consists of a tensile stress in the weld metal itself, falling away parabolically to zero a short distance from the weld boundary, with a balancing compressive stress in the outer part of the plate. Figure 7.28 illustrates typical residual stress field cross-sections for a carbon-steel plate after making a butt weld down the centre using coated electrodes. In wide plates, the residual stress in the weld metal is close to the recorded yield value for weld deposits made with coated electrodes. In a narrow plate, the redistribution of stress results in lower values of residual stress in the weld metal but higher values in the plate.

When making a weld between two plates of moderate thickness (for example, 10 mm thick), most of the residual stress is built up during the first run. In thick plate, however, there is a contraction stress at right angles to the plate surface, and consequently the stress field may intensify progressively as the joint is built up.

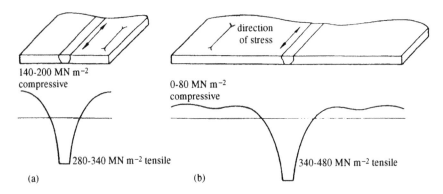

7.28 Typical residual stress fields in longitudinally welded plate; carbon steel welded with coated electrodes. (a) Narrow plate (100–200 mm); (b) wide plate (over 300 mm).

Figure 7.29 shows residual stresses along the centreline of a submerged arc weld in 165 mm thick plate Mn–Mo steel (yield stress nominally 480 MN m^{-2}). The tests were made on a sample 650 mm long cut from a larger plate, and there may have been some mechanical stress relief, particularly of the longitudinal residual stress. In all three directions – longitudinal, transverse (at right angles to the line of weld and parallel to the plate surface) and short transverse (at right angles to the plate surface) – there is a maximum tensile stress close to each plate surface and a maximum compressive (negative) stress in the centre. Details of the residual stress pattern vary with the weld procedure and may, for example, show a tensile region in one half of the cross-section and a compressive region in the other. If it is assumed that the three stresses measured represent principal stresses, the effective residual stress is

$$\bar{\sigma}_r = \frac{1}{\sqrt{2}}[(\sigma_T - \sigma_L)^2 + (\sigma_L - \sigma_{ST})^2 + (\sigma_{ST} - \sigma_T)^2]^{1/2} \qquad (7.42)$$

The fracture toughness of the weldment K_{IC} is reduced by the presence of a residual stress, and it has been found experimentally that this reduction (ΔK_c) is proportional to the effective residual stress:

$$\Delta K_c = k\bar{\sigma}_r \qquad (7.43)$$

For a weld in ductile material, ΔK_c is not large enough to affect the behaviour of the joint, but in embrittled material it might well be a significant factor in promoting an unstable fracture.

Residual stress may cause cracking in brittle materials or in materials that have been embrittled by welding. It may also result in stress corrosion cracking, if the alloy is sensitive to this type of corrosion and is exposed to certain environments. In a ductile metal that is not subject to stress corrosion cracking, such as pure

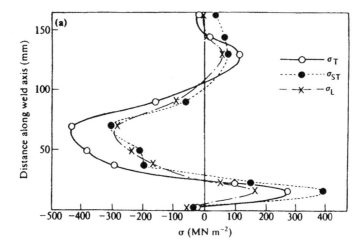
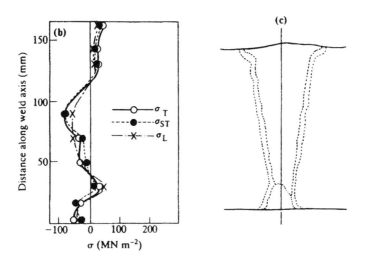

7.29 The distribution of residual stress along the centreline (c) of a submerged arc weld in 165 mm thick Mn–Mo steel: (a) postweld heat treatment 15 min at 600 °C; (b) postweld heat treatment 40 h at 600 °C. Curves are σ_T, stress transverse to weld; σ_{ST}, stress at right angles to plate surface; σ_L, longitudinal stress (from Suzuki et al., 1978).

aluminium, the residual stress due to fusion welding has little or no effect on the properties of the joint.

Residual stresses may be reduced, although not entirely eliminated (see Fig. 7.29) by a **postweld heat treatment**, typically at about 600 °C for a minimum of 1 h per 25 mm thickness. Heating may be local to the weld, as for example in the case of site welds in process pipework, or the whole assembly may be treated in a furnace, as is usually the case with pressure vessels. There is some evidence that carbon and low-alloy steel may be embrittled by such furnace treatment, possibly because of slow cooling from the soaking temperature. It is normal practice in the case of pressure vessels to make a representative welded testpiece, and these samples are given a simulated heat treatment in the laboratory before testing. However, it is rare for testpieces to be subject to the slow cooling typical of a furnace, so that embrittlement that may occur during this period will not normally be detected. It is also unusual to test samples from the shell of a vessel after treatment, but when this is done, impact test results are not infrequently below specification level and on occasions are very low loaded. The damage is probably caused by **temper embrittlement**, which results from the migration of impurity atoms to the prior austenite grain boundaries. With the continued trend towards increased purity in commercial steel, any such problems should be ameliorated, and there is evidence to this effect.

Regardless of any metallurgical change that may occur during postweld heat treatment, the safety record of welded pressure vessels is very good. In particular, brittle failures are rare, and the majority of such failures have occurred during hydrostatic testing, and not in service.

Vibration may also be employed to remove or diminish welding residual stress. A machine which drives rotating eccentric masses, for example, is attached to the structure for a predetermined time. **Vibratory stress relief** is not universally accepted as being effective, and is much less commonly used than thermal stress relief: there is, however, good reason to expect some benefit. It is known that slip takes place in metals subject to fatigue testing, so that vibration would be expected to iron out strain concentration to some degree. The process has been used for machine beds, where residual stress could result in dimensional instability. Residual stresses in welds may be measured by observing the change in dimensions after cutting rectangular or circular sections out of a sample. Alternatively, they may be estimated from the heat flow equations using elastic–plastic analysis and, as a rule, numerical methods of calculation.

The final stage in the construction of pressure equipment is to make a hydrostatic pressure test, or more rarely, a pneumatic or gas leak test. Lifting equipment such as cranes and elevators may be likewise subject to a **proof load** greater than that expected in service. Such tests are intended to prove that the structure is more than capable of bearing its service loading. It also, however, may have a beneficial metallurgical effect.

Most structures contain certain areas of **stress concentration**. In pressure vessels for example, openings such as nozzles and manways constitute such areas. During the hydrostatic test these regions are strained beyond the elastic limit; they stretch plastically and when the pressure is released, relaxation of elastic strain in the surrounding metal puts them into a state of compression. Then when the vessel is in service at a pressure lower than that of the hydrotest, the erstwhile areas of stress concentration will remain in a state of compression, and will not be vulnerable to fatigue or stress corrosion cracking, which could in turn initiate a catastrophic failure.

Very rarely, this effect (sometimes known as **shakedown**) has acted in the wrong direction. The hull of the first Comet aircraft was given repeated pressure tests to simulate the effect of take-offs on the pressurized cabin. Unfortunately, these tests were conducted on a mock-up which had previously been given a static overpressure loading. These tests indicated that the aircraft could be operated safely for the required service life. The operating aircraft were not, however, given an overpressure test; fatigue failures initiated at points of stress concentration and as a result, two of the Comets burst in midair.

In virtually all other cases, the overload test is beneficial. The good service record of pressure vessels, referred to earlier, is in part due to this fact.

References

Bailey, N. and Jones, S.B. (1978) *Welding Journal*, **57** (8), 2–7.
Brandes, E.A. (ed.) (1983) *Smithells Metals Reference Book*, 6th edn, Butterworth, London.
Carslaw, H.S. and Jaeger, J.C. (1959) *Conduction of Heat in Solids*, Clarendon Press, Oxford.
Chew, B. and Willgoss, R.A. (1980) in *Weld Pool Chemistry and Metallurgy*, Paper 25, TWI, Cambridge.
Chubb, J.P. and Billingham, J. (1978) *Metals Technology*, **5** (3), 100–103.
Coe, F.R. (1976) *Welding in the World*, **14** (1/2), 1–7.
Evans, D.B. and Pehlke, R.D. (1965) *Transactions of the AIME*, **233**, 1620–1624.
Ganaha, T. and Kerr, H.W. (1978) *Metals Technology*, **5**, 62–69.
Howden, D.G. and Milner, D.R. (1963) *British Welding Journal*, **10**, 313.
Kasamatsu, Y. (1962) *Journal of the Japanese Welding Society*, **31**, 775.
Keene, B.J. (1983) *A survey of extant data for surface tension of iron and its binary alloys*, NPL Report DM(A)67, National Physical Laboratory, Teddington, Middlesex, UK.
Kobayashi, T., Kuwana, T., Kikuchi, Y. and Kiguchi, R. (1968) International Institute of Welding Document no. XII-461-68.
Kobayashi, T., Kuwana, T. and Kikuchi, Y. (1971) *Journal of the Japanese Welding Society*, **40**, 221–231.
Kobayashi, T., Kuwana, T. and Kiguchi, R. (1972) *Journal of the Japanese Welding Society*, **41**, 308–321.
Kuwana, T. and Kokawa, H. (1983) *Quarterly Journal of the Japanese Welding Society*, **1**, 392–398.
Kuwana, T. and Sato, Y. (1983a) *Journal of the Japanese Welding Society*, **52**, 292–298.
Kuwana, T. and Sato, Y. (1983b) *Quarterly Journal of the Japanese Welding Society*, **1**, 16–21.
Matsuda, F., Nakagawa, H., Kohmoto, H., Honda, Y. and Matsubara, Y. (1983) *Transactions of the Japanese Welding Research Institute*, **12**, 65–80.
Ohno, S. and Uda, M. (1981) *Transactions of the National Research Institute for Metals*, **23**, 243–248.

Pehlke, R.D. and Elliott, J.F. (1960) *Transactions of the AIME*, **218**, 1088–1101.
Suzuki, M., Komura, I. and Takahashi, H. (1978) *International Journal of Pressure Vessels and Piping*, **6**, 87–112.
Taylor, C.R. and Chipman, J. (1943) *Transactions of the AIME*, **154**, 228–247.
Uda, M. (1982) *Transactions of the National Research Institute for Metals*, **24**, 218–225.
Uda, M. and Wada, T. (1968a) *Transactions of the National Research Institute for Metals*, **10**, 21–33.
Uda, M. and Wada, T. (1968b) *Transactions of the National Research Institute for Metals*, **10**, 79–91.
Uda, M. and Ohno, S. (1973) *Transactions of the National Research Institute for Metals*, **15**, 20–28.
Uda, M. and Ohno, S. (1975) *Transactions of the National Research Institute for Metals*, **17**, 71–78.
Uda, M., Dan, T. and Ohno, S. (1976) *Transactions of the Iron and Steel Institute of Japan*, **16**, 664–672.

Further reading

American Welding Society (1977) *Welding Handbook*, 8th edn, Vol. 1: *Welding Technology*, AWS, Miami; Macmillan, London.
Brandes, E.A. (1983) *Smithells Metals Reference Book*, 6th edn, Butterworth, London.
Carslaw, H.S. and Jaeger, J.C. (1959) *Conduction of Heat in Solids*, Clarendon Press, Oxford.
Christensen, N., Davies, V. de L. and Gjermundsen, K. (1965) *British Welding Journal*, **12**, 54–74.
Davies, G.J. and Garland, J.G. (1975) *International Metallurgical Reviews*, **20**, 83–106.
Pfluger, A.R. and Lewis, R.E. (1968) *Weld Imperfections*, Addison-Wesley, Reading, Mass.
The Welding Institute (1980) *Weld Pool Chemistry and Metallurgy*, TWI, Cambridge.

8
Carbon and ferritic alloy steels

8.1 Scope

It is the intention in this chapter to discuss those steels that have a body-centred cubic form at or above normal atmospheric temperature. Included are carbon steels with carbon content up to 1.0%, carbon–manganese steels with manganese content up to 1.6%, and steels containing other alloying elements up to the martensitic type of 12% Cr steel and maraging nickel steel. Higher-alloy ferritic and ferritic–austenitic steels are also included, although they may not be subject to an α to γ transformation. Other than this last group, the common feature about these alloys is that they may all be hardened, to a greater or lesser degree, as a result of passing through the weld thermal cycle, and therefore there may be a change in properties in the region of a fusion-welded joint. General metallurgical questions will be considered first, and a later section will deal with the individual alloys and alloy groups that are used in welded fabrication. Cast iron is included in the material groups since, from a welding viewpoint, it suffers the same type of transformation in the heat-affected zone of fusion welds as steel, albeit in an extreme form.

8.2 Metallurgy of the liquid weld metal

8.2.1 Gas–metal reactions: general

It is probable that the results pertaining to gas absorption from arcs in argon–gas mixtures reported in Chapter 7 are broadly applicable to arc welding processes generally. In particular, gases are absorbed at the arc root more rapidly, and to a higher level, than would be the case in non-arc melting. Excess hydrogen that may consequently be present bubbles out at the rear of the weld pool and under steady welding conditions the surplus gas is purged before the metal solidifies, so that this process does not cause porosity. Nitrogen is less rapidly desorbed, and therefore is likely to be present at an undesirable level in the finished weld if protective measures are not taken. Oxygen, if absorbed to excess, combines with deoxidants and is rejected in the form of slag. If carbon dioxide is used as a shielding gas, it may add both carbon and oxygen to the weld pool. Details of individual gas–metal reactions are described below.

8.2.2 Nitrogen

The amount of nitrogen absorbed in arc welding increases with the partial pressure of nitrogen in the arc atmosphere and with the amount of oxygen. The saturation level for nitrogen is about equal to the equilibrium solubility at 1 atm pressure and liquid metal temperature of 1600 °C, which is 0.045% by mass. The amount dissolved from a 50% N_2–50%; O_2 arc atmosphere rises to 0.16%, so that air is a particularly unfavourable atmosphere for the arc welding of steel. Hydrogen, conversely, reduces nitrogen absorption, so that in a nitrogen–hydrogen mixture the gas content is equal to the equilibrium value for the partial pressure of nitrogen in the arc atmosphere. Figure 8.1 shows the nitrogen content of the transferring drops in self-shielded arc welding using cored wire. Wire A contains only iron powder, so this is effectively bare wire welding, and the nitrogen content falls with increasing current. Wire B has a rutile core, which is not adequate without additional shielding, and the nitrogen content increases slightly with current. Wire C is a typical self-shielded wire, containing a $CaCo_3$–CaF_2 flux core with aluminium and magnesium. The transferring drops are surrounded with a blanket of volatile metal vapour, which keeps the nitrogen content below 0.025%, regardless of current.

Nitrogen is damaging in two ways, first by causing porosity, and second by embrittling the weld deposit. Porosity appears when the nitrogen content exceeds about 0.045 mass %, as shown in Fig. 8.2. This would be expected from the saturation effect discussed earlier. Nitrogen porosity also occurs in the root pass of pipe welds when made with basic coated electrodes. The conditions here are unfavourable since the underside of the weld pool is in contact with air, and the arc atmosphere is oxidizing. Rutile and cellulosic electrodes do not suffer this defect, possibly because of the effect of hydrogen in the arc atmosphere. In self-shielded welding it is necessary to add about 1% aluminium in order to inhibit nitrogen porosity.

Nitrogen in the form of FeN has a severely embrittling effect on weld metal. Deoxidants normally added to welding filler materials are usually adequate to combine with nitrogen to form non-embrittling nitrides. In self-shielded welding, aluminium acts as both deoxidizer and nitride-former. However, aluminium has the effect of closing the gamma loop in the iron–carbon equilibrium diagram (Fig. 8.15) so that if too much is present a coarse, brittle ferritic structure results. This problem is overcome by the addition of austenite stabilizers such as manganese and by restricting the aluminium content. Where improved notch-ductility is required additions of 0.5–2.0% Ni may be made.

Carbon and ferritic alloy steels 213

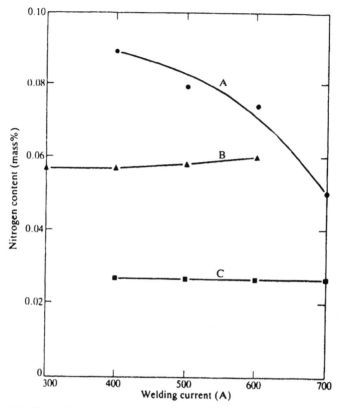

8.1 The nitrogen content of transferring drops in self-shielded flux-cored DC arc welding: curve A, wire with iron powder core; B, rutile core; C, $CaCO_3-CaF_2$ core plus Al and Mg (after Morigaki et al., 1973).

Nitrogen in a carbon or carbon–manganese steel may be responsible for **strain-age embrittlement**. If a nitrogen-bearing steel is subject to plastic strain and simultaneously or subsequently heated at a temperature of about 200 °C, the notch ductility is reduced. Prior to World War II a number of welded bridges were erected in Belgium and Germany using steel made by the air-blown Bessemer process, and therefore of high nitrogen content. Several such bridges failed by brittle fracture, and these failures were ascribed to strain-age embrittlement. Subsequent testing has shown that if a small crack forms adjacent to the weld boundary and is then strained and reheated by successive weld passes, the tip of the crack can be sufficiently embrittled to initiate a brittle fracture. The root passes of a multipass weld are similarly strained and reheated, and this may also lead to a degree of embrittlement, as will be seen later. Strain-age embrittlement may be prevented by the addition of strong nitride-forming elements such as aluminium or titanium. This expedient was used successfully

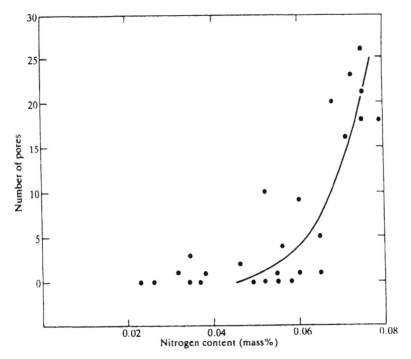

8.2 The number of pores in the fractured surface of a tensile testpiece as a function of nitrogen content; self-shielded flux-cored welding of carbon steel (after Morigaki *et al.*, 1977).

for plate material subsequent to the failures mentioned above, but is not normally possible for arc welding electrodes because Al and Ti burn out in the arc. However, in self-shielded welding the amount of Al in the weld deposit may be sufficient to prevent strain-age embrittlement of the root passes.

8.2.3 Oxygen

Oxygen may be dissolved in the liquid metal either directly from the arc atmosphere or by reaction with a slag or flux. Slag–metal reactions are reviewed in Section 8.2.4.

The arc atmosphere contains substantial amounts of oxygen in the gas metal arc welding of carbon and alloy steel, where it is added to stabilize the arc, in CO_2-shielded welding, and in self-shielded arc welding. Basic electrode coatings and basic submerged arc welding fluxes may also generate oxygen by decomposition of CO_2 but the dominant effect in these cases is the slag–metal reaction.

Oxygen in the arc atmosphere affects the weld metal properties in three ways: through the precipitation of non-metallic inclusions, by oxidizing alloying additions, and by causing CO porosity. Oxygen is present in weld metal as oxides, silicates or other chemical compounds, and in solution. The compounds may be present as macroinclusions (slag inclusions) or microinclusions. Macroinclusions are usually the result of human error, but microinclusions are precipitated either at the rear of the weld pool or in the freshly solidified weld metal because of the fall in oxygen solubility with falling temperature. They are spherical in form, and appear as circular discs in microsections. They are composed of the oxides of silicon, manganese and any other deoxidants that may be present, together with manganese sulphide. Diameters are up to 10 μm, and there are typically 1×10^4–3×10^4 inclusions per square millimetre, so that the volume fraction may be up to 1%. Figure 8.3 shows the composition and size distribution of such inclusions in submerged arc welds made with a flux of basicity $BI = 1.1$ (see Section 8.2.4). These precipitates become coarser as the welding current increases. They may affect the mechanical properties of the weld metal (see Section 8.3).

In processes that employ a consumable electrode, oxidation of the alloying elements takes place in the droplets as they form at the electrode tip and transfer across the arc; there is little reaction with the weld pool. Silicon and manganese are commonly used as deoxidants for carbon steel in arc welding, and their oxidation loss in gas metal arc welding using argon–oxygen shielding is shown in Fig. 8.4. Manganese is lost by vaporization as well as oxidation, silicon by

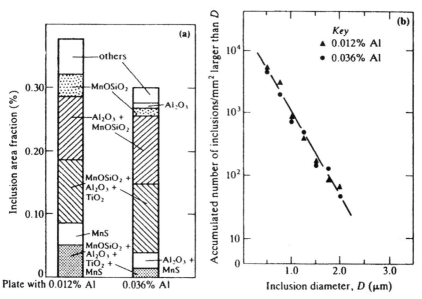

8.3 (a) Inclusion composition and (b) size distribution in submerged arc weld metal made with a flux of basicity $BI = 1.1$ and on plates of different aluminium content (from Hannerz and Werlefors, 1980).

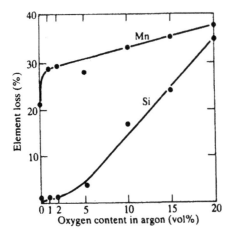

8.4 The loss of manganese and silicon from carbon steel in gas metal arc welding, as a function of the oxygen content of the argon–oxygen atmosphere (from Corderoy and Wallwork, 1989).

oxidation alone. Triple deoxidized wire containing titanium, zirconium and aluminium may be used in order to avoid porosity in the presence of surface contaminants. However, for most practical applications, deoxidization by silicon and manganese is adequate.

The oxygen content of 9% Ni steel weld metal produced by gas metal arc welding with an Ar–O_2 shield is shown in Fig. 8.5. There is virtually no nitrogen pick-up, indicating that the shielding was good. The oxygen content with a 2% O_2–Ar mixture is 0.025 mass%, and from Fig. 7.12 the maximum permissible carbon content to avoid CO blowholes would be 0.1%. For the most part gas metal arc wires have a carbon content below this level. However, in welding with coated electrodes with no added deoxidant (oxide–silicate coating), CO bubbles form in the drop at the electrode tip and eventually burst, generating a fine metal spray. Where deoxidants are present, the final oxygen content of the weld metal will be reduced as indicated by Fig. 8.5.

8.2.4 Slag–metal reactions

The nature of slag–metal reactions is largely determined by the composition of the flux or electrode coating. However, the mechanics (or physics) of the interaction may be important and will be discussed first.

The mechanics of slag–metal interaction

The form of the tip of a coated electrode is illustrated in Fig. 8.6. The coating forms a cone within which a liquid drop forms. Interaction between slag and

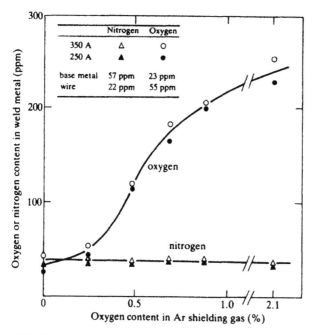

8.5 The oxygen and nitrogen contents in weld metal, as functions of the oxygen content in argon–oxygen shielding gas (from Matsuda et al., 1981).

metal takes place initially at the junction between cone and liquid. If alloying elements are added through the coating, they dissolve at this point and are mixed well enough for the drop, when it detaches, to be almost homogeneous. This indicates rapid circulation in the drop, owing to electromagnetic effects and/or drag from the gas evolved by the coating. At the same time, the liquid metal is heated to a high temperature. In all probability the more important chemical interactions, particularly the absorption of oxygen, occur at this point. In practice, the drop profile is by no means as regular as shown in Fig. 8.6: it is in a state of constant movement, as is the arc root. This does not, however, affect the basic geometry of the arrangement.

With flux-cored wire, the situation is very different, in that the metal sheath normally melts before the flux either melts or detaches. Little interaction can take place at the electrode and the flux–metal reactions must take place in the weld pool.

In submerged arc welding, the arc operates in a cavity of liquid slag and the liquid metal drops frequently transfer through the slag cavity or occasionally around the cavity wall. Once again, because of the high temperature and the large surface-to-volume ratio of the drop, it is likely that significant slag–metal reactions take place at this point. The deep penetration characteristic of submerged arc welding may also be relevant to slag–metal reactions. Suppose

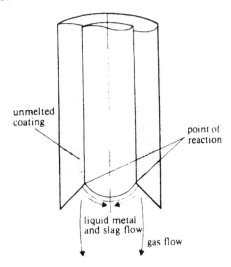

8.6 The slag–metal reaction in submerged metal arc welding.

that it is desired to add an element M through the flux (or through the wire, for that matter). The final concentration of M in the weld pool will depend on the degree of dilution. Thus special additions may, in submerged arc welding, need to be restricted to applications where the degree of dilution is known.

The chemistry of slag–metal interaction

The nature of the chemical reactions between slag and liquid weld metal has been studied largely in connection with the submerged arc process, but in general terms the conclusions that have been reached, particularly in relation to flux basicity, apply also to welding with coated electrodes.

There are several formulae for expressing the basicity of slag, of which the International Institute of Welding (IIW) index BI (due originally to Tuliani, Boniszewski and Eaton) and the Mori index BL are most commonly used:

$$BI = \frac{CaO + CaF_2 + MgO + K_2O + Na_2O + \frac{1}{2}(MnO + FeO)}{SiO_2 + \frac{1}{2}(Al_2O_3 + TiO_2 + ZrO_2)} \quad (8.1)$$

$$BL = 6.05*CaO + 4.8*MnO + 4.0*MgO + 3.4*FeO \\ - (6.31*SiO_2 + 4.97*TiO_2 + 0.2*Al_2O_3) \quad (8.2)$$

where in equation 8.1 CaO, etc., represent the mass per cent of the component in question, while in equation 8.2 *CaO, etc., represent the mol fraction of the component. In terms of the IIW index, the flux is acid when BI is less than 1, neutral when it is 1.0 to 1.5, semibasic when 1.5 to 2.5, and basic when greater than 2.5. Using the Mori index, slags are basic when BL is positive, and acid when it is negative, neutral fluxes having a BL value close to zero.

Carbon and ferritic alloy steels 219

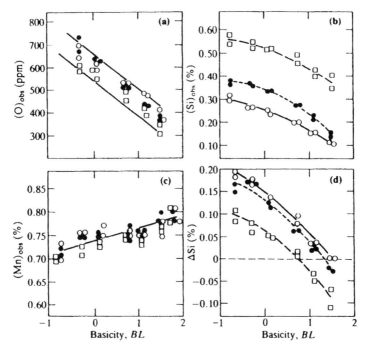

8.7 Relations between slag basicity and element content in weld metal: (a) oxygen; (b) silicon; (c) manganese; (d) silicon increment.

The influence of slag basicity on the content of oxygen, silicon and manganese in submerged arc weld metal is shown in Fig. 8.7. Oxygen and silicon contents are largely governed by the reaction

$$SiO_2 = [Si]_D + 2[O]_D \qquad (8.3)$$

and all these quantities are reduced as the SiO_2 activity in the slag falls with increasing basicity. $(Si)_w$ and $(Si)_p$ in Fig. 8.7 represent the silicon contents of wire and plate, respectively. Increased basicity has a dominant effect in reducing oxygen content, but inceasing silicon content of the wire from 0.05% to 0.82% causes a fall of about 0.01 in the mass per cent of oxygen. Manganese content increases slightly with basicity, but in a somewhat irregular manner.

The test results recorded in Fig. 8.7 were obtained using agglomerated fluxes of the $CaO-MgO-Al_2O_3-SiO_2$ system. By incorporating the stable oxide TiO_2 in the flux, it is possible to obtain lower oxygen contents for any given basicity. Incorporating boron in such fluxes (as B_2O_3) can give a controlled transfer of Ti and B to the weld metal with corresponding improvement in tensile strength and notch ductility.

The sulphur content of the weld metal is reduced by increasing the basicity of the slag. Phosphorus is not so affected; indeed there may be a tendency to increased phosphorus because of the composition of minerals used to manufacture basic fluxes.

The term **neutral** is used in the USA for a flux that does not contain any significant amount of metallic deoxidant or other metallic component, as opposed to an **active** flux where such deoxidants are present and where the amount of manganese and silicon in the weld metal may be affected by welding variables such as heat input rate, voltage and electrode stick-out. Cases have been known where deviations from the prescribed welding conditions led to excessive silicon and manganese contents, resulting in high weld hardness and ultimate failure due to stress corrosion cracking. Similarly, where alloying elements are added via submerged arc flux, variations in the welding parameters may result in failure to obtain the required weld composition. Consequently, some authorities specify 'non-active' or 'neutral' fluxes. In welding with coated electrodes, on the other hand, the weld metal composition is not affected in this way, and it is normal practice to add alloying elements via the electrode coating for low-alloy steels.

Submerged arc fluxes may be **fused, sintered** or **agglomerated (bonded)**. Fused fluxes are melted, cast and ground to the required size. Sintered fluxes are partially fused, while agglomerated fluxes are mixed and then baked at a temperature below that required for sintering. Fluxes may also consist of a blend of fused and bonded types, or of minerals mixed with metallic deoxidizers. All these methods of manufacture may be used to produce a satisfactory flux. One of the more important requirements for a flux is that it should not be hygroscopic, and that for any given moisture content it should produce the lowest possible hydrogen content in the weld metal. Basic fluxes are more sensitive in this respect than semibasic or acid fluxes, and it is thought that the **chevron cracking** (Section 8.5.4) that has been found in weld metal produced with a basic flux may have been due in part to excessive hydrogen content. Basic fluxes are used because the low silicon, oxygen and sulphur contents give better notch ductility in the weld metal. However, slag removal from the deeper weld preparations may be difficult with a fully basic flux, and for some applications it may be preferable to use a semibasic type. Fluxes that absorb moisture need special handling and storage, and may require drying before use.

8.2.5 Hydrogen

From the results presented in Section 7.2.3 it will be evident that the hydrogen solubility at the arc root, s_a, is higher than the equilibrium value s_e, but that hydrogen evolution from the liquid metal surface outside the arc reduces the net amount absorbed to the equilibrium level at about 1600 °C. An attempt has been made to measure s_a by producing an arc-melted pool in thin steel plate, the underside of which was kept solid by cooling. Argon–hydrogen mixtures were used for the arc atmosphere, and the pressure of hydrogen p_A diffusing through the plate was measured and compared with the partial pressure of hydrogen in the arc atmosphere p_E. The arc solubility is then

$$\frac{S_a}{S_e} = \left(\frac{p_A}{p_E}\right)^{1/2} \tag{8.4}$$

Results for carbon steel are given in Fig. 8.8, from which $(p_A, p_E)^{1/2}$ varies from 2.5 up to about 4. In such a test there is necessarily a hydrogen loss from the weld pool surface, so the true figure for S_a/S_e is probably higher.

Both testing and experience show that, under steady welding conditions, it is possible to obtain sound weld metal with up to 50% H_2 in the arc atmosphere. Under unsteady conditions (such as stops and starts), this is not the case, and

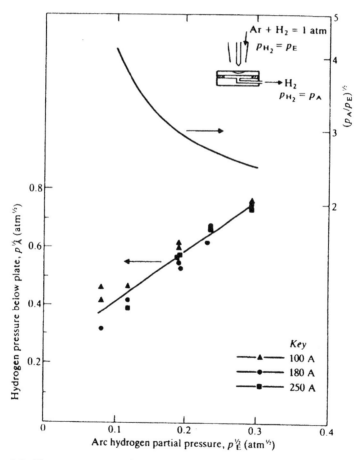

8.8 The pressure p_A of hydrogen diffusing through arc-melted carbon steel, as a function of hydrogen partial pressure p_E in arc atmosphere. Total pressure of Ar + H_2 is 1 atm (after Howden, 1980).

222 Metallurgy of welding

hydrogen porosity is probable if the arc atmosphere is hydrogen-rich. Tunnel porosity can occur even when the hydrogen content is below the equilibrium solubility, as described in Section 7.4.

8.2.6 Solidification and solidification cracking

The primary structure of carbon and low-alloy steel weld metal is similar to that of other metals; it is epitaxial with elongated columnar grains extending from the fusion boundary to the weld surface. The substructure is cellular at low solidification rates and dendritic at higher rates. The phase structure at solidification depends mainly on the carbon and nickel contents. Figure 8.9 shows the upper left-hand corner of the iron–carbon phase diagram. When the carbon content is below 0.10%, the metal solidifies as δ ferrite. At higher carbon contents, the primary crystals are δ, but just below 1500 °C a peritectic reaction takes place and the remainder of the weld solidifies as austenite. The solubility of sulphur in ferrite is relatively high, but in austenite it is relatively low. Consequently, there is a possibility with $C > 0.1$ that sulphur will be rejected to the grain boundaries of primary austenite grains, promoting intergranular weakness and solidification cracking. Manganese tends to inhibit the effect of sulphur, but the higher the carbon content, the higher the manganese sulphur ratio required to avoid cracking, as shown in Fig. 8.10. Sulphur may also segregate to interdendritic regions and promote interdendritic cracks.

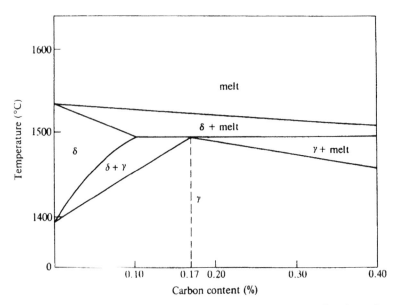

8.9 A section of iron–carbon equilibrium diagram showing the peritectic reaction.

Carbon and ferritic alloy steels 223

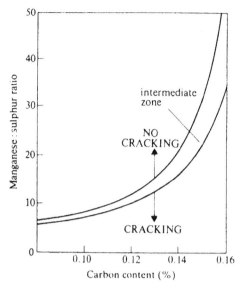

8.10 The effect of manganese/sulphur ratio and of carbon content on the susceptibility of carbon steel weld metal to hot cracking.

Figure 8.11 is part of the Fe–S binary phase diagram, which shows that the equilibrium maximum solubility of sulphur in δ ferrite is 0.18%, while in austenite the maximum solubility is about 0.05%. Under continuous cooling a sulphur-rich liquid will be rejected at somewhat lower bulk concentrations than these, and at worst liquid sulphide can persist down to about 1000 °C. In practice, the presence of sulphur has the effect of increasing the brittle temperature range (see Section 7.8.1). Figure 8.12 shows the effect of sulphur and carbon contents on the brittle temperature range of carbon steel, as determined by the MISO technique (described in Section 7.8.2). This diagram illustrates the cooperative action of increased sulphur and carbon contents in promoting solidification cracking.

It will also be seen from Fig. 8.12 that phosphorus has much the same effect on the brittle temperature range (and hence on solidification cracking). From the Fe–P equilibrium diagram (Fig. 8.13) the maximum solubility of phosphorus in δ ferrite is 2.8 by mass%, and the formation of a phosphide eutectic is improbable. Phosphorus, however, segregates to grain boundaries and could act either by lowering the melting point in the interdendritic regions or by reducing intergranular cohesion.

The effect of other elements on the susceptibility to solidification cracking of carbon steel is shown in Fig. 8.14. Boron increases susceptibility in the same way as phosphorus; nickel, however, acts in the same way as carbon in promoting the formation of austenite as a primary structure. Because of a high risk of

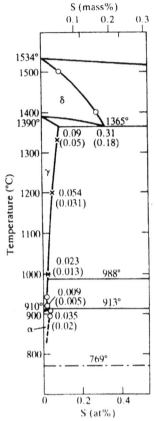

8.11 Part of the Fe–S binary equilibrium diagram (from Hansen and Anderko, 1958).

solidification cracking, manual metal arc welds containing more than 4% Ni are not normally possible; gas metal arc welding with 9% Ni wire is, however, practicable.

The probability of solidification cracking can be assessed for submerged arc welds in carbon–manganese steel by using a cracking index developed by Bailey and Jones:

$$U_{cs} = 230C + 190S + 75P + 45Nb - 12.3Si - 5.4Mn - 1 \qquad (8.5)$$

where C, etc., are in mass%. Cracking of fillet welds is likely if $U_{cs} > 20$ and of butt welds if $U_{cs} > 25$. Bead shape and other factors may alter the cracking risk to some degree.

The risk of solidification cracking is minimized by:

1. maintaining a low carbon content in the weld deposit;
2. keeping sulphur and phosphorus contents as low as possible;

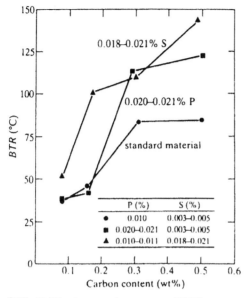

8.12 Brittle temperature range (BTR) versus carbon content as determined by the MISO test (from Matsuda *et al.*, 1983).

3 ensuring that the manganese content is high enough to allow for possible dilution (and ingress of sulphur) from the plate material.

It is normal practice to use low-carbon rimming steel as the core wire for carbon-steel and low-alloy-steel electrodes and a typical carbon-steel all weld metal deposit contains 0.05–0.10 C. In low-alloy (e.g. Cr–Mo) steel, carbon contents are usually about 0.1% but may be reduced below 0.06% by using low-carbon ferro-alloys in the coating or by using a low-carbon alloy core wire.

8.3 Transformation and microstructure of steel

Regardless of the primary phase structure, carbon and low-alloy steels transform to austenite at a temperature not far from the solidification point, and finally to ferrite. The equilibrium structure of iron–carbon alloys is shown in Fig. 8.15. At temperatures where the steel falls within the region designated 'γ phase', the steel is austenitic. On slow cooling, transformation starts when the temperature falls to the point on the lower boundary of this region corresponding to the carbon content of the steel—the **upper critical temperature**. Ferrite containing a small amount of carbon in solid solution is precipitated, leaving austenite grains that become smaller and are progressively enriched in carbon as the temperature falls. At 723 °C (the **lower critical temperature**), the residual austenite, which now contains about 0.8% carbon, transforms into **pearlite** – a laminated

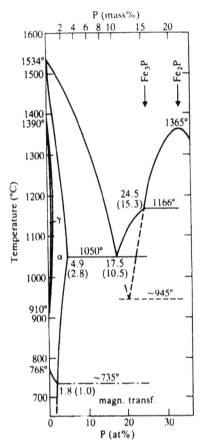

8.13 Part of the Fe–P binary equilibrium diagram (from Hansen and Anderko, 1958).

eutectoid mixture of ferrite and cementite (Fe_3C). The structure so obtained by slow cooling consists of intermingled grains of ferrite and pearlite. Cementite decomposes to iron plus graphite if held at elevated temperature for long periods: this is why the term 'metastable' is used in the caption of Fig. 8.15.

Rapid cooling depresses the temperature at which the γ to α change takes place. As the **transformation temperature** falls, the distance over which carbon atoms can diffuse is reduced, and there is a tendency to form structures involving progressively shorter movements of atoms. Whereas on slow cooling carbon can segregate into separate individual grains of austenite, with more rapid cooling carbides precipitate around or within the ferrite, which now appears in the form of needles or plates rather than equiaxed grains, a structure known as bainite (Fig. 8.16). At even more rapid rates of cooling, the transformation is depressed to a temperature at which **martensite** is formed. This **transformation product** is produced by a shear movement of the austenite

Carbon and ferritic alloy steels 227

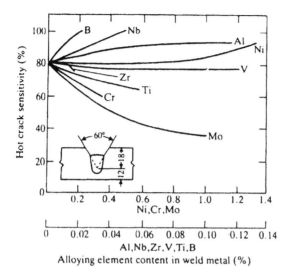

8.14 The relationships between alloying element content and hot cracking sensitivity, measured as a percentage of weld run cracked, with groove preparation as illustrated (from Morigaki et al., 1976).

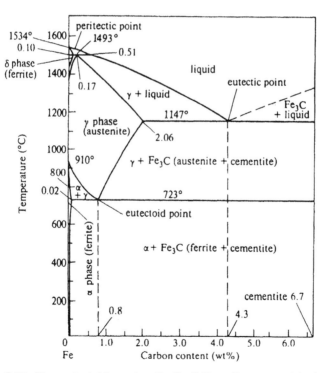

8.15 The metastable system Fe–Fe$_3$C (from Hansen and Anderko, 1958, p. 353).

228 Metallurgy of welding

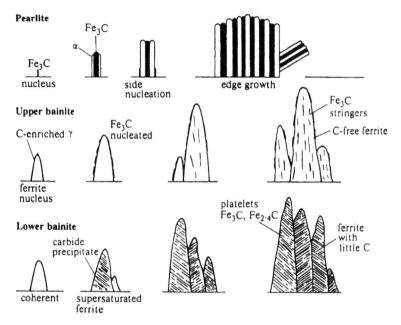

8.16 A diagrammatic representation of the formation of pearlite, upper bainite and lower bainite (from Rollason, 1961).

lattice, the carbon being retained in solid solution in a distorted body-centred cubic lattice. Generally speaking, the transformation product is harder and more brittle the lower the transformation temperature and (particularly for martensite) the higher the carbon content.

The cooling rate in fusion welding increases as the heat input rate q/v decreases, and as the welding speed increases. It is higher for multipass welding in thick plate than for single-pass welding in thin plate, other things being equal. It is reduced by increasing the preheat temperature. The weld metal and heat-affected zone structures form under conditions of continuous cooling, and a **continuous cooling transformation** diagram such as that illustrated later in Fig. 8.20 may indicate the type of structure to be expected. However, there are certain special features to be observed in the microstructure of welds, as discussed below.

8.3.1 Transformation and microstructure of weld metal

The character of the weld metal microstructure depends in the first place on alloy content. In Cr–Mo low-alloy steel, for example, the weld metal transforms uniformly to martensite or bainite. It will then normally be given a postweld heat treatment, which has the effect of tempering and stress-relieving the joint as a

Carbon and ferritic alloy steels

whole. In carbon, carbon–manganese and microalloyed steel, however, the microstructure and weld metal properties are much affected by details of weld procedure and composition. It is this second group of materials, for which the weld metal has a basic carbon–manganese composition, that will be considered here.

Factors that affect the microstructure of this type of weld metal include cooling rate, composition, the presence of non-metallic nuclei and plastic strain.

Cooling rates for weld metal are generally such that diffusion is limited. The overall grain size depends upon heat input rate and composition, the finer structures corresponding with low rates of heat input. Improved productivity, however, is achieved through high heat input rates, and much research effort is devoted to achieving a combination of fine grain structure with high heat input by control of composition. Non-metallics are important in that they may act as nuclei for transformation products; such nuclei may range from relatively coarse inclusions to the micrometre-sized particles discussed in Section 8.2.3. Plastic strain occurs during cooling of the weld and may have an effect on the weld metal properties.

The major microconstituents of carbon–manganese steel weld metal are shown in Fig. 8.17, using a nomenclature proposed by Abson *et al.* (1988). Essentially these structures are a mixture of primary (grain boundary and polygonal) ferrite, here designated PF(G) and PF(I), and Widmanstätten ferrite. The latter may take the form of side plates, which are nucleated at the austenite grain boundaries and form the lamellar structure marked FS(A) in Fig. 8.17(a) and (b). Alternatively the interior of the grain may transform wholly or in part to a fine acicular ferrite, as shown in Fig. 8.17(c) and (d). In some instances, ferrite–carbide aggregates more similar to equilibrium forms may appear, as in Fig. 8.17(e). Acicular ferrite is shown in more detail in Fig. 8.18 and ferrite with aligned martensite, austenite and carbides in Fig. 8.19. Alternative designations for the microstructural constituents shown in Fig. 8.17 are listed in Table 8.1.

Good notch-ductility of the fused zone of carbon–manganese steel is generally associated with a high proportion of acicular ferrite, whilst lamellar structures such as FS(A) and coarse grain boundary ferrite are, from this point of view, undesirable.

Figure 8.20 is a schematic continuous cooling transformation diagram for a carbon–manganese steel weld deposit. On cooling below the critical range, transformation starts at the austenite grain boundary (GB) with the formation of proeutectoid ferrite, followed by transformation of the interior of the grain to either side plates or to acicular ferrite or both. Carbides and martensite appear between the ferrite grains with some retained austenite.

The nature of the transformation products is much influenced by the size and distribution of the nucleating particles. Acicular ferrite is a fine Widmanstätten constituent which has been nucleated by an optimum intragranular dispersion of oxide–sulphide–silicate particles. If these particles are relatively large, as is the

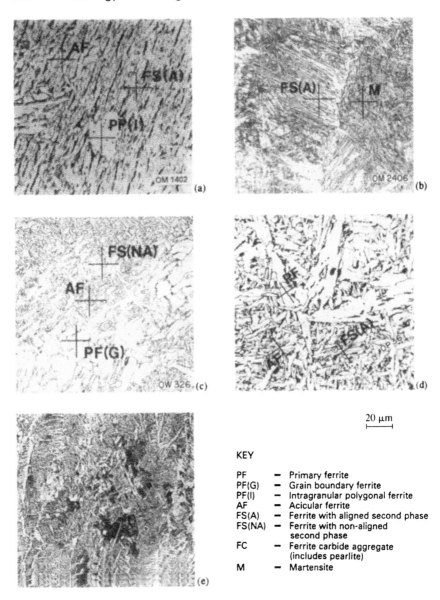

8.17 Typical microstructural constituents of weld metals associated with carbon–manganese steels (×500) (from Dolby, 1983).

case in submerged arc welding with an acid flux, ferrite is nucleated at the grain boundaries early in the transformation and grows into the grain in the form of side plates, giving rise to the undesirable lamellar structures described earlier. The non-metallic particles become larger with increasing oxygen content, and there is an optimum oxygen content for submerged arc weld metal where (other things being equal) the impact strength is a maximum (Fig. 8.21). At low oxygen levels,

Carbon and ferritic alloy steels 231

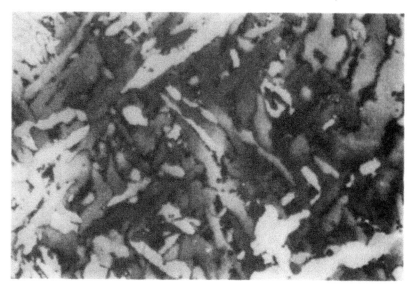

8.18 Acicular ferrite plus carbides in submerged arc weld metal (×1600, reduced by one-fifth in reproduction) (from Almquist *et al.*, 1972).

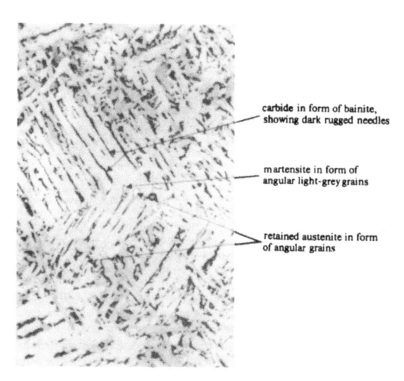

carbide in form of bainite, showing dark rugged needles

martensite in form of angular light-grey grains

retained austenite in form of angular grains

8.19 Lamellar structure in submerged arc weld metal: ferrite with aligned martensite, austenite and carbides (×1000, reduced by one-third in reproduction) (from Almquist *et al.*, 1972).

Table 8.1. Microstructure of ferrite steel weld metal

Designation proposed by Abson et al. (1988)	Code	Earlier or alternative designation
Grain boundary ferrite	PF(G)	GF, proeutectoid ferrite, ferrite veins, blocky ferrite, polygonal ferrite
Intragranular polygonal ferrite	PF(I)	PF, ferrite islands
Acicular ferrite	AF	No change
Ferrite with aligned second phase	FS(A)	AC, ferrite side plates, upper bainite, feathery bainite, lamellar product
Ferrite with non-aligned second phase	FS(NA)	None
Ferrite–carbide aggregate (includes pearlite)	FC	FC, pearlite, ferrite and interphase carbide
Martensite	M	No change

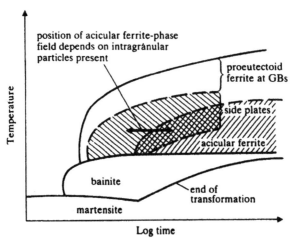

8.20 A schematic continuous cooling transformation diagram for a weld deposit showing the relationship of the acicular ferrite-phase field to those of other constituents (from Dolby, 1983).

there are insufficient nuclei to produce a fine-grained structure and there is an increase in hardness. With high oxygen contents, the structure again becomes coarse and the amount of oxide present may be an embrittling factor.

The oxygen content of submerged arc weld metal is determined primarily by the basicity of the flux. Acid fluxes are based on silica or silicates, and the dissociation constant of these compounds is such that they produce a relatively high level of dissolved oxygen. Basic fluxes are more stable, and the corresponding oxygen content of weld metal is close to the optimum level. The improved

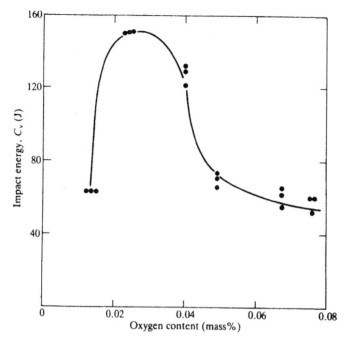

8.21 The impact strength at $-15°C$ of submerged arc weld metal of tensile strength $785\,MN\,m^{-2}$, as a function of oxygen content.

notch-ductility that results accounts for the increasing use of basic fluxes in recent years.

Other than controlling oxygen content, it is possible to improve the notch-ductility of submerged arc welds by the addition of boron and titanium, as recorded in Section 8.2.4. Titanium promotes the formation of acicular ferrite, possibly through the formation of TiN or TiO_2 nuclei, while boron inhibits the nucleation of ferrite at the austenite grain boundaries. Figure 8.22 shows the effect of boron addition via B_2O_3 in the flux. There is an optimum boron content, and too large an addition could do more harm than good. However, titania–boron fluxes can give good results over a wide range of welding currents in submerged arc welding.

In high-current flux-cored wires, it is not practicable to control oxidizable elements with such precision, and other means of refining the weld metal structure, such as the addition of molybdenum, may be used.

8.3.2 Transformation and microstructure in the heat-affected zone

Whereas in the weld metal we are concerned mainly with solidification and cooling, in the heat-affected zone the complete thermal cycle is effective in

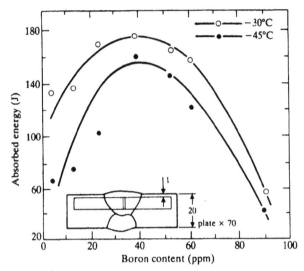

8.22 The effect of boron content on submerged arc weld metal toughness: B introduced through flux; C–Mn microalloyed base steel (from Dolby, 1983).

determining microstructure. On heating the steel through the temperature range between the upper critical temperature and somewhere about 1200 °C, the austenite grains form and grow relatively slowly, but above a specific point (the **grain-coarsening temperature**) the rate of growth increases sharply. Below the grain-coarsening temperature, grain boundary movement is impeded by the presence of certain particles such as aluminium nitride. However, these particles go into solution at a specific temperature level and, above this level, grain growth is unimpeded. Aluminium and other elements are added to steel in order to produce fine grain, and they do so by impeding the growth of austenite grains during the various thermal cycles to which the steel is subjected during processing. The effectiveness of grain-refining additives such as aluminium, titanium, niobium or vanadium is greater the higher the solution temperature of the nitride or carbonitride particles formed.

Titanium nitride may have two effects. First, it enhances intragranular ferrite nucleation in the manner described for weld metal in the previous section, thereby producing an acicular ferrite structure with improved notch toughness. Second, it is a grain growth inhibitor. Optimum austenite boundary pinning is obtained with a TiN particle size less than 0.05 nm, and a titanium content of about 0.015%. The relative effectiveness of various addition elements in restricting austenite grain growth is illustrated in Fig. 8.23.

From a metallurgical viewpoint, the heat-affected zone of a fusion weld in steel may be divided into three zones; supercritical, intercritical and subcritical. The supercritical region may in turn be divided into two parts: the **grain growth**

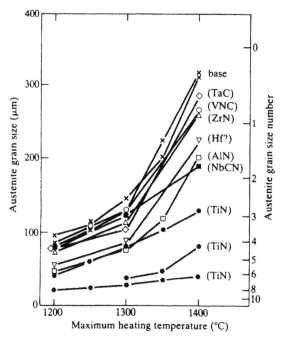

8.23 Variation of austenite grain size with austenitizing temperature for a variety of precipitate types in C–Mn base steel (from Dolby, 1983).

region and the **grain refined region**. Grain growth occurs where the peak temperature of the weld thermal cycle exceeds the grain-coarsening temperature. Below this temperature, the thermal cycle will usually produce a grain size that is smaller than that of the parent metal.

The microstructure of the grain growth region is dominated by two features: the austenite grain size and the transformation structure within the grain. The grain size is in turn governed by two factors: the nature of the weld thermal cycle and the grain-coarsening temperature of the steel. For any given steel, the greater the heat input rate, the longer the time spent above the grain-coarsening temperature, and the coarser the grain. Thus electroslag welding generates a large grain size near the fusion boundary (Fig. 8.24). Also, for any given weld thermal cycle, the higher the grain-coarsening temperature, the shorter the time for grain growth. Aluminium-treated carbon steels contain aluminium nitride particles, which inhibit the growth of austenite grains up to a temperature of about 1250 °C. Above this temperature, the aluminium nitride particles go into solution and grain growth is rapid, such that the grain size at the fusion boundary is not much different from that of steels that do not contain aluminium.

In the coarse-grained region of the heat-affected zone, the steel has been raised to temperatures that are in the **overheating** ranges. Overheating is a phenomenon

8.24 Macrosection of an electroslag weld in 75 mm thick carbon steel in the as-welded condition, showing a typically coarse-grained structure (photograph courtesy of TWI).

that is troublesome in forging technology: if steel is held at a temperature above about 1200 °C for a significant period of time and then cooled slowly, the material is embrittled and impact test specimens may have a coarse intergranular fracture. The embrittlement is due in most instances to the solution of sulphides at high temperature, followed by their reprecipitation at grain boundaries on cooling through the austenite range (see Fig. 8.11). In low-sulphur aluminium-treated steel, aluminium nitride may dissolve and reprecipitate in a similar way. Because of higher cooling rates, the weld heat-affected zone is less subject to this type of embrittlement than forgings. Nevertheless, sulphide precipitation may occur and may contribute to the embrittlement of the coarse-grained region. It also may promote intergranular failure in the weld metal or heat-affected zone at elevated temperature (see Section 8.5.6).

Manganese is generally beneficial in reducing embrittlement due to overheating, since it reduces the solubility of sulphur in austenite and promotes the formation of spherical intergranular precipitates, particularly at high cooling rates.

The type of microstructure in the coarse-grained region depends on the carbon and alloy content of the steel, on the grain size and on the cooling rate. In low-carbon steel proeutectoid ferrite separates first at the prior austenite grain boundaries, and inside the grain a ferrite–pearlite and/or a ferrite–bainite structure develops. The pearlitic constituent may consist of parallel arrays of small globular cementite particles. With increasing cooling rates and/or increasing carbon and alloy content, the proeutectoid ferrite disappears and the austenite grains transform completely to acicular structures: upper bainite, lower bainite or martensite, or a mixture of these components. The coarser the prior austenite grain size, the coarser the microstructure: for example, the wider the spacing between martensite laths, or the larger the bands or blocks of proeutectoid ferrite. Also, a coarse austenitic grain size may promote the formation of harder micro-structures than would result from the transformation of a finer-grained material; e.g. upper bainite may form instead of a ferritic–pearlitic structure.

Partial melting (liquation) may occur near the fusion boundary in the case of certain alloys. Niobium may form patches of eutectic, and in the case of medium (0.4%) carbon high-tensile steel there may be grain boundary liquation, associated with sulphur and phosphorus, leading to liquation cracking. The latter type of cracking is controlled by maintaining very low levels of sulphur and phosphorus: increasing the manganese sulphur ratio does not appear to have much effect.

With increasing distance from the weld centreline, both peak temperature and cooling rate decrease. Consequently, the grain refined region may have a range of microstructure, being similar to that of normalized steel on the outside.

In the intercritical region, which is relatively narrow, partial transformation may take place. For example, in a carbon steel having a ferrite–pearlite structure prior to welding, the pearlite islands, which are of eutectoid carbon content, transform to austenite on heating and to martensite or bainite on cooling. In the 'as-welded' condition, the region may therefore consist of hard grains embedded in a relatively soft, untransformed ferritic matrix (Fig. 8.25).

This type of structure, which must be a normal feature in fusion welds in pearlitic carbon steel, has never, so far as is known, given rise to any failure in service or, indeed, in laboratory tests. This situation is of interest because it happens not infrequently that the acceptability of similar structures, where hard regions are embedded in a relatively soft matrix, comes into question. While each such case must be decided on its own merits, the example of the intercritical region of the heat-affected zone shows that this condition need not be damaging.

The subcritical region does not normally undergo any observable micro-structural change except that a small region of spheroidization may occur: this is generally difficut to detect. Strain-ageing (see Section 8.2.2) is a possibility where the metal has been strained and heated within the temperature range 100–300 °C. This phenomenon is associated with a fine precipitation ($\sim$10 nm) or clustering located at dislocations within the ferrite grains. The precipitates are too fine for

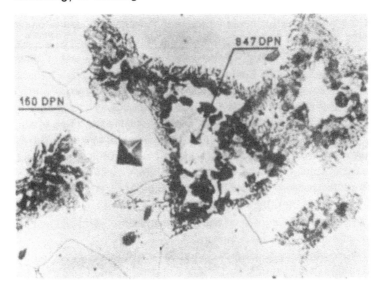

8.25 The region of partial transformation in the heat-affected zone of a carbon steel weld. Solution of carbide and transformation to austenite has occurred preferentially in the centre of the pearlite grains. On cooling, martensite has formed in the transformed areas, as shown by the microhardness test results (×820, reduced by one-third in reproduction).

analysis by energy-dispersive X-ray techniques but are observable under favourable conditions by transmission electron microscopy. The known association of strain-age embrittlement with nitrogen suggests that the particles are nitrides or carbonitrides. Their effect is to shift the Charpy V impact transition temperature upwards by (typically) about 50 °C. The impact fracture morphology remains transgranular, as would be expected from the intragranular nature of the precipitation.

Non-metallic inclusions may have an effect on the hardenability of the heat-affected zone. Sulphide, oxide and silicate inclusions nucleate ferrite within the transforming austenite grains and this produces a lower hardness than in the absence of inclusions. It follows that the heat-affected zone of clean steel is likely, for any given welding procedure, to be harder than that of less refined material. This effect is most apparent in the moderately hardenable carbon–manganese steel, but not so evident in low-carbon or hardenable low-alloy steel. Figure 8.26 illustrates how decreasing the S content of a carbon–manganese steel increases hardenability. Lowering the sulphur level has the greatest effect in aluminium-treated and vacuum-degassed steel, where most of the non-metallics are sulphides.

The increased hardenability of clean steel may be a problem in the fabrication of offshore structures and line pipe, where specifications require a maximum

Carbon and ferritic alloy steels 239

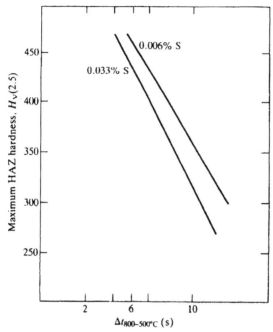

8.26 Variation of heat-affected zone hardenability of 0.22 C–1.4 Mn steel with steel sulphur content (from Hart, 1978).

hardness limit in the heat-affected zone of welds in order to minimize the risk of stress corrosion cracking. Its effect on the incidence of hydrogen-induced weld cracking, once a matter of some concern, has proved not to be significant.

With multi-run welding, fresh heat-affected zones are formed in previously deposited weld metal and previously formed heat-affected zones (HAZ). Transformed regions are retransformed: coarse grains may be refined and vice versa. The effect of multi-run welding is usually beneficial in reducing hardness and increasing notch ductility, and this may be exploited in the temper-bead technique, where a final capping run is made and then ground off, thus avoiding the problem of leaving the final run in as the as-welded condition. The actual properties of the joint, however, depend on details of the welding procedure, and must be determined by procedure testing.

8.4 The mechanical properties of the welded joint

It will be evident that there must be variations in properties from point to point across the welded joint. For example, the root pass in a multi-run weld may pick up nitrogen and be embrittled thereby, or its properties may be altered by admixture with the parent metal or by strain-age embrittlement. The properties of

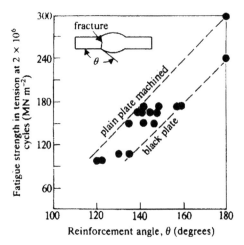

8.27 The effect of joint profile on the fatigue strength of transverse butt welds in carbon steel (from Newman, 1960).

the joint as a whole may also be affected by the form of the weld profile and the way in which the weld is disposed relative to the principal applied stress. To take the simple case of a butt weld in plate, the angle between the reinforcement and the plate surface is a measure of the stress concentration at the fusion boundary. The greater this stress concentration, the lower the fatigue strength (Fig. 8.27). Likewise, the probability of cracking due to other embrittlement mechanisms such as hydrogen embrittlement is increased as the reinforcement wetting angle is reduced.

8.4.1 The mechanical properties of weld metals

For any given composition (and particularly for carbon and carbon–manganese steels), one of the important factors governing mechanical properties is the ferrite grain size. The term 'grain size' is used here in a general sense and refers to a characteristic dimension for the structure in question. For example, in carbon steel weld metal as shown earlier in Fig. 8.18 the width of the acicular ferrite grains may be a good measure of grain size, but the properties of the weld metal may also be influenced by the size and morphology of the proeutectoid ferrite. In bainitic or martensitic structures, the width of the needles provides a measure of grain size. However, it must be recognized that a precise correlation between grain size alone and mechanical properties is not possible since these may be influenced by the morphology and distribution of carbides, the presence of grain boundary precipitates and other features.

With these reservations, it may be said that in general the yield strength of weld metal increases and the impact transition temperature falls as the ferrite grain size is reduced. The ferrite grain size is largely determined by the size of the prior austenite grains and this, as pointed out earlier, is a function of the weld thermal cycle. Thus high heat input rate processes such as electroslag generate a relatively coarse-grained weld metal, while that produced by other processes is relatively fine. Weld metal usually contains a relatively high density of dislocations, which also contributes to increased yield strength in carbon and carbon–manganese steels. The final result is that weld metal normally has higher tensile and yield strength than the equivalent plate material, even when the carbon and/or alloy content is lower. These higher properties are maintained after subcritical heat treatments. In carbon steel, there is no grain growth and little softening at such temperatures, probably because of the carbide network surrounding the ferrite grains (Fig. 8.18). Indeed there is more concern about the danger of excessive yield strength in the weld metal, which may generate high residual stresses and render the joint susceptible to stress corrosion cracking (Chapter 11). The hardness of weld metal correlates with ultimate strength in the same way as for wrought steel, and hardness testing is used as a method of controlling weld metal strength.

The notch ductility of weld metal, as measured for example by impact testing, depends mainly, for any given composition, on microstructure, grain size and inclusion content. Notch ductility is reduced, other things being equal, by the presence of proeutectoid ferrite in the structure, and particularly when it is coarse and block-like in form. Ideally, the weld metal should consist entirely of fine acicular ferrite. The notch ductility is also reduced as the amount of martensite in the structure increases. The grain size of weld metal deposited by the submerged arc process is such that the impact properties are adequate for the common run of applications, but where more stringent requirements prevail (such as impact testing at $-30\,°C$), it is necessary to use basic or semibasic fluxes to reduce oxygen content down to an optimum level, to use a special flux such as the titanium–boron types discussed earlier, and/or to use a nickel alloy wire.

Additions of Ni to the basic carbon–manganese composition up to 2.5% Ni can improve notch ductility as measured by Charpy V or crack-tip opening displacement (CTOD) testing (Chapter 11), but there is a cost penalty. In manual metal arc welding adequate notch ductility is normally attainable with basic coated, unalloyed rods, although the addition of 1% Ni may be necessary for some more stringent requirements. Self-shielded (no-gas) welding using a flux-cored wire is a special case. The microstructure of carbon–manganese and low-Ni alloy steel deposited by this process consists largely of proeutectoid ferrite with side plates, with smaller areas of acicular bainite and lath martensite. Nevertheless it is possible to obtain acceptable Charpy-V results with such welds using specially developed proprietary wires. In gas metal arc welding the problem is different because of the need to add CO_2 or oxygen to the argon shielding gas in order to stabilize the arc. The use of oxidizable elements such as titanium to obtain a refined structure may not therefore be possible and best

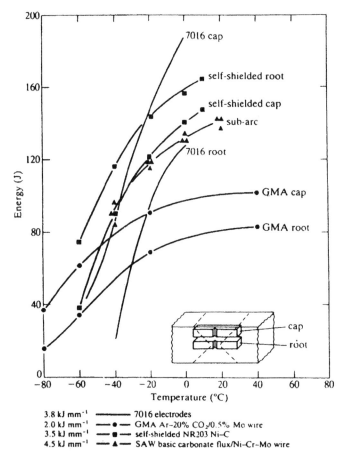

8.28 Charpy V transition curves for various weld metals (sources: Keeler, 1981; George et al., 1981).

results are obtained with 0.5% Mo, which lowers the transformation temperature and increases the proportion of acicular ferrite in the weld metal. Some Charpy V transition curves for these processes are plotted in Fig. 8.28. The results are those achievable under optimum conditions for a plate thickness of 30–50 mm.

For manual metal arc and gas metal arc welds the notch ductility of the subsurface regions (cap) is higher than that of the root. This may in some cases be partly due to dilution with parent metal, but it results mainly from strain-age embrittlement of the root passes during the completion of the joint, the root being subject to a combination of plastic strain and heating at 200 °C. In welds made by the self-shielded process the position is reversed, probably because sufficient aluminium remains in the weld metal to inhibit strain-age embrittlement, and because the shielding of the capping passes is less effective than in the root.

Carbon and ferritic alloy steels 243

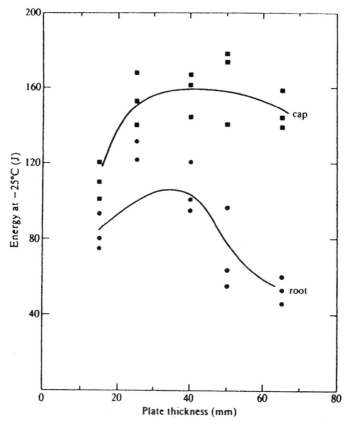

8.29 Variation of Charpy toughness at $-25\,°C$ for double-sided welds made in the vertical position with E7016 electrodes at a heat input rate of $3.8\,kJ\,mm^{-1}$ (after Keeler, 1981).

Figure 8.29 shows how plate thickness may affect the impact properties of weld metals.

The elevated-temperature properties of weld metal have been studied to a rather limited extent. Generally speaking, the short-time elevated-temperature properties of manual metal arc weld metal are similar to those of wrought material of the same composition, although there is considerable scatter in the results obtained. Creep-rupture properties are equal to or lower than those of wrought steel of the same composition.

The effect of increasing carbon and alloy content of weld metal is, in general, to increase strength and hardness. Carbon in particular has a strong effect on these properties. However, for the best combination of fracture toughness and cracking resistance, it is desirable to keep the carbon content within the range 0.05–0.12%. For alloy steel in particular, maintaining about 0.06% C in the deposit helps to avoid excessive hardness values.

The effect of postweld heat treatment in the sub-critical range has been mentioned in Chapter 7, where it is stated that where there is a long, slow cool from the soaking temperature, carbon and low-alloy steels may suffer some degree of embrittlement. When samples are cooled in still air, as is usually the case in laboratory testing, the effect of such treatment is to reduce the level of residual stress, to temper any hard transformation products that may be present, and in most cases to improve the fracture toughness. The yield and ultimate strengths of carbon and carbon–manganese steel weld deposits are not reduced to any significant extent provided that the temperature does not exceed 650 °C.

In weld procedure testing it is essential, in view of the possible embrittlement due to slow cooling, that test samples be subject to a precise simulation of the post-weld heat treatment thermal cycle in all cases where such treatment is specified.

8.4.2 The mechanical properties of the heat-affected zone

In describing the mechanical properties of weld metal, it has been assumed that these are uniform across the fused zone, as would be more or less the case for a single-pass weld. In practice, most welds are multipass, and the weld metal, as well as the unmelted parent metal, contains one or more heat-affected zones. In this section we are concerned with the properties of the heat-affected region as compared with those of the unaffected weld metal or parent plate. The yield and ultimate strength of the heat-affected zone in steel are almost always higher than those of the parent material and the properties of main interest are fracture toughness and hardness.

The hardness of the heat-affected zone

The hardness of the heat-affected zone is a measure of the tensile strength of the steel and, for any given alloy type, gives an indication of the degree of embrittlement. For carbon–manganese and some low-alloy high-tensile steels, a hardness of over 350 BHN in the heat-affected zone would be considered excessive, indicating a susceptibility to cracking. Conversely, the heat-affected zone of nickel alloy steels may be adequately tough with a hardness of 400 BHN.

Hardness depends on the hardenability of the steel, on the cooling rate and, to a lesser degree, on the prior austenite grain size. The hardenability of a steel may be generally correlated with the **carbon equivalent** (CE). In a carbon equivalent formula, the hardening effect of each alloying element is compared with that of carbon, and the relevant alloy content in mass% is divided by a factor that gives the carbon equivalent of that element. A formula that was originally devised by Dearden and O'Neill, and adopted by IIW in 1967 in slightly modified form, is

$$\text{CE} = \text{C} + \frac{\text{Mn}}{6} + \frac{\text{Cu} + \text{Ni}}{15} + \frac{\text{Cr} + \text{Mo} + \text{V}}{5} \tag{8.6}$$

This formula is applicable to plain carbon and carbon–manganese steels, but not to microalloyed high-strength low-alloy steels or low-alloy Cr–Mo types.

The formula adopted by the Japanese Welding Engineering Society, which was due to Ito and Bessyo, is

$$P_{cm} = \text{C} + \frac{\text{Si}}{30} + \frac{\text{Mn} + \text{Cu} + \text{Cr}}{20} + \frac{\text{Ni}}{60} + \frac{\text{Mo}}{15} + \frac{\text{V}}{10} + 5\text{B} \tag{8.7}$$

where, as in the IIW CE, all elements are expressed in mass per cent. P_{cm} is based on a wider range of steels than the IIW formula.

Carbon equivalents may be used to calculate the hardness of the heat-affected zone using a **hardness equivalent**, which takes account of the cooling rate, and also a **weldability equivalent**, which determines the maximum permissible cooling rate for the avoidance of hydrogen-induced weld cracking (see Cottrell, 1984). A maximum carbon equivalent is often specified for structural steel to minimize the risk of excessive hardenability and/or hydrogen cracking. The effect of non-metallics on the hardenability of the heat-affected zone is noted in Section 8.3.2, and in future carbon equivalent formulae may need to take this factor into account. Where figures for carbon equivalent are quoted elsewhere in this book, they are calculated in accordance with the IIW formula (equation 8.6).

For alloy steel, a **hardness traverse** may be taken across transverse sections of procedure test specimens in order to demonstrate that no excessive hardening or softening of the joint has occurred.

The fracture toughness of the heat-affected zone

In some cases all the regions of the heat-affected zone (coarse grain, grain refined, intercritical and subcritical) are embrittled to some degree compared with the parent material. However, if the fracture toughness of the parent material is relatively low, the heat-affected zone may have better properties, particularly in the grain refined region. The factors affecting heat-affected zone toughness are the nature of the weld thermal cycle, grain-coarsening temperature, transformation characteristics, alloy content and non-metallic content. As would be expected, low heat input rate processes which give relatively high cooling rates generate a finer-grained heat-affected zone and less embrittlement in low-carbon steel. In more hardenable steels (including carbon–manganese steels), this effect may be offset by the formation of bainitic or martensitic microstructures. The higher the carbon content, the more brittle the transformed structure. High-carbon twinned martensite is the most brittle of the structures found in the weld heat-affected zone of steel, and is particularly susceptible to hydrogen-induced cold cracking. For any given martensite content, the toughness is improved by a reduction of the width of martensite colonies. However, **autotempered martensite** (low-carbon martensite that forms at a high enough temperature for tempering to occur during further cooling) is a relatively tough product and has better properties in general than bainite. Thus, in a low-carbon low-alloy steel, high cooling rates may, owing to the formation of martensite rather than bainite, generate a more notch-ductile heat-affected zone than low cooling rates.

The addition of aluminium or niobium to produce grain refinement of carbon–manganese steel may or may not be beneficial to heat-affected zone toughness. There is in most instances an optimum content of microalloying elements such as Al, Nb or V. For example, for aluminium additions to carbon–manganese steels, the optimum content is about 0.01%. Higher additions produce an equally fine-grained steel but the fracture toughness deteriorates until at about 0.06% Al it may be little better than for an Al-free steel (Fig. 8.30). As pointed out, even with an optimum Al addition, the coarse-grained region may suffer just as much grain growth as for an Al-free steel. Nevertheless, the average grain size is likely to be lower, and the overall fracture toughness correspondingly higher. Niobium has the effect of suppressing the formation of proeutectoid ferrite and promoting the bainite transformation, as a result of which the coarse-grained region of the heat-affected zone has lower toughness than that of a plain carbon–manganese steel welded at the same heat input rate. The formation of niobium eutectics near the weld boundary may further embrittle this region.

Nickel additions improve the toughness of the heat-affected zone and of steel in general. The lowering of the ductile–brittle transition temperature is greater with increasing nickel content, such that with a 9% Ni addition the heat-affected

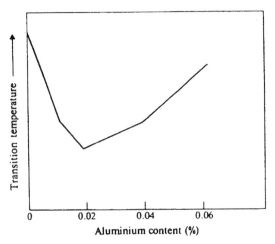

8.30 A typical relationship between aluminium content and Charpy V-notch transition temperature for a quenched and tempered low-alloy steel.

zone has acceptable impact properties down to liquid-nitrogen temperatures in the 'as-welded' condition. Other alloying elements such as chromium and molybdenum affect the toughness of the heat-affected zone primarily by modifying the transformation characteristics. Carbon increases hardenability and also decreases the toughness of any transformation products that are formed, particularly martensite.

The relative degree of embrittlement of the various heat-affected zone regions depends on the hardenability of the steel. Carbon or carbon–manganese steel with a carbon equivalent of around 0.3 will usually form a pearlite–bainite structure in the supercritical region and this may be tougher than a strain-age embrittled subcritical region (strain-age embrittlement is discussed in Sections 8.2.2 and 8.3.2). However, where martensite forms in the supercritical region, as may happen with CE of 0.45 and above, the subcritical region is tougher than the supercritical.

The toughness of the heat-affected zone may be tested using either **Charpy V-notch** impact specimens or by the **crack-tip opening displacement** (CTOD) test. In both cases it is necessary to locate the notch in the correct part of the heat-affected zone. Even so, a plastic zone forms below the notch during testing and this may include other regions of the heat-affected zone or even the parent plate. Such tests therefore indicate how the heat-affected zone as a whole is likely to behave when notched in a particular region. Alternatively, the steel in question may be given a simulated weld thermal cycle so that tests can be performed on a homogeneous microstructure. Both types of test have their place in the study of this complex problem.

The effect of postweld heat treatment on the fracture toughness of the heat-affected zone of a Mn–Mo steel has been investigated using the techniques of linear elastic fracture mechanics. Tests were made on the full section thickness of 165 mm and also on 25 mm × 25 mm and 10 mm × 10 mm specimens, in which any residual stress would be partially or completely removed. Two heat treatments were used, one of 15 min at 600 °C, giving minimal stress relief, and one of 40 h at 600 °C, after which residual stresses were low. Tests were conducted at $-1976\,°C$ so that failures were in the elastic condition and the tests were valid to ASTM E-24. The results are shown in Table 8.2.

Table 8.2. Fracture toughness ($MN\,m^{-3/2}$) of weld heat-affected zone in ASTM A533 (Mn–Mo) steel at $-196\,°C$

	15 min postweld heat treatment	40 h postweld heat treatment
Full section	32.7	40.2
25 mm × 25 mm	57.8	44.2
10 mm × 10 mm	75.4	50.3

Source: from Suzuki *et al.* (1978).

It will be evident that stress relief due to postweld heat treatment has improved the fracture toughness of the complete joint. In the case of the 10 mm × 10 mm specimens, however, which are both free from residual stress, the postweld heat treatment has *reduced* the fracture toughness.

8.5 Stress intensification, embrittlement and cracking of fusion welds below the solidus

The embrittlement (or toughening) of the weld metal and the heat-affected zone due to transformations occurring during the weld thermal cycle has been described above. In exceptional cases, the degree of embrittlement so caused may be sufficient to result in cracking either during the welding operation or in service. Normally, however, some additional factor is required, either to augment the applied or residual stress or to increase the degree of embrittlement before any cracking will occur.

8.5.1 Stress concentration

The effect of weld profile on the stress concentration at the fusion line has been discussed generally in Section 8.4. Figure 8.31 shows the **stress concentration factor** for the fatigue of carbon steel butt welds as a function of reinforcement angle. This curve is derived from Fig. 8.27 and gives some indication of the stress concentration under static loading conditions. For the toe of a fillet weld the stress concentration factor is usually assumed to be about 3, while for the root of a partial penetration butt weld or the root of a fillet weld it may be as high as 7 or 8.

The strain concentration may be further augmented if, because of weld distortion, the joint is subject to bending. For example, consider a butt weld that is welded from both sides. If one side is welded out completely, followed by backgouging and complete welding of the second side, shrinkage of the second side weld will cause bending around the weld centreline. If the angle of bend is θ radians and the bending takes place over a length equal to nw where n is a number and w the plate thickness, it may be shown that the average strain on the weld surface is θ/n and the strain at the fusion boundary is $(\theta/n) \times$ stress concentration factor. More complex strain fields due to distortion are, of course, quite possible.

8.5.2 Embrittlement of fusion welds

Strain-age embrittlement may occur during the welding operation, as already discussed, but there are a number of mechanisms that may result in postwelding embrittlement. The most important of these are hydrogen embrittlement,

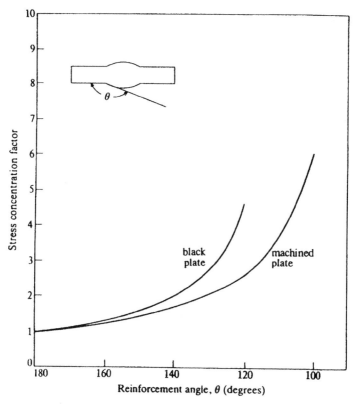

8.31 The stress concentration factor for fatigue of butt welds as a function of reinforcement angle.

secondary hardening and temper embrittlement. Cracking may result from poor transverse ductility in the plate material, and propagation of cracks in service may be due to a number of mechanisms, including stress corrosion and strain-age cracking. The mechanisms that may generate subsolidus cracks during or shortly after welding are discussed below, while those that result in cracking during service will be considered in Chapter 11.

8.5.3 The hydrogen embrittlement and cracking of welds in steel

Steel may suffer two types of embrittlement due to the presence of hydrogen. The first type occurs at elevated temperature and affects carbon and low-alloy steel. It results from chemical reaction between hydrogen and carbides, and causes permanent damage, either decarburization or cracking or both. The second type occurs at temperatures between $-100\,°C$ and $200\,°C$. This embrittlement is due to physical interactions between hydrogen and the crystal lattice, and is reversible in that after removal of the gas the ductility of the steel reverts to normal.

Hydrogen attack

The elevated-temperature effect, known as **hydrogen attack**, has been observed mainly in petroleum and chemical plant where hydrogen forms part of the process fluid. Initially the metal is decarburized, but at a later stage intergranular fissures appear and the metal is weakened as well as being embrittled. Dissolved hydrogen reacts with carbides to form methane (CH_4), which precipitates at the grain boundaries. There is an incubation period before any damage can be detected in normal mechanical tests. This incubation period may be very long (sometimes of the order of years) at low temperature, but it decreases sharply with increasing temperature. Increasing the chromium and molybdenum content of the steel raises the temperature above which hydrogen corrosion occurs, and austenitic chromium–nickel steels of the 18Cr–10Ni type are immune from attack. Decarburization and fissuring that resembles hydrogen attack has been observed in laboratory tests of high-tensile steel welded with cellulosic electrodes. Small areas around defects such as pores and inclusions were decarburized and cracked. It would appear, however, that the properties of the joint were not significantly affected, probably because of the small dimensions of the corroded areas. Of much greater importance in welding is the temporary form of hydrogen embrittlement, since this may cause **hydrogen-induced cold cracking**.

The solution of hydrogen

The equilibrium solubility of hydrogen in ferritic iron below the melting point is

$$s = 47.66 p^{1/2} \exp(-2.72 \times 10^7 / RT) \tag{8.8}$$

where s is in ppm by mass, p is in atmospheres, R is the gas constant (8.134 J mol^{-1}) and T is temperature (K). For $p = 1$ atm and $T = 293$ K, the solubility according to equation 8.8 is 6.7×10^{-4} ppm (7.6×10^{-4} ml/100 g).

The solubility is higher in austenite than in ferrite (refer to Fig. 7.1) and is also modified by the presence of solute elements in the iron. It is also higher in a strained than in an unstrained lattice. If steel is strained to the yield stress σ_{ys}, then according to Oriani and Josephic (1974) the solubility is increased by a factor c/c_n

$$c/c_n = \exp(A\sigma_{ys}^{1/2} - B\sigma_{ys}) \tag{8.9}$$

where

$$A = \frac{2(1+v)\bar{V}}{3RT}\left(\frac{2E}{\pi}\right)^{1/2} \qquad (8.10)$$

$$B = \frac{2(1+v)\bar{V}}{RT}\left(\frac{2}{\pi}\right) \qquad (8.11)$$

and v is Poisson's ratio, $\bar{V}$ is the partial molar volume of hydrogen in iron, E is elastic modulus and other symbols are as before. At 20 °C, $A = 6.926 \times 10^{-2}$ (MN m^{-2})$^{-1/2}$ and $B = 3.242 \times 10^{-4}$ (MN m^{-2})$^{-1}$. For a high-tensile steel where $\sigma_{ys} = 1 \times 10^3$ MN m^{-2}, the solubility is increased by a factor of 6.5. At the tip of a crack, the strain may be higher than that corresponding to uniaxial yield, and the solubility may be still further augmented.

Hydrogen separates out to grain boundaries and to the interfaces between phases in steel. Since hydrogen-induced cracking is intergranular at low and transgranular at high stress intensity factors, the degree of segregation is probably of the same order of magnitude as the increase of solubility due to moderate strain. Hydrogen recombination poisons such as P, S, As and Sb also separate out to grain boundaries and are thought to increase the segregation of hydrogen, thus acting in a cooperative way and enhancing intergranular embrittlement.

Where steel is strained, and particularly when it is subject to plastic strain, the dislocation density increases. Hydrogen diffuses preferentially to these discontinuities in the crystal lattice, and if present in sufficient quantity, precipitates to form microvoids.

Such microvoids immobilize part of the hydrogen content and act as barriers to diffusion, particularly at low levels of mass hydrogen flow. The reduction of apparent diffusivity of hydrogen in steel at lower temperatures shown in Fig. 7.18 is due to the presence of traps. Supersaturation of the metal by hydrogen also generates microvoids.

In fusion welding the mean hydrogen content of steel weld metal immediately after solidification is in the range 1–50 ppm, which is four orders of magnitude higher than the equilibriium value at room temperature. Most of this hydrogen (**diffusible hydrogen**) diffuses out of the weld at room temperature, while the remainder can only be removed by vacuum fusion techniques, to give a measure of the **total hydrogen** content. Diffusible hydrogen is measured by plunging a weld sample into mercury and collecting the evolved gas in a burette. Traditionally, diffusible hydrogen contents have been expressed as ml per 100 g or ppm of *deposited* metal. This method may cause difficulties in comparing submerged arc welding tests with those made using coated electrodes because the amount of dilution is higher with submerged arc. To overcome this problem results may be expressed as ppm or g/ton of *fused* metal (1 ppm = 1 g/ton = 1.12 ml/100 g).

Hydrogen embrittlement

The embrittlement of steel by hydrogen only manifests itself during processes that lead to the fracture of the metal; it does not, for example, increase the hardness as measured by an indentation test. The simplest way of assessing embrittlement is by comparing the reduction of area of a hydrogen-charged specimen δ_H with that of a hydrogen-free specimen δ_0. The degree of embrittlement E may then be expressed as

$$E = \frac{\delta_0 - \delta_H}{\delta_0} \tag{8.12}$$

Alternatively, a V-notch may be machined around the circumference of a cylindrical specimen. If such a testpiece is charged with hydrogen and then subjected to a sufficiently high constant tensile load, it will fail after a lapse of time. The type of result obtained is illustrated in Fig. 8.32. The reduction of strength and the step in the strength–time curve both increase with increasing hydrogen content.

Other techniques employ precracked specimens of various geometry to obtain a measure of the stress intensity factor K_H for cracking of steel that has been

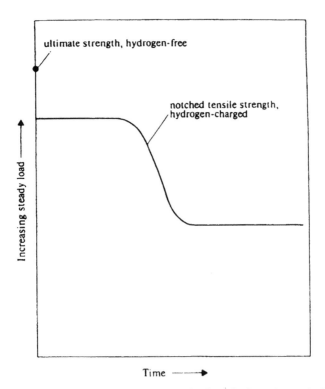

8.32 The notched tensile strength of steel when charged with hydrogen.

charged with hydrogen or is exposed to a hydrogen atmosphere. The most direct method is to use a standard ASTM compact tension specimen and to load it in a hydrogen atmosphere so as to determine the critical stress intensity factor for crack propagation. At atmospheric and lower pressure, the solution of hydrogen in steel is inhibited by very small traces of oxygen, and to obtain reproducible results extreme measures must be taken to purify the hydrogen. In the tests carried out by Oriani and Josephic (1974), which are discussed in the next section, the hydrogen pressure was increased to the point where the crack started to run, and then decreased to the point where it stopped. A small difference between these two pressures indicates that the degree of purification has been effective.

Other measurements have been made with cantilever beam specimens and with centre-cracked sheet metal testpieces. Hydrogen may be introduced by soaking in a hydrogen atmosphere at elevated temperature or, in a less controlled way, by cathodic charging. In such tests it is necessary to allow sufficient time for hydrogen to reach equilibrium level in the strained region at the crack tip. Not many investigators have observed this need, so that results quoted in the literature may be too high, particularly at low levels of stress intensity.

The effect of temperature is well illustrated in Fig. 8.33. This shows the notched tensile strength of a high-tensile material that has been precharged with hydrogen. Fracture occurs when a critical hydrogen content has accumulated in the strained region at the root of the notch. Above about 200 °C the material properties are such that hydrogen concentration and/or embrittlement are no longer possible. Below -100 °C the rate of diffusion of hydrogen is too low for the required amount of hydrogen to accumulate during the period of the

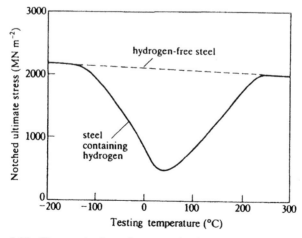

8.33 The notched tensile strength of quenched and tempered low-alloy steel containing hydrogen in the quenched condition, as a function of testing temperature.

experiment. The most severe embrittlement, it will be noted, is close to normal atmospheric temperature.

In general, the presence of hydrogen in steel causes a reduction of strain to failure and of the energy of fracture. This is reflected in the fracture morphology. For example, a ductile microvoid coalescence mode can alter to a brittle intergranular mode, or the extent of cleavage fracture in a mixed fracture mode may increase. Hydrogen-assisted crack growth can also occur by microvoid coalescence alone, and here the hydrogen affects the rate of nucleation of voids and reduces plasticity at the later stages of fracture.

There has been much speculation as to the mechanism of hydrogen embrittlement. An early hypothesis suggested that hydrogen gas accumulated in voids within the metal and built up sufficient pressure to cause fracture. Although **hydrogen blistering** does indeed occur as the result of hydrogen penetration due to corrosion, the mechanism is not now considered to be responsible for hydrogen embrittlement. It was suggested by Petch that the surface energy of steel would be reduced by adsorption of hydrogen, thereby reducing the fracture stress, and there is some evidence that such a mechanism may be effective when steel is exposed to low-pressure hydrogen. The decohesion theory of Oriani and Josephic supposes that the cohesive force between atoms in the metal lattice is reduced in proportion to the hydrogen concentration, and takes account of the increase of hydrogen solubility in the strained region at the tip of a crack. Both these theories relate to the atomistics of the fracture process, whereas in real fractures of bulk materials the main contribution to energy of fracture is from plasticity. There are nevertheless features of the decohesion theory that match with empirical observation: the important role of hydrogen concentration at a crack tip, in particular. Therefore, following van Leeuwen (1976), suppose that the reduction of fracture stress $(\sigma_o - \sigma_H)$ in hydrogen is a simple function of the hydrogen concentration c at the root of a pre-existing crack:

$$\sigma_o - \sigma_H = \beta c^\gamma \tag{8.13}$$

where β and γ are constants. It is further assumed that the radius at the crack tip varies, and is given by the Burdekin–Stone expression:

$$\rho = \frac{K_I^2}{2E\sigma_{ys}} \tag{8.14}$$

where K_I is the stress intensity factor, E is elastic modulus and σ_{ys} is yield strength. If the solubility of hydrogen at the crack tip is augmented in accordance with Equation 8.9 it may be shown that

$$\ln\left(\frac{K_{IC} - K_H}{K_H}\right) = \ln\left(\frac{p}{p_0}\right)^{\gamma/2} - \ln\left[\frac{2}{\beta}\left(\frac{2E\sigma_{ys}}{\pi}\right)^{1/2}\right] + A\sigma_{ys}^{1/2} - B\sigma_{ys} \tag{8.15}$$

where K_{IC} is the fracture toughness of the material in the absence of hydrogen, K_H is the stress intensity factor for initiating hydrogen-assisted cracking and p/p_o is the hydrogen pressure in atmospheres. From the slope and intercept of this equation, the following semi-empirical expression is obtained:

$$\ln\left(\frac{K_{IC} - K_H}{K_H}\right) = \ln\left(\frac{p}{p_0}\right)^{0.132}$$
$$- \ln\left(4.79 \times 10^{-2} \sigma_{ys}^{1/2}\right) + 6.926 \times 10^{-2} \sigma_{ys}^{1/2}$$
$$- 3.242 \times 10^{-4} \sigma_{ys} \tag{8.16}$$

where σ_{ys} is in MN m^{-2} (in this instance $\sigma_{ys} = 1724$ MN m^{-2}).

Equation 8.15 may be rearranged to give

$$\frac{K_H}{K_{IC}} = \frac{1}{1 + (p/p_0)^{\gamma/2} \left(\delta \sigma_{ys}^{1/2}\right)^{-1} \exp\left(A\sigma_{ys}^{1/2} - B\sigma_{ys}\right)} \tag{8.17}$$

where $\delta = (8E/2\pi\beta)^{1/2}$, equation 8.17 is now used to replot Oriani's data and the result is shown in Fig. 8.34. It will be seen that the embrittling effect of hydrogen is relatively insensitive to hydrogen pressure.

As a general rule, and in particular during arc welding, hydrogen cracking occurs when the metal is supersaturated with the gas. Under these circumstances hydrogen is evolved rapidly at the crack tip. It seems reasonable therefore to assume that $p/p_0 = 1$ in which case equation 8.17 may be replotted as a function of yield strength using the relationship between K_{IC} and yield strength given by equation 1.41. Values for K_H are obtained as shown in Fig. 8.35. Also plotted here are data for K_H taken from the literature. These data scatter over a wide band. Hydrogen embrittlement is a time-dependent process, and for its maximum effect, time must be allowed for hydrogen to diffuse into the cracked region and saturate it. Testing is such that the equilibrium condition is rarely achieved; hence the upward scatter. In any event, the theory provides a curve

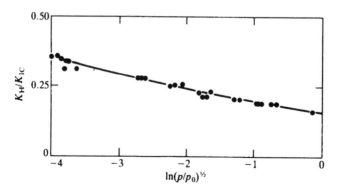

8.34 K_H/K_{IC} as a function of hydrogen pressure required to sustain crack growth.

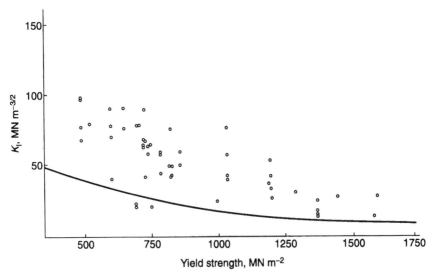

8.35 K_H as a function of yield strength, quenched and tempered steels.

that may equally serve as an empirical lower bound for experimental values of K_H.

The Hydrogen-induced cracking of welds

The three most important factors that determine the probability of hydrogen-induced embrittlement and cracking of welds are hydrogen content, fracture toughness of the weld and heat-affected zone, and the stress to which the joint is exposed as a result of the weld thermal cycle. These factors interact in a complex manner so that a quantitative treatment of the subject is difficult. There remains an element of unpredictability about the hydrogen-induced cracking of welds, and the problem continues to afflict the fabrication industry in spite of much accumulation of knowledge.

The physical appearance of hydrogen cracks in welds is illustrated in Fig. 8.36 and 8.37. In Fig. 8.36 the crack has initiated in the small region of lack of fusion at the root of the fillet and has travelled through the heat-affected zone, mainly in the coarse-grained region. In fillet welds subject to longitudinal restraint, hydrogen cracks tend to be transverse to the weld axis, running through the weld itself. The same applies to hydrogen cracks in a root pass made in thick steel using coated electrodes. In both these cases the restraint is such as to produce the maximum strain in the longitudinal direction. The location of cracks will also depend on the relative hardenability and the carbon contents of the weld metal and parent metal. If the compositions are the same and the longitudinal restraint not too severe, centreline cracks may appear in the weld metal, since the maximum cooling rate is along the weld axis. Many high-tensile steels, however, are welded using a filler metal of low carbon and alloy content, and in such cases

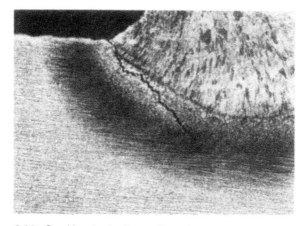

8.36 Cracking in the heat-affected zone of a fillet weld in higher-tensile steel (photograph courtesy of TWI).

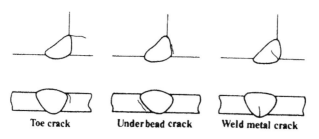

Toe crack   Underbead crack   Weld metal crack

8.37 Typical forms of cold cracking.

heat-affected zone cracking is more likely. Transverse cracks tend to be straight, but longitudinal and heat-affected zone cracks are jagged. The cracking may be intergranular, transgranular or both, relative to the prior austenite grains.

In most instances there is an incubation period before cracks become visible, and hydrogen cracks in welds may take days or even months to develop. More often, they appear immediately or a short time after welding. Because of the incubation time required for crack development, hydrogen cracking is also known as **delayed cracking**.

Fracture toughness of the weld and heat-affected zone has been covered in earlier sections: hydrogen content and stress are considered below.

Hydrogen in arc welding

Hydrogen is derived from hydrogenous chemical compounds that are dissociated in the arc column. These compounds are for the most part either hydrocarbons or water. In both cases they may be introduced into the arc by contamination of gas lines, electrodes, flux on the workpiece, while water vapour may be drawn in from the air. However, the most important source of hydrogen is electrode coatings or fluxes. Electrode coatings consist of minerals, organic matter, ferro-alloys and iron powder bonded with, for example, bentonite (a clay) and sodium silicate. The electrodes are baked after coating, and the higher the baking temperature the lower the final moisture content of the coat. Rutile and cellulosic electrodes contain organic matter and cannot be baked at more than about 200 °C, whereas basic coatings, being all mineral, may be baked at 400–450 °C. The relationship between moisture content of the coating and hydrogen content of the weld deposit for a typical basic-coated rod is shown in Fig. 8.38. In the case of rutile and cellulosic electrodes, hydrogen is generated by both residual water and cellulose, to give a hydrogen content in the standard test that is typically 20–30 ppm.

Submerged arc fluxes may be fused, sintered or agglomerated. The first two contain no water, but agglomerated fluxes are made in a similar way to electrode coatings and may have a residual moisture content. Basic submerged arc fluxes may be of this type and Fig. 8.28 shows the relationship between residual water and hydrogen in the weld metal. Note that for the same water content, basic submerged arc flux generates about twice as much hydrogen in the deposit as

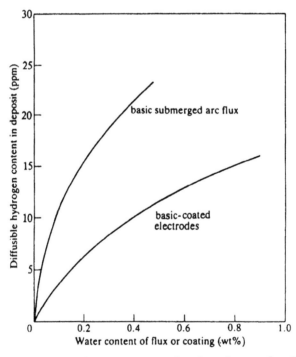

8.38 The hydrogen content of a deposit as a function of the water content of the flux or electrode coating (after Evans and Baach, 1976).

Carbon and ferritic alloy steels

basic electrode coatings. Flux dried at 800 °C can, however, produce hydrogen contents below 10 ppm.

Basic-coated electrodes and basic fluxes may pick up moisture if exposed to the atmosphere. The susceptibility to water absorption depends on baking temperature and on the type of binder that is used, as illustrated in Fig. 8.39 and 8.40. In setting up shop procedures for the use of basic rods and fluxes, it is good practice to determine the maximum exposure time of the consumables by making a series of exposure tests and determining the hydrogen content of weld deposits produced using the exposed electrodes or flux. The IIW recommended terminology for such hydrogen contents is:

1 very low, 0–5 ml/100 g deposit;
2 low, 5–10 ml/100 g deposit;
3 medium, 10–15 ml/100 g deposit;
4 high, over 15 ml/100 g deposit.

Where hydrogen cracking is a risk, the hydrogen content is maintained in the low or very low range. In US practice, the moisture content of the coating of basic low hydrogen electrodes is controlled, the upper limit according to American Welding Society (AWS) standards being as shown in Table 8.3.

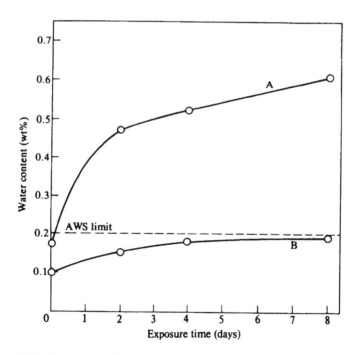

8.39 The effect of binders A and B on the amount of water absorbed by basic coated electrodes after exposure; 50% RH, 23 °C (after Evans and Baach, 1976).

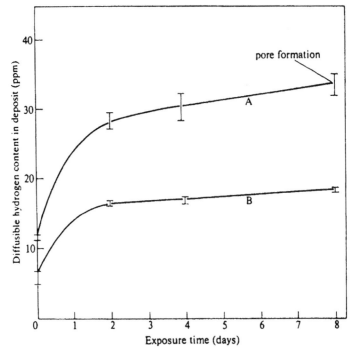

8.40 The effect of exposing a basic submerged arc flux (A and B previously baked at 500 °C and 800 °C, respectively) on the subsequent weld-metal hydrogen content; 90% RH, 23 °C (after Evans and Baach, 1976).

Table 8.3. AWS requirements for moisture content of hydrogen-controlled electrode coatings

Specified minimum ultimate strength of weld deposit		Maximum moisture content (mass %)
(lb in^{-2})	(MN m^{-2})	
70 000	482.6	0.6
80 000	551.6	0.4
90 000	620.5	0.4
100 000	689.5	0.2
110 000	758.4	0.2
120 000	827.4	0.2

Carbon and ferritic alloy steels

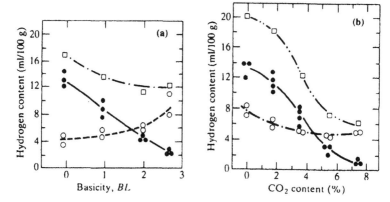

8.41 The flux composition in submerged arc welding and weld-metal hydrogen: (a) effects of flux basicity on hydrogen in weld metal and slag; (b) effect of flux CO_2 content on hydrogen content of weld metal and slag. The curves show: ● diffusible hydrogen, $(H)_D$; ○ hydrogen in slag, $(H)_S$; and □ their sum, $(H)_D + (H)_S$.

Other things being equal, and the flux being dry, the diffusible hydrogen content of submerged arc weld metal is reduced by increasing the basicity. This appears to be due to hydrogen transfer from molten metal to molten slag, as shown in Fig. 8.41 for the $CaO-MgO-Al_2O_3-SiO_2$ system discussed earlier in relation to oxygen and silicon partition. The same figure shows the effect of incorporating carbonates in the flux. Generating CO_2 in the flux cavity reduces the partial pressure of hydrogen in the arc atmosphere, and hence the dissolved hydrogen content.

Note that the diffusible hydrogen contents quoted in this section were all measured using standard techniques that require quenching the specimen immediately after welding. In actual welded joints, where the weld metal is allowed to cool in still air and the hydrogen has the opportunity to diffuse out between passes, the hydrogen content may be much lower. Figure 8.42 shows a comparison between pipeline welds made with cellulosic and basic-coated electrodes. When air cooled, the cellulosic weld metal gives a hydrogen content of less than 4 ml/100 g, as compared with 20–30 ml/100 g for the same electrodes as tested by a standard method. Clearly, the hydrogen content of welds may depend upon details of the welding procedure as well as on the type of electrode coating.

The results shown in Fig. 8.42 are for the joint as a whole. In pipe welding the diffusible hydrogen content is higher in the root pass than in the bulk of the weld because of higher cooling rates, combined with the relatively small weld bead size.

The hydrogen content that is effective in promoting cracking is the amount present below 200 °C, when the embrittlement mechanism becomes operative. Preheat can reduce the effective hydrogen content by reducing the cooling rate

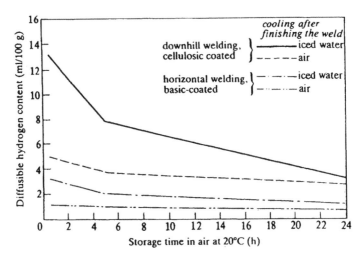

8.42 The content of diffusible hydrogen in a 100 mm V-joint, depending on the cooling rate and the storage time. Cellulosic and basic-coated electrodes; steel API5LX60 (from Rabensteiner, 1976).

and allowing more hydrogen to diffuse out of the weld. In the extreme, if a preheat temperature of 200 °C is maintained until the joint is given a postweld heat treatment, then the risk of hydrogen-induced cracking is virtually eliminated. Such a treatment is not usually compatible with production requirements, however.

Stress

The stress and/or strain that develops in the joint during cooling may result from self-restraint or external restraint, or from a combination of the two. Self-restraint is that caused by local contraction of the weld relative to its immediate surroundings. Local strains may also be generated by the γ–α transformation, but clear evidence as to the importance of this factor is lacking. External restraint is that due to the surrounding structure, which may be stiff or compliant according to the design. One method of testing for hydrogen-induced cracking (to be discussed later) is the **rigid restraint cracking** test, in which a testpiece of variable width is clamped in a rigid frame and then welded. The **intensity of restraint** R is measured as

$$R = Et/l \qquad (8.18)$$

where E is the elastic modulus, t is plate thickness and l is the restrained width. This quantity is expressed in N/mm.mm or equivalent. In some structures the intensity of restraint may be calculable and may be a factor to consider in setting up the welding procedure. Excessive restraint is, of course, a feature that should be avoided in the design of welded structures.

Self-restraint may result in stresses of yield point magnitude in the fused zone, and if the external structure is rigid, there may be a significant amount of plastic strain, particularly in earlier passes. Such plasticity increase the risk of hydrogen-induced cracking, and may cause microcracks to develop into a macroscopic running crack.

Apart from the stress field at the root of any crack that may be present, stress concentrations may form internally around non-metallic inclusions. The degree of stress concentration depends on the shape of the inclusions, and it has been found in practice that cerium-treated steel, which contains rounded inclusions, is less susceptible to hydrogen cracking than steel containing elongated inclusions (this factor may also be significant in lamellar tearing: see Section 8.5.5). Inclusions may also generate **fish-eyes**. These are bright circular areas surrounding an inclusion and found on the rupture face of a tensile test that failed in the weld metal. The bright area is a region of brittle failure that occurred due to hydrogen accumulation around the inclusion during the test. The fact that fish-eyes are not found on impact fracture surfaces indicates the need for hydrogen diffusion.

Tests for susceptibility to hydrogen-induced cracking

Laboratory tests for hydrogen cracking in welds were developed for a number of reasons: for example, to determine the variables that govern the cracking phenomenon, to compare the susceptibility of different steels, or to develop procedures for preventing cracks.

Cold cracking tests fall into two categories: self-restraint tests such as the Lehigh slit groove test and TWI's controlled thermal severity (CTS) test, and tests in which a known external load is applied to a specimen.

The Lehigh test simulates conditions in a butt weld. A plate 12 × 8 in (305 × 203 mm) of the thickness to be tested is prepared with a central slot in the form of a double U preparation. Slits are made laterally along the long edges of the testpieces, the depth of the slit determining the restraint (Fig. 8.43). A weld run is made in the root of the joint, and the depth of slit required to prevent cracking is determined. Alternatively, the percentage length of crack is recorded.

In the controlled thermal severity test a square plate of the required material is fillet-welded to the baseplate along two opposite sides. The remaining two sides are then welded up, the second weld being laid down immediately after the first. The assembly is allowed to stand for a period of time, after which the welds are sectioned and examined for cracks. The severity of the test may be varied by altering the thickness of the plates, the hydrogen level in the test welds and the composition of the weld metal (Fig. 8.44). The cooling rate is designated by means of a **thermal severity number** (TSN). TSN 1 is the thermal severity corresponding to heat flow along a single steel plate 6 mm thick. TSN 2 is obtained in a butt weld between 6 mm plates, while in a 6 mm T joint, where

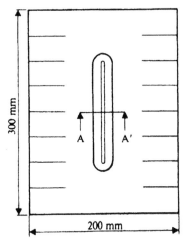 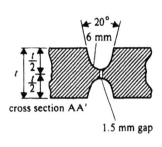

8.43 The Lehigh cracking test.

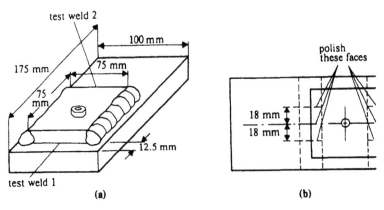

8.44 The controlled thermal severity cracking test.

there are three heat flow paths, the thermal severity number is 3. The TSN number is also increased in proportion to the plate thickness, so that the second test weld in a CTS testpiece (which also has three heat flow paths) in 12.5 mm plate would have TSN 6.

After the CTS test had been in use for some time, it was found that setting a gap (typically 1.5 mm) at the root of the test weld gave more severe cracking conditions: this variant is known as the **modified CTS test**.

In more recently developed tests, it is customary to apply a known load or strain to the test specimen. The **implant test** uses a cylindrical specimen of the steel to be tested. This specimen is notched and then inserted in a hole in a plate made from similar or compatible material. A weld run is then made over the

Carbon and ferritic alloy steels

specimen (or row of specimens), which are located so that the notch lies in the heat-affected zone. After welding and before the weld is cold, a load is applied to each specimen and the time to failure is determined. The plot of stress against time to failure gives an assessment of the relative susceptibility of different steels. Alternatively, the critical cooling time between 800 °C and 500 °C below which cracking occurs may be recorded and used to predict required welding procedures.

In Japan much use has been made of the Y-groove or Tekken cracking test. In the self-restraint form (illustrated in Fig. 8.45), a single weld bead is laid at the root of a gapped Y preparation (as shown) at a standard heat input rate of 1.7 kJ mm^{-1}. The intensity of restraint is a function of plate thickness h, and is approximately

$$R_y = \begin{matrix} 70\ h & \text{for } h \leq 40 \text{ mm} \\ 2800 \text{ kgf/mm.mm} & \text{for } h > 40 \text{ mm} \end{matrix} \qquad (8.19)$$

The critical hydrogen concentration (diffusible hydrogen content) for cracking may be determined as a function of plate thickness and carbon equivalent.

The Y-groove test is also performed with plates clamped rigidly in a frame, in which case the weld is made right across the plate. This is the rigid restraint cracking (RRC) test discussed earlier, and it allows restraint to be varied

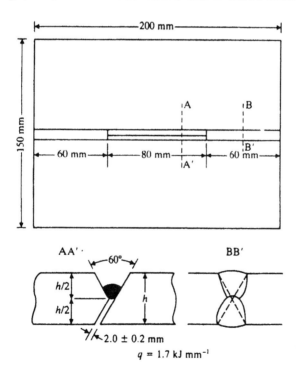

8.45 The JIS Y-groove cracking test (Tekken test).

independently of plate thickness. In this case the restraint intensity for cracking is determined as a function of carbon equivalent for a given hydrogen content and cooling rate.

A test developed at Rensselaer Polytechnic Institute (RPI) employs small specimens (51 × 13 mm), which are stacked together so that a weld bead can be laid along one edge. These are then held in a fixture that imposes a uniform stress or a uniform plastic strain, and the development of cracks is observed through a microscope on a previously polished surface. For stresses below the yield point, the curve of strength against time to initiate microfissures has the same form as that for notched tensile specimens shown in Fig. 8.32. Such curves permit different steels to be compared. In the RPI test, cracks do not propagate unless the specimen is notched or plastically deformed.

Tests such as those described above have been used to determine the effect of variables such as steel composition on the hydrogen crack susceptibility. The CTS and the Y-groove tests have also been used to develop recommendations for minimum preheat temperatures, as indicated below.

Measures to avoid hydrogen-induced cold cracking

The risk of hydrogen-induced cold cracking is minimized by reducing the hydrogen content of the weld deposit, by developing a non-sensitive microstructure, by avoiding excessive restraint and, where necessary, maintaining the temperature above the cold cracking range. These objectives may be accomplished by:

1 material selection;
2 design to avoid restraint;
3 selection of welding process;
4 control of welding procedures, in particular preheat, heat input rate and postweld heat treatment.

Material selection is generally a compromise between the need to obtain maximum design strength at minimum cost on the one hand, and weldability on the other.

In general, weldability is improved by reducing the carbon content and carbon equivalent, by minimizing alloy content and by aiming for low yield strength and high ductility. Fine-grained steel is usually less sensitive to cracking than a coarse-grained steel of equivalent yield strength. As carbon and alloy contents increase, so it becomes necessary to restrict the content of sulphur and phosphorus and to control residuals that might promote temper embrittlement. Likewise, more stringent precautions need to be taken in welding. In high-strength alloy steel for land-based operations, the carbon content should be maintained at the lowest practicable level, preferably below 0.18%. Specifying an

Carbon and ferritic alloy steels

upper limit to the carbon equivalent is an important means of material control for carbon, carbon–manganese, microalloyed or HSLA steels, and steels conforming to such limits (typically in the range 0.4–0.5) are commercially available.

Much can be done at the design stage to avoid details that have too much self-restraint. In structural work it may be possible to calculate restraint factors and establish an upper limit. In many instances, however, calculations of restraint are not practicable, and empirical rules must be used. For example, in pressure piping it is customary to place welds not closer than 50 mm. For alloy-steel pressure vessels, nozzles and attachments should be sited so that the distance between the toes of adjacent welds is, preferably, more than three times the plate thickness. Rigid box-like structures should be avoided in crack-sensitive materials.

The freedom of choice of welding processes is necessarily limited. The application must be practicable and economic. Further constraints may be applied by the job specification: for example, a requirement for impact testing may make it necessary to use coated electrodes instead of submerged arc welding, or it may necessitate the use of a basic flux with submerged arc welding, which in turn increases the risk of hydrogen contamination of the weld. In general, automatic processes (electroslag, submerged arc) or semi-automatic welding are a lower risk than manual metal arc because their higher heat input rates reduce the cooling rate. In welding high-tensile or alloy steel, it is normal practice to use basic-coated electrodes, but for plain carbon steel cellulosic or rutile-coated electrodes are often a better selection.

Very often the material, design and process selection are determined by others and the only freedom of action lies in the control of welding procedures. Here the essential steps include control of hydrogen content of consumables, heat input rate to the workpiece, preheat and postwelding heat treatment. The hydrogen content of basic-coated electrode coatings and basic submerged arc flux is controlled by storage under dry conditions, usually in a heated or air-conditioned store, or by baking before use. Procedures are set up to ensure that rods or flux are not exposed to atmosphere beyond a certain length of time, surplus material being returned to store for rebaking. Periodic checks may be made of moisture content and/or diffusible hydrogen in welds. Basic-coated electrodes are frequently kept in a heated canister after issue from the stores and until they are used.

Preheat is effective in reducing cooling rate, and thereby modifying the transformation products, and in reducing weld and heat-affected zone hardness. It may also take the metal out of the region of maximum sensitivity to hydrogen embrittlement (see Fig. 8.33) and may allow hydrogen to diffuse out of the weld. Preheat is in most instances applied locally to the weld and it is important that a sufficient width of plate (say three times thickness minimum) is uniform in temperature. In joining sections of dissimilar section thickness, it may be necessary to apply more preheat to the thicker part. Heating may be by flame or by electric elements: both are satisfactory but electric heating lends itself more readily to automatic control. Control is by temperature-indicating crayons, optical pyrometer or thermocouples. Although local heating is the norm, it may be desirable for severely restrained parts to heat the whole component. In such cases it would be good practice to maintain the preheat until the component is given its postweld heat treatment. In most cases, however, it is undesirable or impracticable to maintain preheat, and instead the weld is cooled slowly to room temperature, or the preheat is maintained for a period of, say, 30 min after completion and before cooling. Postweld heat treatment finally reduces the hydrogen content to a low level so that further cold cracking is improbable.

Various methods have been used to obtain formulae to enable preheat temperatures to be calculated from the carbon equivalent and other variables. In the UK, BS 5135 has been developed based on CTS testing and practical experience. The standard contains tabulations and charts that enable preheat temperatures to be specified. For a given combination of carbon equivalent and diffusible hydrogen, preheat temperatures are indicated as a function of arc energy (volts × amps/welding speed) and the combined thickness of plates at the joint. There is no quantitative allowance for the restraint intensity, but there is a warning that under conditions of high restraint higher preheat may be necessary.

In many instances it is also necessary to control the **interpass temperature**. This temperature is measured immediately before starting another weld run, and either a minimum or maximum may be specified. Where a preheat is required to avoid hydrogen cold cracking, the interpass temperature is a minimum, and is at least equal to the preheat temperature. A **maximum** interpass temperature may be required for one of several reasons. If the alloy is susceptible to liquidation or super-solidus cracking, this limitation will minimize the time spent in the hot brittle range. Austenitic chromium–nickel steels may be subject to this mode of cracking but not, as a rule, carbon or low-alloy steel. An upper limit may be specified for carbon or carbon–manganese steel in order to minimize grain growth and thereby achieve required levels of notch-ductility.

An upper preheat limit may also be called for in the welding of quenched and tempered steels. These materials are used, for example, in the manufacture of heavy goods vehicle bodies. Cooling rates obtained in manual or high-speed submerged arc welding are similar to those used in quenching the parent metal, so it is practicable to achieve an acceptable strength in an as-welded joint. To obtain optimum cooling rates it is, however, necessary to specify maximum preheat and interpass temperatures.

Recommendations for the avoidance of hydrogen-induced cold cracking, and details of preheat and interpass temperatures for structural steels will be found in the American Welding Society's structural welding code AWS D.1.1. Those seeking further guidance on this subject, and on the welding of steel structures in general, are strongly recommended to consult this document. Figures for low-alloy heat and corrosion-resistant steels are given in the code for pressure piping, ANSI B.31.3. A list of typical minimum preheat and interpass temperatures, together with postweld heat treatment ranges, is given in Table 8.4.

Table 8.4. Preheat, interpass and post-welding heat treatment temperatures for various steels

Steel type	Minimum preheat and interpass temperature (°C)	Postweld heat treatment temperature range (°C)
Carbon steel <19 mm	None	None
Carbon steel >19 mm	100	580–650
C$\frac{1}{2}$Mo	100	650–690
1Cr$\frac{1}{2}$Mo	150	650–700
2$\frac{1}{4}$Cr1Mo	200	690–740
5Cr$\frac{1}{2}$Mo } 9Cr1Mo } 12CrMoV }	200	700–760
3$\frac{1}{2}$Ni	None	580–620
9Ni	None	None
Austenitic Cr–Ni steel	None	Normally none (see text)

8.5.4 Chevron cracking

The use of basic submerged arc fluxes has been attended in earlier times by a type of cracking known as **staircase, 45°** or **chevron cracking**. Figure 8.46 shows a longitudinal section from a weld so affected. Sometimes the cracks are relatively straight, but in many cases they have a zig-zag shape, whence the term 'staircase'. The cracks were at first ascribed to hydrogen, since the early basic fluxes were baked at 500 °C. As will be seen from Fig. 8.40, such a flux may generate a high weld-metal hydrogen content if it has been exposed to the atmosphere for a relatively short time. Increasing the preheat temperature overcame this problem. At the same time, the baking temperature of the flux was increased to 800 °C. Cracks have nevertheless been found in welds made with the improved flux.

The circumstantial evidence favours hydrogen embrittlement as a cause of chevron cracking. However, the morphology of the cracking is not characteristic of a normal hydrogen crack. Figure 8.47 shows a typical staircase in which there are a series of open, intergranular cracks joined by fine transgranular cracks. The open cracks are staggered so as to form a line at 45° to the weld axis, hence the typical form. The intergranular surfaces (which are intergranular relative to prior austenite grains) show thermal facets, indicating that they have been exposed to high temperature.

270   Metallurgy of welding

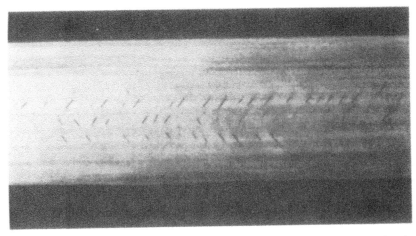

8.46 A longitudinal section through a submerged arc weld showing chevron cracks (photograph courtesy of Cranfield Institute of Technology).

8.47 A chevron crack located in the recrystallized region of a weld metal (photograph courtesy of Cranfield Institute of Technology).

It has been shown that under laboratory conditions chevron cracks form in a progressive manner. No cracks appear until the later weld passes have been made; at this stage microfissures form in the lower runs, and as the weld is completed these extend to form a staircase crack. Two hypotheses have been advanced to explain the experimental evidence. The first suggests that the weld metal is subject to **ductility dip cracking**. Steel weld metal may lose ductility as it cools from 1200 to 1000 °C, after which it recovers and is normal at about 800 °C. This loss of ductility is thought to be due to segregation of impurity atoms (such as sulphur and phosphorus) to grain boundaries, and may be promoted by some constituent of the flux. Thus intergranular ductility dip cracks form at elevated temperature, and at low temperature these cracks join by a hydrogen cracking mechanism. The second hypothesis suggests that the cracking is all due to hydrogen, but that once again it occurs in two stages. In the first stage intergranular microcracks form in earlier passes, and later these join by transgranular hydrogen cracking.

Although this defect has been found mainly in submerged arc welds, it is also known to occur in welds made with basic-coated electrodes. Only a small number of such cases have been recorded.

Although the incidence of chevron cracking has diminished, it is impossible to rule out the recurrence of any hydrogen-induced defect, at least in the case of arc welding processes that employ a flux.

8.5.5 Lamellar tearing

Lamellar tearing is a form of cracking that occurs in the base metal of a weldment due to the combination of high localized stress and low ductility of the plate in the through-thickness direction. It is associated with restrained corner or T joints, particularly in thick plate, where the fusion boundary of the weld is more or less parallel to the plate surface. The cracks appear close to or a few millimetres away from the weld boundary, and usually consist of planar areas parallel to the surface joined by shear failures at right angles to the surface. Figure 8.48 illustrates a typical case.

The susceptibility to lamellar tearing depends upon the type of joint and the inherent restraint, on sulphur and oxygen contents, on the type and morphology of inclusions (which affect the **through-thickness ductility**) and on the hydrogen content of the weld.

Lamellar tearing has affected weld fabrication in the machine tool industry, where T and corner joints in heavy plate are required for frames and bed plates. It is also a hazard in the fabrication of offshore oil platforms, and in welded-on attachments to boilers and thick-walled pressure vessels.

Lamellar tears initiate by separation or void formation at the interface between inclusions and metal, or by shattering of the inclusion itself. The voids so formed

272 Metallurgy of welding

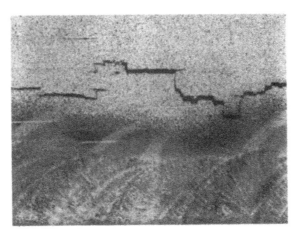

8.48 A lamellar tear under T butt weld in C–Mn steel. Note the steplike morphology of the crack illustrating its tendency to run in planes parallel to the plate surface (×5) (photograph courtesy of TWI).

link together in a planar manner by necking, microvoid coalescence or cleavage. Subsequently these planar discontinuities, when they exist at different levels, are joined by vertical shear walls. It would be expected that susceptibility to lamellar tearing would correlate with the number of inclusions as counted using a Quantimet apparatus, but this does not appear to be the case except in very broad terms. It is possible that submicroscopic inclusions play a part in generating this type of crack.

Silicate and sulphide inclusions both play a part in initiating lamellar tearing. Testing does not always show a clear correlation between sulphur and silicon content on the one hand and tearing susceptibility on the other, but reduction of sulphur content is generally regarded as one of the methods of control. Cerium or rare-earth metal (REM) treatment is another means of control. Hydrogen has a significant effect on lamellar tearing. In high-strength steels that form martensite in the heat-affected zone, hydrogen-induced cold cracks will generally form preferentially, but in plain carbon steels of low hardenability, hydrogen increases the susceptibility to lamellar tearing quite markedly. There is little or no correlation between heat input rate and the incidence of lamellar tearing, but in the presence of hydrogen a low heat input rate might tip the balance towards hydrogen cracking.

Lamellar tearing may, in principle, be avoided by ensuring that the design does not impose through-thickness contraction strains on steel with poor through-thickness ductility. Some possible design modifications are illustrated in Fig. 8.49. Such changes will usually entail an increase in cost and therefore need to be justified by experience. It is also possible to grind or machine away the volume of metal where tearing is anticipated, and replace the cut-away portion with weld metal, a process known as **buttering**. In severe cases the assembly is then stress-

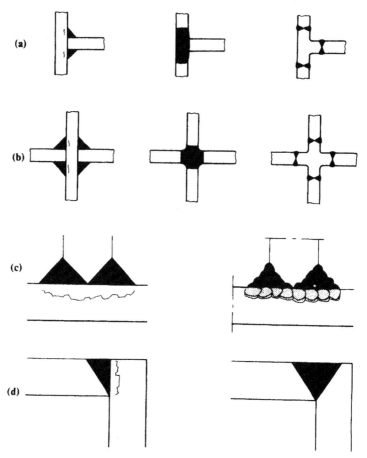

8.49 Redesign to avoid lamellar tearing: (a) and (b) replace fillets with solid weld metal or forged sections; (c) buttering; (d) modify preparation of corner joints (in part after Dorn and Lai Choe Kming, 1978).

relieved before welding on the attachment. The risk of tearing may be further reduced by specifying a material of high through-thickness ductility, which is usually achieved by limiting the sulphur content to a low value, say less than 0.007%. Preheating may also reduce the risk of lamellar tearing in some cases.

The most widely used test for susceptibility to lamellar tearing is the through-thickness ductility test. Plates are welded at right angles to and on opposite sides of the plate to be tested, or round bar may be friction welded thereto. Specimens are then cut out of this assembly and machined to a round test bar so that the original plate forms the central part of the gauge length. If the plate is thick enough the whole testpiece may be machined from it.

The ductility in a tensile test made on such a specimen is taken as a measure of susceptibility; material having a through-thickness ductility less than, say, 25% is regarded as susceptible. Other tests employ restrained or externally loaded welded specimens. Attenuation of an ultrasonic beam was at first considered to be a possible means of testing for lamellar testing susceptibility, but this proved not to be the case.

After the loss of the accommodation platform *Alexander L. Kielland* in 1980 (see Section 11.2.2) it has become normal practice to specify that plates used in critical locations in North Sea offshore structures should have a minimum through-thickness ductility of 20 or 25%. This practice, combined with continued improvements in the cleanliness of steel, would appear to provide a reasonable security against such accidents.

8.5.6 Reheat cracking

Reheat or **stress relaxation cracking** may occur in the heat-affected zone of welds in low-alloy steel during postweld heat treatment or during service at elevated temperature. The factors that contribute to reheat cracking are:

1. a susceptible alloy composition;
2. a susceptible microstructure;
3. a high level of residual strain combined with some degree of triaxiality;
4. temperature in the strain relaxation (creep) range.

Most alloy steels suffer some degree of embrittlement in the coarse-grained region of the heat-affected zone when heated at 600 °C. Elements that promote such embrittlement are Cr, Cu, Mo, B, V, Nb and Ti, while S, and possibly P and Sn, influence the brittle intergranular mode of reheat cracking. Molybdenum–vanadium and molybdenum–boron steels are particularly susceptible, especially if the vanadium is over 0.1%. The relative effect of the various elements has been expressed quantitatively in formulae, due to Nakamura (8.20) and Ito (8.21):

$$P = Cr + 3.3Mo + 8.1V - 2 \tag{8.20}$$

$$P = Cr + Cu + 2Mo + 10V + 7Nb + 5Ti - 2 \tag{8.21}$$

When the value of the parameter P is equal to or greater than zero, the steel may be susceptible to reheat cracking. The cracks are intergranular relative to prior austenitic grains (Fig. 8.50) and occur preferentially in the coarse-grained heat-affected zone of the weld, usually in the parent metal but also sometimes in the weld metal. There are two distinct fracture morphologies: low-ductility intergranular fracture and intergranular microvoid coalescence. The former is characterized by relatively smooth intergranular facets with some associated

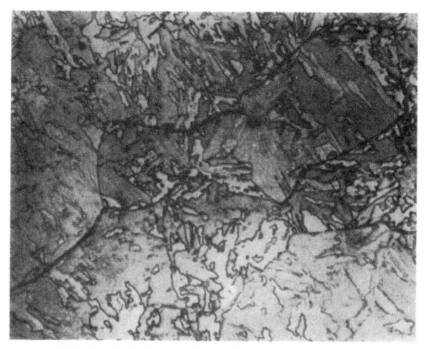

8.50 A typical reheat crack in Cr–Mo–V coarse-grain heat-affected zone (from Glover *et al.*, 1977).

particles, and occurs during heating between 450 and 600 °C, whereas the latter shows heavily cavitated surfaces and occurs at temperatures above 600 °C (Fig. 8.51). The brittle intergranular mode is initiated by stress concentrators such as pre-existing cracks or unfavourable surface geometry; in the absence of stress intensifiers, the intergranular microvoid coalescence type of fracture is dominant. In the latter case, particles within cavities are either non-metallic inclusions containing sulphur or Fe-rich M_3C-type carbides. Microcracks that form during postweld heat treatment are likely to extend during service at elevated temperature.

The Nakamura formula relates to the Japanese steel HT 70, which is a low-carbon Ni–Cr–Mo low-alloy type. Compositions that have suffered reheat cracking in practice are Mo or Cr–Mo steels with more than 0.18% V, all of which have parameter values greater than zero, and a Mo–B steel, which proved to be particularly subject to this type of defect. Another susceptible material is the 0.5Cr–0.5Mo–0.25 V steel used by the steam power industry in the UK. Cu and Sb are detrimental in this steel, while titanium deoxidation is superior to aluminium deoxidation. ASTM steels that are known to be subject to reheat cracking in thick sections are A508 Class 2, A517 Grades E and F, A533B, A542 and A387 Grade B.

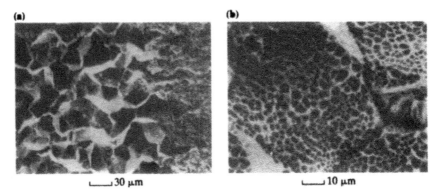

8.51 Typical fracture morphologies of reheat cracks in alloy steel: (a) low-ductility intergranular fracture; (b) intergranular microvoid coalescence (from Hippsley, 1985; micrographs reproduced with permission of the author and the UKAEA, which retains the copyright).

There are indications that a structure having poor ductility (such as upper bainite) will be more susceptible to elevated-temperature embrittlement. Likewise coarse-grained material is more likely to crack than fine-grained. It follows that the use of low heat input processes will be better than submerged arc welding. The elevated-temperature strength may also be important. If the coarse-grained region of the heat-affected zone is stronger than the parent metal at the postweld heat treatment temperature, then relaxation takes place outside the heat-affected zone and the risk of cracking is reduced. The degree of restraint and the yield strength of the weld metal are important factors, as with hydrogen cracking. However, reheat cracking generally affects only thick sections (over about 50 mm), suggesting that a higher level of residual stress is required to cause failure. This would indeed be expected since the cracks form above 400 °C where the residual stress has already been reduced. High-pressure steam drums with closely spaced nozzles have failed owing to reheat cracking in the nozzle and plate material. Cracking of the same type may occur below stainless-steel weld-deposit cladding if the backing steel is susceptible and is given a heat treatment after cladding. The stress here is due to the differential expansion between austenitic and ferritic steel. The cracks generally occur during the heating cycle before reaching soaking temperature, probably in the 450–700 °C range. The heating and cooling rates do not appear to have any significant effect on the result.

There is evidence from Auger analysis of crack surfaces that, in the brittle intergranular mode, sulphur separates out to the crack tip. Two alternative models have been proposed to describe this behaviour. In the first model, it is assumed that solute atoms are driven to the crack-tip vicinity by elastic interaction with the crack-tip stress field. When sufficient concentration is reached, local embrittlement occurs and the crack jumps forwards by brittle fracture into a fresh region of unsegregated grain boundary, so that growth occurs in a stepwise manner

(Hippsley, 1985). This model is similar to that for hydrogen-induced cracking, except that sulphur atoms occupy substitutional, not interstitial, locations and the detailed mechanism of migration is not the same.

The second proposal assumed that intergranular sulphides, which are originally precipitated by quenching from high austenitizing temperatures, dissolve when they are exposed on the crack surface. Elemental sulphur so formed then diffuses across the surface to the crack tip, enabling brittle fracture to progress at a steady rate. For both models, the rate of crack growth is governed by the rate of diffusion of sulphur.

Intergranular microcracking associated with sulphur segregation has been observed in gas tungsten arc 2.25Cr–1 Mo weld metal in the as-welded condition. The fracture surfaces were stepped, and there was evidence of sulphide precipitation during cooling through the austenitic range. Such a fracture appearance is consistent with the sulphur segregation mechanism discussed above, but in this instance the cracks must have formed on cooling and not during reheating. This type of cracking is rare, and is thought to be associated with an unusually low oxygen content (Allen and Wolstenholme, 1982).

Reheat cracks may also form or extend in service if the welded component is operating at elevated temperature and if joints are exposed to tensile stress, due to either inadequate stress relieving or service loads.

Reheat cracking tests may be divided into three types: self-restraint tests, high-temperature tensile tests and stress relaxation tests. One technique is to make up butt welds with about two-thirds of the weld completed. The samples are cut into strips, and the strips welded to an austenitic stainless-steel bar. This assembly is then heated and held for 2 h at the postweld heat treatment temperature. The greatest length of sample in which no cracks are observed is a measure of susceptibility.

Hot tensile tests are made after first subjecting the specimen to a simulated weld thermal cycle. Subsequently a tensile test is made at 600 °C and both strength and reduction of area are measured. A combination of strength below that of the base metal and reduction of area below 20% indicates susceptibility to reheat cracking. Stress relaxation testing is carried out using a bar that is notched in the region of interest. The bar is loaded in four-point bending and maintained at constant radius during heating up to, say, 700 °C. Load relaxation is measured, and the load–temperature curve can indicate the initiation and growth of cracks. The specimen is finally broken at low temperature for fractographic examination.

Reheat cracking is avoided and/or detected by the following means.

1 Material selection: for heavy sections, limit alloy content as indicated by the Japanese formulae and limit vanadium to 0.10% maximum.
2 Designing to minimize restraint: where restraint is unavoidable, consider making a stress relief treatment after the vessel is part welded.
3 Using a higher preheat temperature; dressing the toes of fillet and nozzle attachment welds; using a lower-strength weld metal.

4   Carrying out ultrasonic and magnetic particle testing after postweld heat treatment.

Austenitic chromium–nickel steels and some nickel-base alloys may also suffer reheat cracking (see Chapters 9 and 10).

8.5.7 Temper embrittlement

If an alloy steel is held for a period of time within the temperature range 375–575 °C or, for the more susceptible compositions, is cooled slowly through this range, it may suffer an increase in the impact transition temperature. The susceptibility of a steel to temper embrittlement is normally measured by the temperature shift of either the 55 J impact energy or the 50% fracture appearance transition temperature (FATT) after exposure to a standard combination of time and temperature. Step cooling, in which specimens are held for increasing periods of time at a series of decreasing temperatures within the temper embrittlement range, has been much used in studying this phenomenon. Step cooling is essentially a control test, and for a quantitative result it is necessary to expose samples at constant temperature over a long period of time.

Straight chromium steels are very susceptible to temper embrittlement, but the addition of 0.5% Mo greatly reduces the susceptibility. Elements that markedly increase the degree of embrittlement are Sb, P, Sn and As, while Mn, V, B and Si increase susceptibility slightly. Weld metal of the same composition as the alloy plate is also subject to temper embrittlement. The effect is reversible, and steel may be 'de-embrittled' by heating at temperatures of 600 °C and above. It is caused by segregation of tramp elements to the prior austenite grain boundaries, and fractures are typically intergranular.

Embrittling and alloying elements act synergistically, and for example Mn and Si increase the effect of P and Sn. For quenched and tempered 2.25Cr–1Mo steel with As below 0.02% and Sb less than 0.004%, the susceptibility to embrittlement may be measured by the Watanabe J factor:

$$J = (Mn + Si)(P + Sn) \times 10^4 \qquad (8.22)$$

Figure 8.52 shows the correlation between the J factor and the fracture appearance transition temperature for long-term isothermal heating. The J factor can be reduced by lowering the silicon content and controlling P, As, Sn and Sb. The manganese level cannot be reduced without sacrificing tensile properties.

Temper embrittlement is of concern in the operation of heavy-wall pressure vessels for nuclear power or petrochemical plant. If, for example, cracking is found during an inspection of the shell, then in order to assess the integrity of the vessel it is necessary to know the fracture toughness. From the initial properties and the service life, and using a correlation such as that shown in Fig. 8.52, it may be possible to calculate the fracture risk. Vessels used for hydrogenation such as

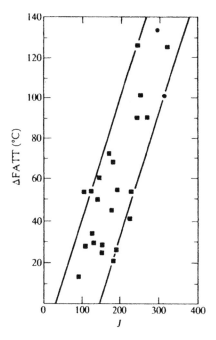

8.52 The correlation between *J* factor and ΔFATT results from long-time isothermal embrittlement studies (from Viswanathan and Jaffee, 1982).

hydrocrackers are especially vulnerable because on cooling from operating temperature they contain supersaturated hydrogen, which is a further embrittling factor.

8.6 Steelmaking

8.6.1 General

The past few decades have seen a remarkable development – amounting to almost a revolution – in methods of producing steel. Originally designed to improve productivity, these changes have made possible radical improvements in the quality of steel plates, almost all of which have been beneficial to welding operations. The steelmaking developments are described below, following which the various types of ferritic carbon and alloy steels will be reviewed in relation to their weldability. The subject-matter of this section is covered in greater detail by Lancaster (1997).

8.6.2 The evolution of steelmaking processes

The production of steel on a large scale first became possible as a result of the invention of the bottom-blown converter by Henry Bessemer in the middle of the nineteenth century. At that time the market was dominated by three main ferrous products: wrought iron, cast iron and steel. Wrought iron was made by the direct reduction of iron ore in a puddling furnace, and in plate form was used in the construction of ships. Cast iron was also employed for structural purposes, in buildings, bridges and to some extent in shipbuilding. Steel was made on a relatively small scale by melting carburized bar in crucibles, and found application almost entirely in tools and instruments.

Bessemer steel was not an immediate success. Initially there were technical problems and some of the early products were brittle. Also, the price of steel plate was at first higher than that of iron. It was not until the last decade of the nineteenth century that Bessemer steel displaced iron as a constructional material in shipbuilding.

In Britain and the USA Bessemer steel had a relatively short life. After the beginning of the twentieth century it was overtaken by the open-hearth or Siemens–Martin process, which allowed larger tonnages to be made in a single melt, and in which the steel was not contaminated by nitrogen. The basic Bessemer, or Thomas process, which employed a Bessemer converter with a basic lining, remained the dominant steelmaking process in much of continental Europe, however. The high nitrogen content associated with an air-blown converter was known to be damaging and in particular to promote strain-age embrittlement, and attempts were made to produce 'improved' Thomas steels by, for example, blowing with a mixture of air and steam. Blowing with oxygen was not possible because of severe erosion of tuyères. It would appear that the nitrogen problem continues to give rise for concern because in European Standard EN 10 025 for structural steels, published in 1990, nitrogen limits are specified for other than fully killed grades.

In Austria after World War II there were concerns of a different kind: namely, how to improve output and productivity, as compared with the capability of existing open-hearth furnaces. Based on some promising tests, it was decided to develop a top-blown converter process, using oxygen. At first the oxygen lance was plunged into the melt to obtain circulation but further work showed that this was not necessary; simply by blowing a jet of oxygen on to the surface of the liquid iron a vigorous circulation was obtained and the carbon content was reduced from a typical 4% to 0.05% in a 20-minute blow. Figure 8.53 shows the principle of bottom-blown and top-blown converters respectively, and Fig. 8.54 compares the refining curves for top-blown, basic Bessemer, acid Bessemer and open-hearth processes. The top-blown converter process is known in Austria as the L–D (Linz–Donawitz), and elsewhere as the basic oxygen process.

Carbon and ferritic alloy steels 281

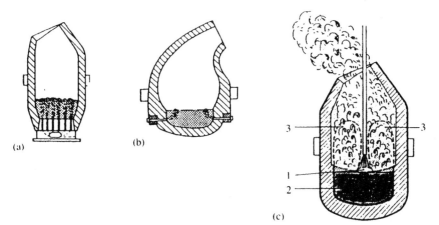

8.53 Steelmaking converters: (a) bottom-blown; (b) side-blown; (c) L-D (basic oxygen). In the basic oxygen converter refining takes place in three stages: 1, oxygen penetrates the slag and reacts with carbon, producing temperatures up to 2500 °C, 2, liquid steel circulates; 3, metal reacts with slag (from Wallner, 1986).

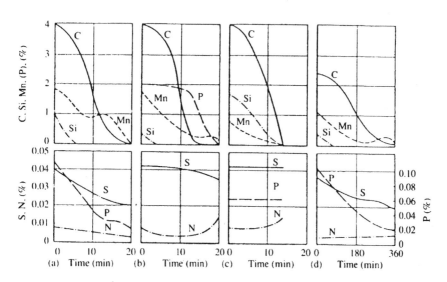

8.54 Typical refining curves for different steelmaking processes: (a) basic oxygen; (b) basic Bessemer; (c) acid Bessemer; (d) open-hearth (from Wallner, 1986).

So far as productivity is concerned, the top-blown process combines the best features of open-hearth (large bath capacity) and those of other converter processes (short refining time). Comparative production rates are:

open hearth   40 t h^{-1}
basic Bessemer   70 t h^{-1}
basic oxygen   600 t h^{-1}

The economic advantage of the new process was therefore overwhelming and the bulk of steelmaking, other than in the USSR (as it then was) and Eastern Europe, converted rapidly to the basic oxygen technique. Moreover there were technical advantages: the nitrogen content was low and by various means, including bottom blowing with an inert gas, the carbon content could be brought down to very low levels. Numerous modifications of this kind have been introduced but top blowing with oxygen remains the common feature in all but a very few processes.

The basic oxygen process can absorb a relatively small amount of scrap. Some of the modifications of the process have been designed to increase the scrap capacity but their overall effect has been small. Scrap is the preferred charge for electric furnaces, so the development of basic oxygen steelmaking has been accompanied by an increase in the amount of electric steel.

The other steelmaking development of importance to weldability is the removal of sulphur, phosphorus and other impurities. This is accomplished upstream of the converter, in the torpedo car carrying the unrefined hot metal from the blast furnace, or alternatively in an intermediate vessel. Sulphur is removed by injection of magnesium or of calcium in the form of calcium carbide, lime, soda ash or calcium carbonate. In this way sulphur can be reduced to 0.01%. Phosphorus may be removed by reducing the silicon content to 0.2% and treating with a basic oxidizing slag to produce an iron with 4.3% carbon, 0.05% silicon, 0.015% phosphorus and 0.005% sulphur. Downstream of the converter it is possible to reduce sulphur by treatment with a basic slag. Additions of calcium or misch metal (rare earth metals) may also be used to globularize any residual sulphides.

By such means it is possible to produce **clean steel**. In bulk steel production this is primarily required in order to obtain good through-thickness ductility in rolled plate. Classification societies and other statutory authorities usually require such properties for critical locations on offshore structures. Even higher levels of cleanness, obtained for example by electroslag remelting, may be specified for automobile and aircraft applications.

8.6.3 Casting

Prior to the introduction of the oxygen converter process, steel was normally cast into ingot moulds. The economics of this procedure is much affected by the amount of ingot that is cut off and recirculated as scrap. This in turn depends on the degree of deoxidation. If no deoxidants are added the oxygen in the liquid steel combines with carbon to form CO, which bubbles out with a spectacular display of sparks. The outer rim of the ingot solidifies as almost pure iron, and impurities segregate to the centre. Adding small amounts of deoxidant produces **semi-killed** or **balanced** steel. The ingot is porous and fills the mould but impurities segregate at the top. When the steel is fully deoxidized (**killed**) there is no porosity but there is a relatively large shrinkage pipe. The amount of metal that must be cropped from the ingot prior to rolling is least for rimming steel, greatest for killed steel, and intermediate for semi-killed. Rimmed steel when rolled into plate has an augmented carbon and impurity content in the centre, and the weldability may be affected. Semi-killed steel is more or less uniform in composition but may have laminar discontinuities that result from porosity in the ingot. Such laminations do not have any significantly damaging effect on the integrity of structures, but they have a considerable nuisance value in fabrication work; in the need for repair for example, or rejection in bad cases. Killed steel is acceptable for welded fabrication without reservations.

This situation has been radically changed for the better by the introduction of continuous casting. The layout of a continuous slab casting machine is illustrated in Fig. 8.55. Liquid metal from the ladle is bottom-poured into a tundish, whence it feeds into a water-cooled copper mould. Here a solid skin is formed, the mould being given a reciprocating motion to prevent sticking. The partly solidified strand is bent through 90° by rolls to emerge horizontally, and is eventually cut into slabs. Such equipment requires a high rate of steel production, a

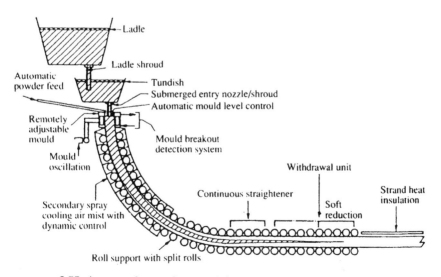

8.55 Layout of a continuous slab casting machine (Islam, 1989).

requirement that can be met by the basic oxygen process. The advantage in productivity and in the reduction of scrap will be self-evident.

So far as weldability is concerned the advantage of continuous casting is that the steel is fully killed. There may still be some segregation to the centre of the strand and although this may be significant in some applications it is unimportant for welding. The lamination associated with semi-killed practice is eliminated, and although there may be other causes of laminating in rolled products, this problem is greatly minimized.

8.6.4 Controlled rolling

In the period immediately after 1945 it became evident that there was a need for improved notch-ductility in structural steel. Traditionally the means of obtaining improved properties in a hot-rolled steel was by normalizing: that is to say, heating to a temperature above the ferrite–austenite transition and cooling in air. Normalizing produces a fine equiaxed ferritic grain structure of good notch-ductility, but it is a relatively costly procedure. Attempts were therefore made to find other means of grain refinement.

The ferrite grain size of a rolled plate is largely a function of the austenite grain size and the way it is deformed. At rolling temperatures above 1000 °C the austenite is coarse-grained and transforms to a coarse ferrite–bainite structure which has low notch-ductility. When deformed between 1000 and 900 °C the austenite recrystallizes continuously and a moderately fine-grain steel is obtained. However, if the rolling is performed below 900 °C (or below the recrystallization temperature, which may vary), elongated austenite grains containing slip bands are formed. On cooling through the transformation temperature fine grains of ferrite nucleate on the slip bands and at the austenite grain boundaries, resulting in a steel that has improved notch-ductility combined with higher yield strength.

The addition of small amounts of niobium assists this process by raising the austenite recrystallization temperature and increasing the temperature range in which elongated grains are obtained. Also, the finishing temperature (completion of rolling) may be raised, which improves productivity.

Originally, controlled rolling (which is also known as **thermomechanically controlled rolling**) was continued below the austenite–ferrite transition in order to refine the ferrite grains still further. This practice resulted in a lamellar weakness that showed up as longitudinal splits in impact specimens, and currently the trend is to rely on deformation below the austenite recrystallization temperature. Figure 8.56 is a diagrammatic illustration of hot rolling and the two controlled rolling processes noted above. A combination of controlled rolling with water quenching or normalizing is also possible, but these elaborations go beyond the original aims of the procedure.

Carbon and ferritic alloy steels

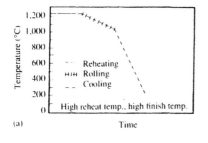
(a)

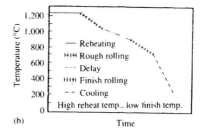
(b)

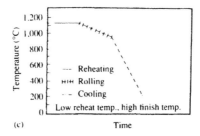
(c)

8.56 Rolling schedules: (a) hot rolling; (b) controlled rolling to low finishing temperature; (c) recrystallization-controlled rolling (from Paules, 1991).

Controlled rolling is only applicable to the rolling of slabs in a reversing mill. In a continuous mill, where the steel is rolled continuously through a train of rolls to produce strip, with a finishing temperature of, say, 650 °C, such control is not possible. However, improvements in properties have been obtatined by quenching down to the 650 °C finishing temperature prior to coiling. This results in a very fine grain size; if microalloying elements are present, an improvement in tensile properties may be obtained because of precipitation during the cooling of the coiled strip.

The relatively low finishing temperature in a strip mill means that the steel is subject to a degree of cold rolling. While in some respects this treatment is beneficial, it can reduce the resistance to **hydrogen-induced cracking** quite severely. Hydrogen-induced cracking is caused by H_2S corrosion, which results in the release of atomic hydrogen at the metal surface and its diffusion into the interior. Such hydrogen will recombine at planes of laminar weakness in the steel, and may cause disintegration of the material.

Laminar weakness may be caused by the presence of elongated sulphide inclusions. The effect is to elongate such inclusions still further, and thus provide the conditions for hydrogen-induced cracking. Controlled rolling that is continued to low finishing temperatures may have the same effect. This problem is avoided by desulphurizing the steel and by adding calcium or misch metal to globularize the sulphides.

Metallurgy of welding

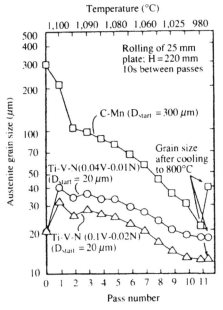

8.57 The effect of titanium–vanadium–nitrogen treatment on austenite grain size (from Paules, 1991).

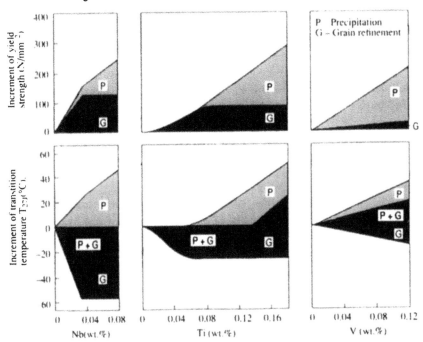

8.58 The effect of niobium, titanium and vanadium on the properties of controlled-rolled steel (from Stuart, 1991).

8.6.5 Microalloying

The use of small additions of niobium to produce fine grain in conjunction with controlled rolling was noted above. Titanium and vanadium are also added in small amounts to carbon steel. Titanium combines preferentially with nitrogen, and TiN particles inhibit austenite grain growth during the reheating of slabs prior to rolling (Fig. 8.57), so giving a better start point for controlled rolling. Vanadium acts chiefly by precipitation hardening. The way in which the three elements act in isolation as a function of their concentration is shown in Fig. 8.58. However, in microalloyed steels it is common practice to add two or more of these elements, sometimes in combination with molybdenum, and their interaction may be complex; for example there is a competition between Ti and V for nitrogen and this may reduce the precipitation hardening effect.

8.7 The welding of iron and steel products

8.7.1 Cast iron

The grades of cast iron that are welded include **grey iron** and **spheroidal graphite cast iron**. Grey iron is the most common and least costly of all cast materials; it is a 2.5–3.5% carbon iron in which much of the carbon is present as graphite flakes. The distribution of graphite in grey iron causes it to be brittle, and consequently the standard set for welds in this material is not very high. Spheroidal graphite (SG) iron, however, is cast with magnesium, nickel or rare-earth additions, and as a result the graphite is in the form of speroids, with a ferritic or pearlitic matrix. Unlike ordinary grey cast iron it has some ductility in the 'as-cast' state, up to about 4% elongation in a tensile test, and after annealing this is increased to 15–25%.

The weldability of SG iron is somewhat better than that of grey iron, because the sulphur and phosphorus contents are generally at a lower level, so that the risk of hot tearing in the weld metal is reduced. The metallurgical changes that take place in the heat-affected zone of fusion welds in these two materials are, however, basically the same. In the region that is heated above the eutectoid temperature, the ferrite is transformed to austenite. Above about 800 °C, graphite starts to go into solution and simultaneously cementite is precipitated, first at the grain boundaries, and at higher temperatures, when more graphite is dissolved, within the austenite grains. At still higher temperatures, some melting occurs. On cooling, the cementite network remains but the austenite transforms: high-carbon regions to martensite and low-carbon regions to pearlite. Thus the heat-affected zone of fusion welds in cast iron has a complex structure comprising remelted regions, undissolved graphite, martensite, fine pearlite, coarse pearlite and some ferrite. Needless to say this structure is very hard and brittle and, if such a weld is tested in tension or bending, it fails through the weld boundary zone (Fig. 8.59).

There are various ways of mitigating these effects. Preheat (combined with the use of low-hydrogen electrode coatings) may be used to minimize the danger of hard-zone cracking, although it is not always practicable, particularly in repair welding. An optimum preheat temperature is 300 °C. The hardness of the heat-affected zone may be reduced by postwelding heat treatment at 650 °C. It may be still further reduced by means of a full anneal. The effect of annealing, however, is to decompose the cementite; the graphite forms in a fine chain-like pattern, which impairs the ductility of spheroidal graphite iron and weakens grey iron (high preheat temperatures over 450 °C have a similar effect). Therefore 650 °C is the optimum postwelding heat treatment temperature. Where preheat is not used, it is desirable to reduce the heat input rate to the lowest practicable level by making short runs and allowing the metal to cool after each run. This minimizes the width of the heat-affected zone and reduces the extent to which graphite is dissolved and reprecipitated as carbide. Correctly applied, this technique is capable of producing sound load-bearing joints.

Various filler alloys have been used for cast iron. For the repair of casting by gas welding, a cast-iron filler rod may be applied. Castings so treated are

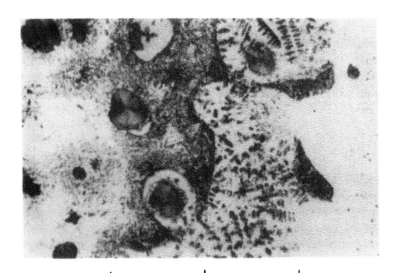

| Graphite, ferrite and low carbon transformation products | Partly dissolved spheroidal graphite, cementite and transformation products | Remelted cast iron: graphite, cementite and martensite | Nickel-alloy weld metal |

8.59 The heat-affected zone of a fusion weld on spheroidal graphite cast iron (×300, reduced by one-third in reproduction) (photograph courtesy of International Nickel Co. (Mond) Ltd).

preheated to between 400 and 475°C and cooled slowly under insulation. Cast iron is not suitable as a filler material for arc welding, however, and for this purpose coated electrodes depositing a 55% nickel–45% iron alloy are the most successful. These electrodes have the advantages of relatively low melting point and low yield strength, which minimize the hardening effect and stress due to weld metal shrinkage. They are also relatively tolerant of sulphur, and the metal transfer is of the large droplet type, which reduces the amount of dilution of the weld deposit. A suitable technique for joining spheroidal graphite iron with the nickel–iron type of electrode is to preheat at 300 °C, butter the edges (i.e. apply a surface deposit on the edges of the joint to be welded) with 55%Ni–45%Fe and then heat treat at 650 °C. Where distortion must be avoided, the preheat and postheat of the **buttering run** are omitted. Subsequently the joint may be completed using 55%Ni–45%Fe electrodes without preheat or postheat. It is characteristic of cast iron that, even when welded in the manner described above, the joint strength and ductility vary erratically. Coated electrodes with a nickel or monel core wire are also used for welding cast iron.

Braze welding, discussed in Chapter 4, is a time-honoured method of repairing iron castings, and should be considered if an operator skilled in the art is available.

8.7.2 Carbon steel for structural applications

Broadly speaking, unalloyed steels used in structural work (bridges, buildings, structures and shipbuilding) fall into two categories: low-carbon steel (up to 0.25% C) in the form of plates and sections, used for welded constructions, and higher-carbon steel (0.5–0.7% C) for reinforcing bar and rails. The second category constitutes about one-third of the total tonnage in the UK although this figure varies from time to time. Rails are joined by flash-butt welding into transportable lengths and field joints are made with low hydrogen coated electrodes or by thermit welding. Thermit welding is also used for joints that do not lend themselves to flash-butt welding, for example at points and on bends.

Reinforcing bar is welded with low hydrogen electrodes and with a preheat of 100–250 °C. Guidance as to procedure is given in AWS Standard D12-1.

Plain carbon steel is the preferred material for the bulk (about 90%) of structural work. Higher-tensile grades are required when stress is a governing consideration – for example, at the node sections of offshore constructions – but where deflection is the limiting factor increased tensile strength offers no advantage. The same applies to welded parts where fatigue loading is the main design consideration.

Both carbon and higher-tensile structural steels are specified in BS 4360 and in numbers of ASTM specifications, some of which are grouped according to applications, such as ASTM A709 for bridging. BS 4360 was replaced in part by the European Standard EN 10 025, which covers unalloyed steel having tensile strengths which range from $310\,\text{N}\,\text{mm}^{-2}$ to $690\,\text{N}\,\text{mm}^{-2}$ and which was published in 1990. Table 8.5 lists the specified compositions and properties of typical EN 10 025 grades, while Table 8.6 is a similar listing of those BS 4360 grades that have not been superseded. The mechanical properties are for sections up to 150 mm thick.

It is characteristic of structural steel that tensile and other tests are carried out on samples that represent a cast or batch, whereas in steel for boilers and pressure vessels tests are made on each plate.

Other standards cover through-thickness properties and weathering resistance. BS 6870 specifies three acceptance classes for through-thickness ductility: Z15, Z25 and Z35, where the number represents the minimum average percentage reduction of area for three transverse tests. Steels with such properties are marketed commercially as **Hyzed** steels.

Weathering steels are used for bridges and steel-framed buildings, particularly in the USA and to a lesser extent in the UK. These steels contain a small amount of copper and sometimes chromium, and when exposed to moderate atmospheric conditions develop a protective layer of rust on the surface. They have the advantage of reduced maintenance, but the rusty appearance is not universally acceptable.

Preheat requirements for non-alloyed structural steels (and this term includes microalloyed and controlled rolled plate) are specified in the UK in BS 5135. In this document four variables are used to determine the preheat: the hydrogen content of the weld deposit, the carbon equivalent of the steel, the combined thickness of the joint, and the heat input rate in $\text{kJ}\,\text{mm}^{-1}$. The categories of hydrogen content are in accordance with the IIW recommendations (Section 8.5.3), likewise the carbon equivalent (equation 8.6). The combined thickness is the sum of the thickness of plate being joined by the weld. Based on these variables, graphs or tabulations give a minimum preheat and interpass temperature.

The AWS Code D1.1 for structural welding has a simpler approach to the problem. Here the variables are ASTM standard and grade, welding process and plate thickness. Coated electrodes are divided into two categories: low hydrogen and others. The required preheat is tabulated as a function of these variables. For example, the preheat for plain carbon steel up to 30 ksi ultimate strength and thickness less than 19 mm (3/4 in) is nil, and for thicknesses between 19 and 38 mm, 66 °C, and so forth.

Preheating is an onerous and costly requirement in welding large structures, so the processes and procedures that reduce or eliminate preheat requirements are much favoured.

Carbon and ferritic alloy steels

Table 8.5. Selected grades of structural steel to European Standard EN 10 025

Grade	Type of deoxidation*	Chemical composition (max %)						Ultimate stress (N mm^{-2})	Yield stress (N mm^{-2})	Elongation	Charpy V impact	
		C	Mn	Si	P	S	N				Temp (°C)	J
Fe 360 B	FU	0.23			0.055	0.055	0.011	340–470	195	22	20	27
Fe 360 D2	FF	0.19			0.045	0.045	–	340–470	195	22	–20	27
FE 430 B	FN	0.25			0.055	0.055	0.011	400–540	225	18	20	27
Fe 430 D2	FF	0.21			0.045	0.045	–	400–540	225	18	–20	27
Fe 510 B	FN	0.27	1.7	0.6	0.055	0.055	0.011	470–630	295	18	20	27
Fe 510 D2	FF	0.24	1.7	0.6	0.045	0.045	–	470–630	295	18	–20	27
Fe 510 DD	FF	0.24	1.7	0.6	0.045	0.045	–	470–630	295	18	–20	40

*FU = rimming steel; FN = any other than rimming steel; FF = fully killed.

Table 8.6. Structural steel to BS 4360: 1990

Grade	Chemical composition (%)						Ultimate strength (N mm^{-2})	Yield strength		Elongation (%)	Charpy V impact		Supply condition*
	C max	Si	Mn max	P max	S max	Nb	V		Thickness limit (mm)	N mm^{-2}		Temp (°C)	J
40 EE	0.16	0.1–0.5	1.5	0.04	0.03	—	—	340–500	150	205	25	−50	27	N
43 EE	0.16	0.1–0.5	1.5	0.04	0.03	—	—	430–580	150	225	23	−50	27	N
50 EE	0.18	0.1–0.5	1.5	0.04	0.03	0.003–0.1	0.003–0.1	490–640	150	305	20	−50	27	N
50 F	0.16	0.1–0.5	1.5	0.025	0.025	0.003–0.08	0.003–0.1	490–640	40	390	20	−60	27	Q & T
55 C	0.22	0.6 max	1.6	0.04	0.04	0.003–0.1	0.003–0.2	550–700	25	430	19	0	27	AR or N
55 EE	0.22	0.1–0.5	1.6	0.04	0.03	0.003–0.1	0.003–0.2	550–700	63	400	19	−50	27	N
55 F	0.16	0.1–0.5	1.5	0.025	0.025	0.003–0.08	0.03/0.1	550–700	40	415	19	−60	27	Q & T

* AR = as rolled; N = normalized; Q & T = quenched and tempered.

8.7.3 Higher tensile structural steels

Table 8.7 lists structural steels having augmented yield strength. The 'normalized' grades are in some cases controlled rolled microalloyed steels. For example, DW 17100 St 52-3 is an aluminium-treated fine-grained steel, one of a family of such steels. The aluminium addition is intended to combine with any dissolved nitrogen and to produce a fine grain in as-rolled plate. This has the effect of eliminating strain-age embrittlement, increasing yield strength and lowering the ductile–brittle transition temperature.

The standard grade most commonly employed in the past for the most highly stressed members of offshore structures, such as the node sections, was BS 4360 50D. This specification has been superseded, and it may be replaced by Fe 510D or by proprietary types. In any event materials for critical locations in North Sea structures must usually meet additional requirements, such as:

1. minimum CTOD value in the heat-affected zone of weld procedure tests;
2. minimum through-thickness ductility of 25% or 35%;
3. maximum hardness value in sections of a weld run across the plate surface;
4. maximum hardness or minimum CTOD value in a region 5 mm from the weld boundary.

Other extra requirements may include higher impact strength to compensate for deterioration in the heat-affected zone. There is a tendency for specifications in this field to become progressively more stringent with the passage of time. Hardness requirements in the bead-on-plate test may be a problem and Fig. 8.60 shows how increasing the preheat temperature and modifying the composition can reduce maximum hardness values.

8.7.4 Quenched and tempered steel

The steels HY 80 and HY 100 listed in Table 8.7 are high-tensile quenched and tempered grades that are used in submarine hull construction. The combination of high yield strength and relatively thick material places special demands on welding procedures. Consumables are tested for notch-ductility by means of the Pellini explosion bulge test. Coated electrodes are subject to stringent drying procedures and are periodically checked for moisture content during welding operations. The heat input rate is maintained within specified limits. Weld examination is by radiography with ultrasonic testing as a back-up.

HY 80 and HY 100 are developments of the chromium–molybdenum–nickel armour plate steels used by the German Navy during World War I. It may be possible to use microalloying techniques in order to achieve better weldability combined with equal and possibly higher mechanical properties.

Quenched and tempered steels are also used in the construction of heavy road vehicles, earthmoving equipment and cranes. This type of steel is represented in

294 Metallurgy of welding

Table 8.7. Structural steels with augmented yield strength

Steel type	Yield strength min. for 30 mm thickness (N mm^{-2})	Standard	Steel grade	Alloying elements
High-strength normalized steels	345	ASTM	A 537	Mn
	345	DIN 17100	St 52-3	Mn
	345	DIN 17102	StE 355 FG 36	MnNb
	345	BS 4360	50D	MnNb
	380	ASTM	A 572	MnNbV
	430	BS 4360	55E	MnNbV
	450	DIN 17102	StE 460 FG 47 CT	MnNiV/MnCuNiV
High-strength, quenched and tempered steels	414	ASTM	A 678/B	Mn
	430	BS 4360	55F	MnNbV
	500	–	StE 500, XABO 500	CrMo/NiMo
	550	MIL-S	HY 80	NiCrMo
	690	–	StE 690, N-A-XTRA-70/T1	CrMoZr/NiCrMoB
	690	MIL-S	HY 100	NiCrMo

Source: from Baumgardt *et al.* (1984).

Table 8.7 by StE 690, which is the Thyssen designation for the American steels N-A-XTRA 70 and US steel T1. The second-named type has been widely used also for road tankers and rail tank cars. These steels are as a rule welded in relatively thin sections using an automatic process such as submerged arc and do not, in such thicknesses, present any significant welding problems. The properties of the parent metal are easy to match in the weld metal using, for example, a manganese–molybdenum filler alloy. Properties in the heat-affected zone are maintained provided that the cooling rate is high enough. To this end the preheat and interpass temperatures are limited to a maximum of 200 °C and the heat input rate is maintained below the value specified by the steel supplier.

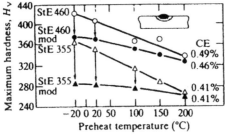

8.60 The effect of preheat temperature on maximum hardness in the heat-affected zone of high-strength steels (IIW bead-on-plate test) (from Baumgardt *et al.*, 1984).

Quenched and tempered steels of this type are specified in ASTM A514, A517 and A709; Tl steel is standardized as A517 grade F. The AWS Bridging Code permits the use of such materials in bridge construction. In the UK designers are not so adventurous and in the postwar period have generally specified BS 4360 grades 50B, 50C or 50D for tension members.

In offshore oil and gas wells the operating depth is normally such that the pressure of circulating fluids is very high. This in turn requires the use of high-tensile steel tubing, which must be welded end-to-end as drilling proceeds. The mud system, for example, which conveys slurry down to the drill tip and removes debris, uses tube in the diameter range 75–125 mm with wall thickness of 17–22 mm. The materials commonly employed in the high-pressure areas are AISI 4130 and 4140, the compositions and mechanical properties of which are listed in Table 8.8. The relatively high carbon contents of these alloys necessitate strict adherence to established weld procedures.

Typical preheat temperatures are 200 °C minimum for 4130 and 250 °C minimum for 4140, with a maximum interpass temperature of, say, 350 °C. Preheat is maintained for 1 h after completion of the weld. It is common practice to make the root and second pass with the gas tungsten arc process using filler rod to AWS ER80SD2, while filler passes are made with coated electrodes to AWS E10018-D2 for 4130 and E11018-D2 for 4140. Electrodes are best vacuum-packed. After removal from the package they are transferred to an oven held at the temperature recommended by the electrode manufacturer, then to a heated quiver immediately before use. Periodic checks of moisture content on electrode coatings may be desirable.

Postweld heat treatment is usually carried out at 620–660 °C, with heating and cooling rates controlled at 150 °C per hour. The holding temperature must not exceed the original tempering temperature.

Ultra-high-strength steels used for welded fabrications include developments of the AISI medium-carbon automotive steels such as AISI 4340 and maraging steels. Maraging steels are low-carbon Ni–Co–Mo alloys with small additions of Ti and Al. They have been developed to combine high proof stress with good fracture toughness. High tensile and proof-stress values are obtained by a martensite transformation followed by age hardening at a temperature of about 500 °C. The steel is solution-annealed at 820 °C and air-cooled, which is normally sufficient to obtain 100% martensite, and is then age-hardened. The combination of low carbon content with 10% Ni promotes good fracture toughness; in addition, the steel is vacuum-melted and Si, Mn, B, Pb and non-metallics are kept at a low level. Table 8.8 includes details of the composition of three grades of maraging steels, together with those of AISI steels 4130, 4140 and 4340. The main application of such materials in the welded condition is in aerospace components, but they are also used for dies, gears, automotive parts and ordnance. Tank-mounted assault bridges have been fabricated from maraging steel. Welding is by inert gas shielded gas metal arc or gas tungsten

Table 8.8. Composition and mechanical properties of high-tensile steels

Designation	Nominal composition (%)							Typical properties				
	C	Ni	Cr	Co	Mo	Ti	Al	0.2% proof ($MN\ m^{-2}$)	Ultimate tensile strength ($MN\ m^{-2}$)	Elongation (%)	Reduction of area (%)	K_{IC} ($MN\ m^{-3/2}$)
4130	0.3	–	1.0	–	0.2	–	–	570	720	–	–	–
4140	0.4	–	1.0	–	0.2	–	–	614	774	–	–	–
4340	0.4	1.75	0.8	–	0.25	–	–	1700	1850	5	22	60
18 Ni 1400	0.03	18.0	–	8.5	3.0	0.2	0.1	1400	1435	12	55	100
18 Ni 1700	0.03	18.0	–	8.0	5.0	0.4	0.1	1700	1735	10	45	90
18 Ni 1900	0.03	18.0	–	9.0	5.0	0.6	0.1	1900	1935	8	40	65

arc welding. Flux-shielded processes are unsuitable owing to cracking and poor toughness in the weld metals. With inert-gas shielding (normally argon) there is no heat-affected zone cracking problem and the steel may be welded without preheat. After welding the mechanical strength of the weld is equal to the parent metal in the solution-annealed condition, but postweld ageing at 480 °C for 3 h gives properties that match those of the fully heat-treated alloy.

The need for postweld ageing at elevated temperature may be a limitation for larger structures such as the assault bridge noted earlier. Cost, of course, is the other factor limiting the use of this weldable super-strength alloy.

8.7.5 Line pipe

In line pipe there are two welding problems: the longitudinal weld that forms the pipe, normally made by submerged arc welding, and the circumferential welds joining lengths of pipe, normally made by stovepipe welding. Stovepiping is a downward welding procedure using cellulosic electrodes, and it has the advantage of being much faster than alternative manual techniques. For both types of weld it is normal practice to specify minimum Charpy impact values as close as possible to that specified for the plate material.

In formulating a flux–wire combination for submerged arc welding it is necessary to bear in mind that the weld metal transferred from the electrode is diluted by about twice the amount of parent metal, and the composition may be modified accordingly. To obtain good impact results it is necessary to aim at a microstructure consisting of acicular ferrite with the minimum amount of coarse grain boundary ferrite (see Section 8.3.1). Even small amounts of martensite are detrimental if, as is normally the case, welds are not postweld heat treated. Therefore the amount of coring (microsegregation) must be kept to a minimum, and this is done by keeping the carbon content down. Alloy additions promote the formation of acicular ferrite but also increase the risk of martensite formation. Further, microalloying constituents in multi-run deposits may cause embrittlement by precipitation in reheated zones or during postweld heat treatment. It will be evident from Fig. 8.20 that the range of cooling rates under which an ideal microstructure forms is quite narrow. For optimum weld-metal notch ductility, it is also necessary to control oxygen content using semibasic or basic fluxes, or possibly the titania–boron type of flux mentioned in Section 8.2.4. The higher the strength of the base metal, the greater the care necessary in control of procedure and consumables.

Achieving impact strength requirements is not so much a problem with the manual circumferential welds because of the lower heat input rate and correspondingly finer structure. Nor is hydrogen-induced cold cracking a serious problem, even though the welds are made with cellulosic electrodes. A possible reason for this is the relatively short time between passes in stovepipe welding such that the interpass temperature is relatively high even under cold ambient conditions. Also, the actual hydrogen content may, as seen earlier, be much lower than might be assumed from the results of diffusible hydrogen tests. In the Alaska pipeline, which was welded under severe Arctic conditions, only 28 cracks were found in 1200 km of pipeline, and these were thought to be due to excessive bending during handling rather than to welding.

Table 8.9 shows the API standard grades of steel for high-strength line pipe. In addition, two special types have been developed for the higher-strength grades, namely **acicular ferrite** and **pearlite-reduced steels**. The composition of acicular ferrite steel is formulated so that 'as-rolled' plate has a microstructure rather similar to that of carbon-steel weld metal. It consists of acicular ferrite grains with islands of martensite and scattered carbides. The proeutectoid ferrite found in weld metal is absent, but there may be some polygonal ferrite. This type of structure may be obtained from a steel with 0.06% C maximum, 1.5–2.2% Mn, 0.1–0.4% Mo and 0.04–0.10% Nb. Acicular ferrite steel of this general type to API 5 LX 70 can achieve impact properties of more than 200 J cm^{-2} at $-30\,°C$.

Pearlite-reduced steel, as the name implies, is a low-carbon ferrite–pearlite steel with a lower proportion of pearlite in the microstructure than with the normal API grades. Some typical proprietary compositions are shown in Table 8.10. Like acicular ferrite steel, pearlite-reduced steel has a good combination of impact and tensile properties, suitable for large-diameter longitudinally welded pipes used in arctic conditions. For the most part, such steels are of the microalloyed type and are subject to controlled rolling in order to obtain the optimum combination of properties.

Submerged arc weld metal for high-strength line pipe steel may contain manganese, molybdenum, nickel, niobium and sometimes small amounts of titanium. A combination of 0.01–0.02% Ti and a basic flux in submerged arc welding can give an impact transition temperature in the weld metal as low as $-60\,°C$. The heat-affected zone, however, is normally lower in impact strength than the parent metal. Cellulosic-coated rods for the circumferential seams are manganese–nickel or Mn–Mo–Ni, the lower heat input rate permitting a lower alloy content.

The root and hot passes of stovepipe welds are made by pairs of welders working from opposite sides of the pipe, starting at the twelve o'clock and finishing at the six o'clock positions. In some circumstances two pairs of welders may work together. This is the critical operation in pipeline work. The rate of pipelaying is largely governed by the productivity of the root and hot pass welders. For long overland lines machine welding is used to an increasing extent, and experience in Canada has shown that available machines can double the number of welds completed per day. However, the equipment is costly and heavy, and special field-machined bevels are required. In addition, pipe welding is a positional operation and this requires the use of gas metal arc welding in the

Table 8.9. High-strength line pipe steel

Spec. no.	Grade	Ladle analyses (%)						Yield stress, min. (N mm^{-2})	Ultimate stress, min. (N mm^{-2})	Elongation, min. (%)
		C, max	Si	Mn	V, min	Nb, min	Ti, min			
API Std 5LX	X42	0.28	–	<1.25	–	–	–	290	410	25
	X46	0.28	–	<1.25	–	–	–	315	430	23
	X52	0.28	–	<1.25	–	–	–	360	450	22
	X56	0.26	–	<1.35	0.02	0.005	–	385	490	22
	X60	0.26	–	<1.35	0.02	0.005	0.03	415	520	22
	X65	0.26	–	<1.40	0.02	0.005	0.03	450	550	20
	X70	0.23	–	<1.60	–	–	–	480	560	20

Table 8.10. Typical composition of pearlite-reduced X70 line pipe steel

Grade	No.	Chemical composition (%)					
		C	Si	Mn	Al	V	Nb
API 5LX 70	1	0.15	0.13	1.35	0.05	0.05	0.03
	2	0.10	0.18	1.17	0.06	0.04	0.03
	3	0.09	0.27	1.71	0.067	0.08	0.049

short-circuiting mode, with the risk of porosity, lack of fusion and spatter. The use of pulsed arc welding could minimize this technical disadvantage. Economically, manual welding still has the advantage for short lines and rugged terrain (Rothwell et al., 1990).

8.7.6 Steels for use at subzero temperature

The selection of steel for low-temperature use is influenced by section thickness as well as by temperature, and guidance on the temperature–thickness relationship may be found in BS 5500 and the ASME pressure vessel code. In general, however, it is possible to employ impact-tested carbon steel down to $-50\,°C$, 3.5% Ni steel down to $-100\,°C$, 5% Ni steel down to $-120\,°C$, and 9% Ni steel, austenitic Cr–Ni steel and non-ferrous metals down to the lowest operating temperatures. Below a particular temperature ($-20\,°F$ in US practice, but somewhat higher in most other countries) the notch-ductility of ferritic steel is controlled by impact testing. Welds are likewise tested in the weld metal and often in the heat-affected zone. For carbon steel and 3.5% Ni steel it is possible when using matching electrodes to obtain impact values in the weld and heat-affected zone that are acceptable to (for example) the ASME code. With the submerged arc process, however, there may be difficulties in obtaining good results with carbon steel, even using a multi-run technique with basic flux, at the lower end of the temperature range. Under these circumstances it is necessary to use an alloy filler, usually 1% Ni or 2.5% Ni.

When the nickel content exceeds a nominal 3.5%, it is no longer practicable to use matching filler material because of the high susceptibility to solidification cracking. Normal practice is to employ a nickel-base coated electrode or wire, with which there is no difficulty in obtaining the required impact results. The nickel-base fillers originally used for 5% and 9% Ni steel were substantially lower in strength than the parent metal, but subsequently higher-strength fillers have been developed. Postweld heat treatment is not required for 9% Ni except (to the ASME code) in sections thicker than 50 mm, nor for 3.5% Ni below 19 mm. The value of postweld heat treatment for nickel-alloy steels is uncertain, but when it is carried out the soaking temperature for 3.5% Ni should be below 620 °C, because the lower transformation temperature is reduced by the Ni content.

Nickel-containing filler alloys (typically 1% Ni) may be used for welding carbon–manganese steel for extreme atmospheric exposure, such as North Sea operations. However, this type of welding consumable may not be accepted on the grounds that the addition of nickel increases the susceptibility to stress corrosion cracking.

Nickel additions may also be used to compensate for the relatively poor notch-ductility of unalloyed carbon-steel weld metal as produced by self-shielded welding.

8.7.7 Low-alloy corrosion- and heat-resisting steels

Carbon–molybdenum and chromium–molybdenum steels are used for their enhanced strength at elevated temperature, for resistance to hydrogen attack (see Section 8.5.3) and for resistance to corrosion by sulphur-bearing hydrocarbons. The two steels most frequently used for elevated temperature strength are 0.5Cr–0.5Mo–0.25V and 2.25Cr–1Mo. Both steels are used in major power stations, and the 2.25Cr steel is also employed for hydrocracker reactors, where it is required to withstand hydrogen attack. The 0.5Cr–0.5Mo–0.25V steel presents major welding problems because it is particularly susceptible to reheat cracking. This alloy is welded with a 2.25Cr–1Mo filler, and the cracking is almost entirely in the coarse-grained heat-affected zone. Cracks appear after postweld heat treatment or after a period of service at elevated temperature. The susceptibility to cracking is reduced by controlling residual elements such as copper, arsenic, antimony and tungsten to a low level, and by using a low heat input rate process to limit the grain growth in the heat-affected zone. How far these precautions are adequate to prevent reheat cracking in 0.5Cr–0.5Mo–0.25V steel welds is not certain.

2.25Cr–1Mo steel is welded using matching electrodes and, as would be expected from the composition, has a lower susceptibility to reheat cracking. Provided that the correct preheat temperature is maintained, this steel does not present any outstanding welding problems. The same applies to the 1.25Cr–0.5Mo, 5Cr–0.5Mo and 9Cr–1Mo steels that are used for corrosion resistance in the petroleum and petrochemical industries (Table 8.11). These steels are welded with matching electrodes, and are normally given a postweld heat treatment, with a possible exemption for thin-walled pipe. Pressure vessels are fabricated from alloy plate up to 5Cr, but the higher alloys are usually in the form of pipe. There is also a 12Cr–1Mo steel with similar weldability that is used to a limited extent for steam lines and superheaters.

Possible needs for chromium–molybdenum steels with enhanced elevated temperature properties has motivated the development of a 9Cr–1Mo alloy

302 Metallurgy of welding

Table 8.11. Heat-resistant and corrosion-resistant chromium-molybdenum steels

Designation	Nominal composition (%)				Mechanical properties			Preheat (°C)	PWHT (°C)
	C, max	Cr	Mo	V	0.2% proof (MN m^{-2})	UTS (MN m^{-2})	Elong. (%)		
0.5Mo	0.30	–	0.5	–	200	380	22	80	595–720
0.5Cr–0.5Mo–0.25V	–	0.5	0.5	0.25	–	–	–	–	–
1.25Cr–0.5Mo	0.15	1.25	0.5	–	200	400	22	150	700–725
2.25Cr–1Mo	0.15	2.25	1.0	–	200	400	22	175	700–750
5Cr–0.5Mo	0.15	5.0	0.5	–	200	400	22	175	700–750
9Cr–1Mo	0.15	9.0	1.0	–	200	400	22	175	700–750

modified by additions of vanadium and niobium, and in 1986 the improved alloy was accepted by ASME. This material has been used to a limited extent in power boilers. Parallel work has been conducted on 21/4Cr–1Mo steel with vanadium, titanium and boron additions, and this alloy has also been accepted by ASME in Code Case 1960. Potential applications are for superheaters and in the petrochemical industry (Lundin, 1990). It must be borne in mind that the compositions so developed are likely to be susceptible to reheat cracking and may therefore not be applicable to heavy-wall pressure vessels or piping.

8.7.8 Ferritic and austenitic–ferritic chromium stainless steels

In this section we are concerned with steels containing over 11% Cr and with a primarily ferritic matrix. Such steels are resistant to chloride stress corrosion cracking so that, combined with the advantage of lower cost, they may be preferred to austenitic chromium–nickel steels for certain applications.

There are four main categories of ferritic Cr steels: the 12Cr, the 18Cr, the 27Cr and the ferritic–austenitic types such as 26Cr–4Ni. All these basic types may be modified by additions of Mo, Ti and Nb. Against their advantages must be set three major disadvantages: poor weldability, brittleness and susceptibility to temper embrittlement.

These problems may in part be related to the Fe–Cr phase diagram. This is shown for a 0.05% C alloy in Fig. 8.61. The higher Cr alloys solidify entirely as δ ferrite and below 20% Cr transform to $\gamma + \delta$ on cooling below about 1400 °C. At lower temperatures, the austenite is either retained or transforms to martensite. The martensite appears at grain boundaries and may cause intergranular brittleness.

If the steel is held in the δ ferrite region for a significant length of time (as during the weld thermal cycle), rapid grain growth takes place. The grain boundary area is reduced and the grain boundaries are correspondingly enriched in impurities. This further aggravates grain boundary embrittlement. In addition, carbides or carbonitrides may precipitate at the grain boundaries causing the steel to be susceptible to intergranular corrosion. Sensitization of ferritic stainless steel to intergranular attack is caused by heating to the austenitizing temperature (above 925 °C) and rapid cooling. The susceptibility is removed by slow cooling or by annealing at 775–800 °C.

To minimize the intergranular corrosion problem, niobium or titanium may be added to the steel, so that, 'as rolled', the interstitial carbon is very low. The effect of so reducing carbon in solution is to reduce the extent of the γ field (the **gamma loop**) shown in Fig. 8.61 to lower levels of Cr. Consequently, on heating through the weld thermal cycle, even a 12Cr steel may be ferritic as it passes the nose of the gamma loop. Initially grain growth is inhibited by the presence of

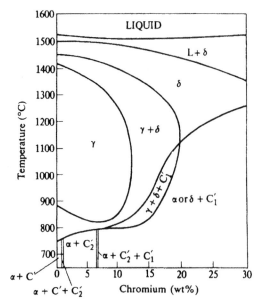

8.61 The chromium–nickel phase diagram for a carbon content of 0.05%.

carbonitride particles but at a higher temperature these go into solution, and grain growth (and possibly some austenite formation) may occur. On cooling, the carbides reprecipitate as a fine dispersion and may harden and embrittle the heat-affected zone, or they may precipitate in an intergranular form and render the steel susceptible to corrosion. Adding stabilizing elements does not necessarily prevent intergranular corrosion in ferritic stainless steel, and limitation of carbon and nitrogen to very low levels is necessary for immunity.

In spite of these complex and somewhat intractable problems, commercial 12Cr alloys may be welded in the form of sheet material and good properties obtained. Two such alloys are shown in Table 8.12. Welding is by gas-shielded processes using a type 309 (23Cr–12Ni) or 310 (25Cr–20Ni) filler, and with a heat input rate preferably in the range $0.8–1.2 \text{ kJ mm}^{-1}$. Welded 12Cr steel is used in automobile exhaust systems, for mildly corrosive food industry applications such as sugar, and for seaborne containers. AISI types 405 and 410 steels, which have no stabilizing elements added, are employed extensively in the petroleum industry for lining towers, drums and heat exchangers handling sulphur-bearing hydrocarbons at elevated temperature. The material used is carbon or low-alloy steel plate roll-clad with 12Cr, but since the cladding is not required to take any load, local weakening or embrittlement is not a problem. The filler metal on the clad side is AISI type 309 or 310. The welding of solid-wall 12Cr steel for pressure parts is not normal practice in hydrocarbon processing. Note that type 410 steel has sufficient carbon to be martensitic in the

Table 8.12. Chemical composition of ferritic and ferritic–austenitic stainless steels

Designation	Type	Nominal composition (%)							
		C, max	Cr	Ni, max	Mo	Al	Ti	Nb	N
AISI 405	Ferritic 12Cr	0.08	11.5–14.5	0.6	–	0.1–0.3	–	–	–
AISI 409	Ferritic 12Cr	0.08	10.5–11.75	0.5	–	–	6×C, 0.75 max	–	–
AISI 410	Martensitic 12Cr	0.15	11.5–13.5	0.75	–	–	–	–	–
NSS 21-2	Ferritic 12Cr	0.06	11.5	–	–	0.5	–	–	0.02
409 D	Stabilized 12Cr	0.01	11.4	–	–	–	0.22	–	0.02
AISI 430	Ferritic 17Cr	0.12	16.0–18.0	0.75	–	–	–	–	–
1803 T	Stabilized 17Cr	0.015	17.5	–	2.0	–	0.5	–	0.02
19:2 Nb	Stabilized 17Cr	0.018	19.0	–	2.0	–	–	0.5	0.03
18:2 Ti	Stabilized 17Cr	0.018	18.0	–	–	–	0.3	–	0.03
AISI 446	Ferritic 27Cr	0.20	23.0–27.0	0.6	–	–	–	–	0.25
ASTM A669	Ferritic–austenitic	0.03	18.0–19.0	4.25–5.25	2.5–3.0	–	–	–	–

heat-affected zone: for this reason a low-carbon variety, type 410S, is sometimes preferred. It is possible to weld type 410 steel with matching electrode, but this is not a common procedure.

18% Cr steels are used where higher corrosion resistance is required, and the standard AISI type 430 and some commercial alloys are listed in Table 8.12. Type 430 is not often welded, being more susceptible to embrittlement due to grain growth and the presence of interstitial elements than the 12Cr type. This embrittlement is minimized by electron-beam refining or other sophisticated steel-making techniques, and/or by the addition of stabilizers (Ti and Nb) and molybdenum. Addition of molybdenum also increases the corrosion resistance to a level similar to that of the austenitic chromium–nickel steels. Thus 18Cr–2Mo steel has potential advantages for sheet metal fabrication and particularly where chloride stress corrosion cracking is a known hazard. The metallurgical problems are similar to those for 12Cr types; in particular Ti must be limited to avoid precipitation hardening in the heat-affected zone, while excessive niobium may cause liquation cracking. The high-chromium ferritic steels as specified by AISI (e.g. type 446, Table 8.12) are not suitable for welding except for lightly loaded welds or emergency repairs. These steels are mainly used for heat-resisting duties, where their intrinsic brittleness is accepted. Welds are made with coated electrodes using a low heat input rate technique as for cast iron with no preheat and a type 309 or 310 filler.

A 26Cr–1Mo steel (E-brite) made by electron-beam re-melting and having very low C and N has been developed for special corrosion-resistant duties. This steel can be welded but the corrosion resistance of the joint is not equal to that of the parent metal.

The ferritic–austenitic stainless steels are normally designed to resist chloride stress corrosion cracking and consist of islands of austenite in a ferritic matrix. These steels, one of which is listed in Table 8.12, have the advantages of the 12Cr and 18Cr ferritic steels (resistance to chloride stress corrosion cracking) together with some of their disadvantages (susceptibility to temper brittleness). They do not, however, suffer excessive grain growth in the heat-affected zone and consequently their weldability is much better. The Mo-bearing types may have a corrosion resistance equal to that of type 316 SS. Welding is possible using most processes with austenitic Cr–Ni filler. There are a number of proprietary steels similar to A669, for example, UHB 44L and Sandvik 3 Re 60.

The ferritic and ferritic–austenitic steels are not very widely used in welded fabrication. In part this is due to the metallurgical disadvantages and in part it is due to the fact that individual alloys are rarely available in all product forms. They are most suitable for thin sections, and find their largest output in the form of sheet (for example, for automobile trim and exhausts) and as tube for heat exchangers. In stressed applications the operating temperature is limited to a maximum of 400 °C because of temper embrittlement. The 27Cr and ferritic-austenitic steels may be further embrittled by σ phase formation at operating temperatures above 600 °C. When such steels are used for heat-resisting applications, the parts must be designed to accommodate the expected embrittlement.

References

Abson, D.J., Duncan, A. and Pargeter, R. (1988) International Institute of Welding document, no. IX-1533-88.
Allen, D.J. and Wolstenholme, D.A. (1982) *Metals Technology*, **9**, 266–273.
Almquist, G., Polgary, S., Rosendahl, C.H. and Valland, G. (1972) in *Welding Research Relating to Power Plant*, CEGB, London, pp. 204–231.
Baumgardt, H., de Boer, H. and Müsgen, B. (1984) *Metal Construction*, **16**, 15–19.
Corderoy, D.J.H. and Wallwork, G.R. (1980) in *Weld Pool Chemistry and Metallurgy*, TWI, Cambridge.
Cottrell, C.L.M. (1984) *Metal Construction*, **16**, 740–744.
Dolby, R.E. (1983) *Metals Technology*, **10**, 349–362.
Dorn, L. and Lai Choe Kming (1978) *Schweissen und Schneiden*, **30**, 84–86.
Evans, G.M. and Baach, H. (1976) in *Metals Technology Conference*, Sydney, paper 4–2.
George, M.J., Still, J.R. and Terry, P. (1981) *Metal Construction*, **13** 730–737.
Glover, A.G. Jones, W.K.C. and Price, A.T. (1977) *Metals Technology*, **4**, 329.
Hannerz, N.E. and Werlefors, T. (1980) in *Weld Pool Chemistry and Metallurgy*, TWI, Cambridge, p. 335.
Hansen, M. and Anderko, K. (1958) *Constitution of Binary Alloys*, McGraw-Hill, New York.
Hart, P.H.M. (1978) in *Trends in Steel and Consumables for Welding*, TWI, Cambridge.
Hippsley, C.A. (1985) *Materials Science and Technology*, **1**, 475–479.
Howden, D. (1980) in *Weld Pool Chemistry and Metallurgy*, TWI, Cambridge.
International Institute of Welding (1982), International Institute of Welding document, no. IX-1232-82.
Islam, N. (1989) *Metals and Materials*, **5**, 392–396.
Keeler, T. (1981) *Metal Construction*, **13**, 750–753.
Lancaster J.F. (1997) *Handbook of Structural Welding*, Abington Publishing, Cambridge.
van Leeuwen, H.P. (1976) *Corrosion*, **32**, 34–37.
Lundin, C.D. (1990) in *Advanced Joining Technologies*, ed. T.H. North, Chapman & Hall, London.
Matsuda, F., Ushio, M., Saikawa, S., Araya, T. and Maruyama, Y. (1981) *Transactions of the Japanese Welding Research Institute*, **10**, 25–33.
Matsuda, F., Nakagawa, H., Nakata, K., Kohmoto, H. and Honda, Y. (1983) *Transactions of the Japanese Welding Research Institute*, **12**, 65–80.
Morigaki, O., Matsumoto, T., Yoshida, T. and Makita, M. (1973) International Institute of Welding document, no. XII-B-134-73.
Morigaki, O., Matsumoto, T. and Takemoto, Y. (1976) International Institute of Welding document, no. XII-630-76.
Morigaki, O., Matsumoto, T. and Maki, S. (1977) International Institute of Welding document, no. XII-B-217-77.
Newman, R.P. (1960) *British Welding Journal*, **7**, 172.
Oriani, R.A. and Josephic, P.H. (1974) *Acta Metallurgica*, **22**, 1065–1074.

Paules, J.R. (1991) *Journal of Metals*, 41–44.
Rabensteiner, G. (1976) in *Metals Technology Conference*, Sydney, paper 16–5.
Rollason, E.C. (1961) *Metallurgy for Engineers*, 3rd edn, Edward Arnold, London.
Rothwell, A.B., Dorling, D.V. and Glover, A.G. (1990) in *Advanced Joining Technology*, ed. T.H. North, Chapman & Hall, London.
Stuart, H. (1991) *Journal of Metals*, 35–40.
Viswanathan, R. and Jaffee, R.I. (1982) *Trans. ASME*, **104**, 220–226.
Wallner, F. (1986) *Metal Construction*, **18**, 28–33.

Further reading

General

American Welding Society (1982) *Welding Handbook*, 7th edn, Section 4. AWS, Miami; Macmillan, London.
American Welding Society, *Structural Welding Code* AWS D.1.1.
Hrivňák, I. (1992) *Theory of Weldability of Metals and Alloys*, Elsevier, Amsterdam.
Lancaster, J.F. (1997) *Handbook of Structural Welding*, Woodhead Publishing, Cambridge.
Lancaster, J.F. (1997) *Engineering Catastrophes*, Abington Publishing, Cambridge.
Linnert, G.E. (1965) *Welding Metallurgy*, Vol. 1, American Welding Society, Miami.
Linnert, G.E. (1967) *Welding Metallurgy*, Vol. 2, American Welding Society, Miami.
North, T.H. (ed.) (1990) *Advanced Joining Technologies*, Chapman & Hall, London.

Gas–metal and slag–metal reactions

The Welding Institute (1980) *Weld Pool Chemistry and Metallurgy*. TWI, Cambridge.

Solidification cracking

The Welding Institute (1977) *Solidification Cracking of Ferritic Steels During Submerged Arc Welding*, TWI, Cambridge.

Weld metal and heat-affected zone properties

Cottrell, C.L.M. (1984) *Metal Construction*, **16**, 740–744.
Dolby, R.E. (ed.) (1975) *The Toughness of Heat-affected Zones*, TWI, Cambridge.
Sekiguchi, H. (1976) *Fundamental Research on the Welding Heat-affected Zone of Steel*, Nikkau Kogyo Shimbun, Tokyo.
The Welding Institute (1983) *The Effects of Residual, Impurity and Micro-Alloying Elements on Weldability and Weld Properties*, TWI, Cambridge.

Weld-metal microstructure

Pargeter, R.J. and Dolby, R.E. (1984) *Identification and Quantitative Description of Ferritic Steel Weld Metal Microstructures*, International Institute of Welding document, no. IX-1323-84.

The International Institute of Welding (1985) *Compendium of Weld Metal Microstructures and Properties*, TWI, Cambridge.

Hydrogen embrittlement and cracking

British Standard BS 5135: 1984.
Coe, F.R. (1973) *Welding Steels Without Hydrogen Cracking*, TWI, Cambridge.
Suzuki, H. and Yurioka, N. (1982) *Prevention Against Cold Cracking by the Hydrogen Accumulation Parameter P_{HA}*, International Institute of Welding document, no. IX-1232-82.

Materials and weldability

The Welding Institute (1979) *Trends in Steel and Consumables for Welding*, TWI, Cambridge.
The Welding Institute (1979) *2nd International Conference on Pipewelding*, TWI, Cambridge.

The influence of second-phase non-metallic particles on weld properties

Dolby, R.E. (1983) *Metals Technology*, **10**, 349–362.

9
Austenitic and high-alloy steels

9.1 Scope

Austenitic steels considered in this chapter are the **austenitic chromium–nickel corrosion-resistant** of the 18Cr–(10–12)Ni (commonly known as 18/8) type, the **superaustenitic** which have augmented alloy content for higher corrosion resistance, and the **creep-resistant** and **scaling-resistant** containing up to a nominal 25 % Cr. Hardenable high-alloy steels are reviewed briefly; these are arbitrarily classified as those containing more than 20 % of alloying elements but that are capable of being hardened by heat treatment. Reference to austenitic–ferritic chromium–nickel steels will be found in Section 8.7.8 of the previous chapter.

The corrosion and oxidation resistance of these steels results from the formation of a self-healing surface film of chromium oxide. They may be welded by any of the major processes, but chief consideration will be given to the phenomena associated with fusion welding. In addition to metallurgical problems affecting welded joints in austenitic and high-alloy steels, austenitic–ferritic joints in welds and weld overlays will be considered. Corrosion and other problems that occur during service are discussed in Chapter 11.

9.2 The weld pool

9.2.1 Gas-metal and slag-metal reactions

The solubilities of oxygen and nitrogen in austenitic chromium–nickel steels are greater than in iron, largely because of the chromium content. It will be seen from Table 7.1 that the interaction parameter for chromium is numerically large for both nitrogen and oxygen. In gas tungsten arc welding the pattern of absorption of nitrogen is similar to that in iron as shown in Fig. 9.1: at low partial pressures of nitrogen in nitrogen–argon mixtures, the solubility more or less follows Sievert's law but at a higher level than for equilibrium conditions, while at higher partial pressures the net absorption becomes constant. The increased solubility of

Austenitic and high-alloy steels 311

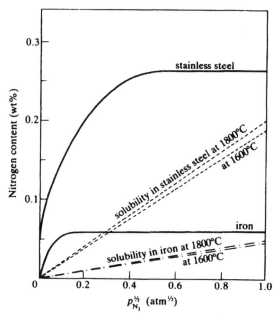

9.1 The nitrogen absorption in gas tungsten arc welding of austenitic chromium–nickel steel, as a function of square root of partial pressure of nitrogen in a nitrogen–argon atmosphere. Conditions: total pressure, 1 atm; 250 A; 20 cm min^{-1} (from Kuwana et al., 1984).

nitrogen in the presence of oxygen is shown for stainless steel in Fig. 7.8, while Fig. 7.9 illustrates how increasing current reduces the amount of nitrogen absorbed.

Submerged arc welding is employed for thick sections of austenitic stainless steel and for cladding. As with ferritic steel, the oxygen content of submerged arc weld metal falls as the basicity index of the flux increases (Fig. 9.2). Low oxygen content is desirable for good notch-ductility at very low temperature, but in other respects the oxygen content has little effect on the properties or microstructure of the weld metal.

The equilibrium solubility of hydrogen in liquid chromium–nickel steel under non-arc melting conditions is slightly higher than for liquid iron; for example, at 1600 °C it is 31.4 ml/100 g as compared with 26.3 ml/100 g for iron. Typical hydrogen contents for austenitic chromium–nickel steel weld metal produced by coated electrodes are compared with those of ferritic steel deposits in Table 9.1. As would be expected, the total hydrogen content is slightly higher in the austenitic welds. However, the diffusion coefficient for austenitic at room temperature is very much lower than that for ferritic steel, such that the amount of hydrogen that diffuses out of the specimen during a test period of 48 h (diffusible hydrogen) is very small. This fact has been used to explain the

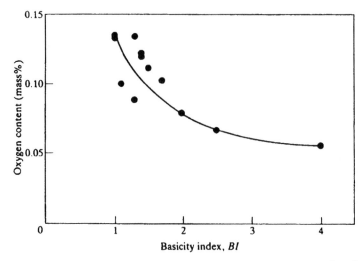

9.2 The oxygen content of stainless steel weld metal, as a function of the basicity index BI in submerged arc welding (from Ekstrom and Olsson, 1980).

Table 9.1. Representative weld-metal hydrogen levels for ferritic and austenitic electrodes and for nickel-base electrodes

Consumable type	Drying conditions	Hydrogen content (ml/100 g deposited metal)		
		Diffusible	Residual	Total
Basic-coated ferritic	150°C/2h	12	0.5	12.5
	450°C/1h	3.5	0.5	4.0
Rutile-coated 20/9/3	as received	0.5	12.5	13.0
	150°C/1h	0.5	11.0	11.5
	350°C/1h	0	5.5	5.5
	450°C/1h	0	4.0	4.0
Basic-coated 23/13	as received	nd*	nd	17.5
	130°C/1h	nd	nd	16
	350°C/1h	0	10	10
	450°C/1h	0	5	5
Rutile 29/9	as received	nd	nd	17
	130°C/1h	nd	nd	13
	350°C/1h	0	12	12
	450°C/1h	0	11	11
Ni-base (AWS E NiCrFe-3)	as received	1	23	24

*nd = not determined.
Source: from Gooch (1980).

successful use of austenitic electrodes for the welding of such materials as armour plate and for repair welding of crack-sensitive steel. Their use for such purposes is discussed later in this chapter.

It should be noted that the formulation of electrode coatings for austenitic stainless steel electrodes commonly differs from that of ferritic electrodes. Basic (AWS type 15), coatings contain carbonates and fluorides; rutile (also called lime-titania, AWS type 16) coatings contain basic components + rutile (TiO_2 mineral). There is not much difference between the oxygen contents of weld metal produced by these two types of electrode, but dissociation of the rutile may add a small amount of titanium to the weld deposit.

It should also be noted that the basic-coated 23/13 electrode listed in Table 9.1 generates a higher hydrogen content in the weld deposit than the rutile-coated 20/9/3. Where hydrogen content is critical, as in the repair welding of medium-carbon low-alloy steel, baking the electrodes at 450 °C may be necessary. With austenitic chromium–nickel steels, a basic-coated electrode is not necessarily a low-hydrogen electrode.

9.2.2 Weld pool shape: cast-to-cast variations in penetration

The problem of variable penetration in the gas tungsten arc welding of austenitic stainless steel first became evident in the 1970s. It affected, in particular, the longitudinal automatic welding of thin wall (up to 3 mm) type 304 and 316 austenitic stainless steels. Poor casts gave wide, shallow welds for which there was no practical margin between penetration and collapse of the weld pool. Similar problems have occurred with nickel-base alloys and ferritic steels, but the main area of concern is the automatic welding of austenitic chromium–nickel steels.

Most authorities agree that variable penetration is associated with flow in the weld pool, and that the dominant factor affecting such flow is surface tension. More specifically, a gradient of surface tension will generate a shear stress at the surface which is balanced by the viscous force associated with a velocity gradient in the liquid metal. This mechanism is discussed generally in Section 1.4.3, where it is noted that surface tension gradients may be set up by temperature gradients. In a pure metal, or in an alloy that is not contaminated by surface-active elements, the surface tension falls with increasing temperature, so that in a weld pool of the type produced in gas tungsten arc welding a force would act from the centre towards the weld pool boundary. Where surface-active agents are present, however, this situation may be reversed. Such agents reduce surface tension, but their concentration at the surface falls with increasing temperature, so that the gradient of surface tension with temperature may become positive. Under such circumstances the surface force, and the corresponding flow, will be inwardly directed. Inward flows, in general, are unstable, and tend to degenerate into a

314 Metallurgy of welding

spin, as in water draining from a bath. Some gas tungsten arc weld pools do, in fact, spin, but where surface-active elements are present the flow is generally inward, but sluggish, such that heat flow is primarily by conduction and the cross-section of the weld pool remains roughly semicircular. Outwardly directed flows, by contrast, are not inhibited in this way, and may be very fast.

The two surface-active agents that may be present in austenitic stainless steel in significant quantities are oxygen and sulphur. The relevant quantity is the amount of surfactant that is in the free, or uncombined state. So far as oxygen is concerned, this quantity is governed by the percentage of aluminium that remains after deoxidation. In the case of a liquid metal of the 18Cr–8Ni type, the amount of free oxygen is, approximately, given by

$$(O)_{free} = (O)_{total} - 0.9(Al)_{total} \qquad (9.1)$$

where all quantities are in parts per million (ppm) by mass.

Figure 9.3 shows the depth/width ratio of gas tungsten arc welds in stainless steel containing 150–200 ppm oxygen. To obtain D/W greater than 0.5 requires an aluminium content of less than 50 ppm. According to equation 9.1 this would leave a free oxygen content of 100–150 ppm, which would be sufficient to give a positive gradient of surface tension with temperature, and correspondingly an inward (convergent) flow. However, it is not practicable to place an upper limit of 50 ppm Al for austenitic stainless steel; a value of 100 ppm maximum is achievable but it is not low enough to obtain the desired control over flow.

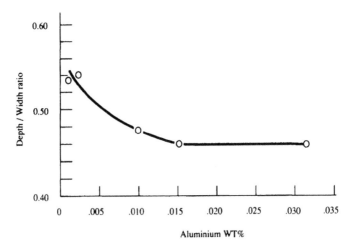

9.3 The effect of aluminium content of austenitic stainless steel on the depth/width ratio of gas tungsten arc welds (from Lancaster and Mills, 1991).

Austenitic and high-alloy steels 315

There has inevitably been some pressure to reduce the sulphur content of stainless steel, first to improve its hot working characteristics, and second to increase corrosion resistance. At the same time the steelmaking developments outlined in Chapter 8 have made it practicable to achieve very low sulphur contents even in structural steel. One means of sulphur control is by calcium treatment; unlike aluminium, calcium reacts strongly with sulphur, as do magnesium and the rare earth metals. A rule for calculating the free sulphur content is

$$(S)_{free} = (S)_{total} - 0.8Ca - 1.3Mg - 0.22(Ce + La) \qquad (9.2)$$

where, as before, Ca etc. represent ppm by mass. The amount of sulphur required to produce a positive gradient of surface tension with temperature is about 50 ppm (Fig. 9.4). Assuming an upper limit of 7 ppm for each of the elements on the right-hand side of equation 9.2, the required minimum sulphur content for a good D/W ratio is 70 ppm.

Based on such considerations, the International Institute of Welding has issued recommendations for control of minor element content in austenitic chromium–nickel steels. These recommendations, which refer to applications where variable penetration in gas tungsten arc welding may be a problem, are as follows:

1 Calcium-treated steels and those with rare-earth additions are not acceptable.
2 Minor element limitations shall be:

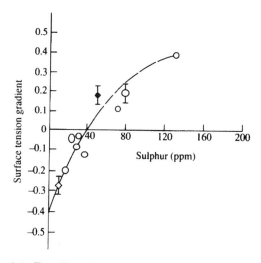

9.4 The effect of sulphur content on the gradient of surface tension with temperature (dγ/dT) in an austenitic stainless steel (from Lancaster and Mills, 1991).

sulphur: not less than 100 ppm (0.01 %);
aluminium: not greater than 100 ppm (0.01 %);
calcium, magnesium, cerium and lanthanum: not greater than 10 ppm (0.001 %) for each element.

In some cases, specifications place an upper limit of 150 ppm on the sulphur content. Where this is so it is recommended that the minimum sulphur level should be 70 ppm.

The document which incorporates these recommendations (Lancaster and Mills, 1991) provides a review of the variable penetration problem, and gives recommendations for methods of testing samples to assess weldability.

9.3 Alloy constitution

Sections of the ternary chromium–nickel–iron constitution diagram are reproduced in Fig. 9.5. Alloys that are rich in nickel solidify as austenite, and any ferrite that may be present forms in the interdendritic regions. With an intermediate range of compositions, dendrites solidify initially as ferrite, but transform by a peritectic reaction to austenite plus ferrite before final solidification (the peritectic reaction being similar to that which occurs in low-carbon steel, illustrated in Fig. 8.9). At low nickel contents, the dendrites again solidify initially as ferrite, and then transform to austenite plus ferrite in the solid state. These solidification modes are illustrated in Fig. 9.6.

The ferritic region below the solidus in Fig. 9.5 is an extension of the δ ferrite region of the iron–carbon diagram, so the ferritic phase in primarily austenitic steels is referred to as δ ferrite. From the equilibrium diagram, it would be expected that such ferrite would be enriched in Cr and lean in Ni; this is indeed the case and ferrite formers such as Mo also segregate preferentially to δ ferrite.

The low-temperature γ–α transformation does not necessarily occur on cooling to room temperature, so that the structure of quench-annealed 18Cr–10Ni steel consists typically of metastable (retained) austenite, sometimes with a little ferrite. Increasing the nickel content has the effect of reducing the transformation temperature of the austenite, while the presence of chromium makes the transformation sluggish. Thus the stability of the austenite is increased by raising the nickel content. With low nickel (about 4 %, for example) it is possible to obtain a martensite transformation by cooling to subzero temperature, and such compositions are used for hardenable high-alloy steel in machine and aircraft construction. When the nickel content is about 6 %, the γ–α transformation may only be achieved by cold-working the steel, and in annealed 18Cr–10Ni the austenite remains untransformed down to the lowest attainable temperature.

Austenitic and high-alloy steels 317

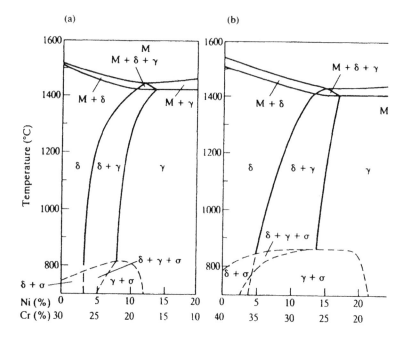

9.5 Sections of the ternary Cr–Ni–Fe constitution diagram: (a) 70% Fe; (b) 60% Fe (from Schafmeister and Ergang, 1939).

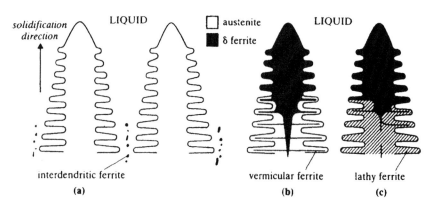

9.6 Solidification modes of austenitic chromium–nickel steels: (a) primary austenite with interdendritic ferrite; (b) peritectic reaction; (c) primary ferrite transforming to austenite plus ferrite below the solidus temperature (from Allen, 1983).

9.3.1 Weld-metal microstructure

The actual structure obtained in austenitic chromium–nickel steel weld metal varies with composition and cooling rate. For manual welding with coated electrodes, the differences in cooling rate may for this purpose be ignored, so that constitution depends primarily on composition. The various alloying elements used may be classified as either **austenite formers** or **ferrite formers** and, depending upon their balance, so will the structure be more or less austenitic. Chromium, molybdenum, silicon, niobium and aluminium are the common ferrite-forming elements, while nickel, carbon, nitrogen and manganese favour the formation of austenite. The combined effects of ferrite and austenite formers on the constitution of weld metal are summarized in Fig. 9.7, which is due to Schaeffler. This diagram has been widely used as a simple means of predicting weld-metal constitution.

The effect of austenite formers and ferrite formers is accounted for by numerical factors in the **chromium equivalent** (Cr_{eq}) and **nickel equivalent** (Ni_{eq}) as shown in Fig. 9.7. These are the Schaeffler equivalents, and are commonly used for predicting room-temperature ferrite content. In assessing solidification mode, the Hammar and Svenson equivalents may give better results:

$$Cr_{eq} = Cr + 1.37Mo + 1.5Si + 2Nb + 3Ti \qquad (9.3)$$

$$Ni_{eq} = Ni + 0.81Mn + 22C + 14.2N + Cu \qquad (9.4)$$

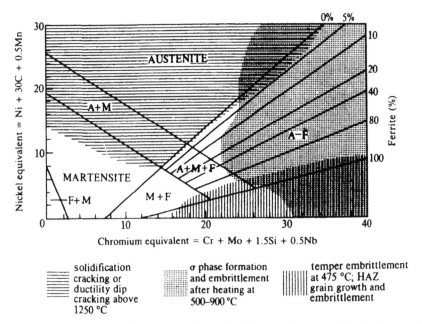

9.7 A constitution or Schaeffler diagram for Fe–Cr–Ni weld metal showing approximate regions in which typical defects may develop (courtesy Professor R. L. Apps).

Austenitic and high-alloy steels 319

In both sets of formulae the symbols Cr, etc., mean per cent by mass.

A slightly different diagram (the DeLong diagram) was developed by the US Welding Research Council (WRC) in order to account for the effect of nitrogen, which is a strong austenite-former, and to relate ferrite content as determined by metallographic analysis with that obtained by magnetic measurements made in accordance with the AWS standard A4.2-74. Figure 9.8 shows the revised diagram, which plots the **ferrite number** as a function of the chromium and nickel equivalents, the latter being modified to take account of nitrogen content. Where the weld metal has not been analysed for nitrogen, its content is taken to be 0.12% for self-shielded arc welding, 0.08% for normal gas metal arc welds, and 0.06% for welds made by other processes.

It is claimed that Fig. 9.8 gives better accuracy in determining a value (the ferrite number) that can be used in specifications designed to avoid solidification cracking problems. In practice, figures obtained from either diagram correlate broadly but not exactly with the results of magnetic and/or metallographic determinations.

9.3.2 Solidification cracking

The susceptibility of an austenitic chromium–nickel steel to solidification cracking depends on the solidification mode, the sulphur and phosphorus

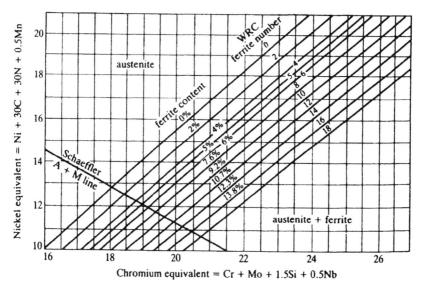

9.8 A revised (DeLong) constitution diagram for austenitic chromium–nickel steel weld metal.

contents, the manganese content and the presence or otherwise of rare-earth elements.

Solidification mode is the most important factor. When Cr_{eq}/Ni_{eq} is less than 1.48 using the Schaeffler formula (or 1.5 using Hammar and Svenson equivalents), solidification is austenitic and the susceptibility to cracking is high (Fig. 9.9). With Cr_{eq}/Ni_{eq} between 1.5 and 1.95, the solidification is ferritic–austenitic; that is to say, the primary dendrites are ferritic and these transform to austenite by the peritectic reaction during solidification. Ferritic solidification, when the austenite nucleates after solidification, occurs with $Cr_{eq}/Ni_{eq} > 1.95$. The ferritic–austenitic mode is the most crack-resistant. This appears to be mainly due to two factors: first, the solubility of S and P in ferrite is higher than it is in austenite, so that these elements are partly trapped in the solidified ferrite prior to transformation; and second, there are no γ–γ boundaries, which are particularly susceptible because they are wetted by sulphide or phosphide films. The presence of small amounts of interdendritic ferrite in austenitic solidification reduces the crack susceptibility to some degree.

Practical control of hot cracking in 18Cr–8Ni-type steels relies mainly on the ferrite content of weld metal at room temperature, and specifications commonly

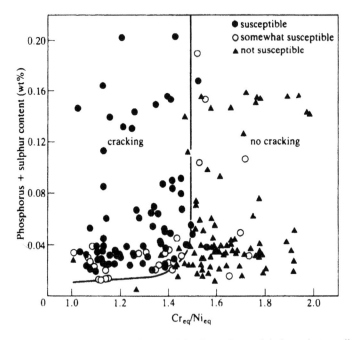

9.9 The susceptibility of austenitic chromium–nickel steels to solidification cracking, as a function of Schaeffler Cr_{eq}/Ni_{eq} and sulphur and phosphorus contents (from Kujampaa et al., 1980).

require the ferrite content to be maintained between 3 and 10%. The line for $Cr_{eq}/Ni_{eq} = 1.5$ crosses the Schaeffler diagram obliquely, indicating that at higher chromium equivalents the room-temperature ferrite content required to inhibit cracking should be greater. This requirement is not apparent in practice.

Fully austenitic (or nearly fully austenitic: 1–2% ferrite is sometimes permissible) grades of 18Cr–10Ni steel may be required for non-magnetic or extremely low-temperature applications, and a fully austenitic grade of either 18Cr–10Ni–2.5Mo or 25Cr–22Ni–2Mo composition is used for cladding urea reactors. Methods of avoiding solidification cracks in welding such material include reducing S and P, increasing manganese content and adding rare-earth elements. Figure 9.10 shows the crack length in a spot weld test for a low-carbon 25Cr–20Ni steel, and indicates that S + P should be maintained at very low levels, preferably below 0.002%. With normal commercial practice, impurity levels are $P + S \approx 0.02$–0.05%, but procedures are available to obtain substantially lower contents. Note, however, that the effect of sulphur is non-linear at high concentrations, and the welding of free-machining stainless steels containing 0.2–0.5% S is not a problem if the ferrite content is maintained within the normal 3–10% limits. Additions of Ta, Nb, Zr, Mo and Mn may give improved cracking resistance but may also have side effects. Mn up to 7% has been used, but may reduce corrosion resistance. Rare-earth metal and lanthanum additions are very effective by combining preferentially with S and P and raising the solidification temperature of sulphides and phosphides. According to Matsuda the required addition of La is

$$La = 4.5P + 8.7S \tag{9.5}$$

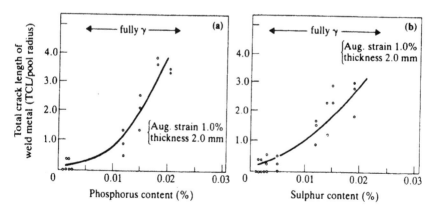

9.10 Crack length in a Transvarestraint test using a spot gas tungsten arc weld as a function of: (a) phosphorus in a 0.003S–25Cr–20Ni steel; (b) sulphur in a 0.002P–25Cr–20Ni steel (from Ogawa et al., 1982).

Using a suitable combination of such expedients, the welding of fully austenitic chromium–nickel steels is quite practicable.

Silicon and carbon affect cracking, but are mainly significant in relation to heat-resisting compositions, which normally contain substantial amounts of both elements. Increasing silicon content promotes cracking, whereas increasing carbon content reduces crack sensitivity. The combined effect of these two elements on a 15Cr–35Ni weld deposit with low sulphur and phosphorus contents is illustrated in Fig. 9.11. For this and other fully austenitic weld deposits, the carbon content should exceed $0.22\sqrt{Si}$ to keep cracking to acceptable levels. This requirement cannot be met by low-carbon deposits, which therefore contain 4–6 % Mn and 0.1–0.15 % N to improve cracking resistance. High-carbon weld metal has low ductility because of the presence of substantial amounts of

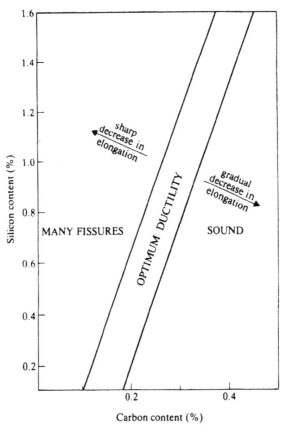

9.11 The effect of carbon and silicon on microfissuring and ductility of 15Cr–35Ni-type weld metal (from Rozet et al., 1948; courtesy of the American Welding Society).

Austenitic and high-alloy steels

primary carbide in the structure, while low-carbon metal has low ductility because of microfissuring. Hence there is a carbon:silicon ratio for optimum ductility. Fissuring due to silicon may be a serious problem in welding heat-resistant castings, where silicon is added to promote castability or, in the wrought 15Cr–35Ni–2Si alloy, where the silicon is added to improve scaling resistance. In both cases, low-silicon welding materials are used and the welding technique is adjusted to minimize dilution of weld metal by the parent metal.

Although the addition of carbon is beneficial in reducing the risk of hot cracking in 25Cr–20Ni welds, it reduces ductility when added to the parent metal, and may cause further embrittlement in service at elevated temperature due to carbide precipitation.

9.3.3 Hot cracking in the heat-affected zone during welding

Cracks sometimes appear in the parent metal close to the weld boundary immediately after welding. This type of defect is very much less common than weld-metal cracking, and is most troublesome under conditions of severe restraint, or when welding relatively thick sections (over about 18 mm) of certain

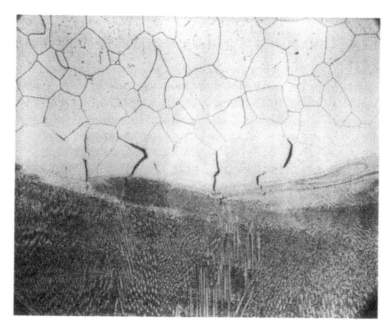

9.12 Hot cracking in the heat-affected zone of an austenitic chromium–nickel steel weld. This type of crack occurs during welding (×200) (photograph courtesy of TWI).

grades of stainless steel, particularly the fully austenitic 18Cr–13Ni–1Nb type. Cracking is intergranular and frequently starts at the plate surface immediately adjacent to the weld boundary and propagates inwards, either at right angles to the surface or following the weld boundary profile (Fig. 9.12).

Various tests have been used to evaluate the susceptibility of a material to heat-affected zone cracking, notably tests that simulate the welding thermal cycle and measure the temperature, during either heating or cooling of the specimen, at which the ductility falls to zero (the **nil-ductility test**). There is a general correlation between the temperature range below the solidus over which ductility is zero or small and the cracking tendency: the greater the nil-ductility range, the greater the tendency to cracking adjacent to the weld. Indications from this type of test are that niobium, zirconium and boron are damaging addition elements.

The mechanism of this type of cracking is not well understood. Cracking starts in high-temperature regions, where there is micrographic evidence of the liquation of low-melting constituents, but may propagate through regions of substantially lower temperature, where there is no liquation. Nor can direct evidence of low-melting constituents be found along the crack boundaries. However, the correlation between nil-ductility range and cracking tendency does suggest that a grain-boundary segregation or liquation mechanism is involved.

9.3.4 Embrittlement and cracking due to liquid metals

Austenitic Cr–Ni steels are susceptible to embrittlement and cracking by a number of liquid metals, but those most important in metal joining are zinc, copper and brass. The last-named will crack stainless steel if it is brazed using common brazing metal, but brazing at low temperatures using silver brazing alloys is satisfactory. Zinc cracking occurs if stainless steel is welded to galvanized carbon steel using an austenitic filler material. Such welds may be severely cracked. This and other cases of cracking by liquid zinc are clearly avoidable but nevertheless happen from time to time.

9.4 Mechanical properties

Austenitic stainless-steel weld metal is inherently tough and ductile, and its properties at room temperature are relatively insensitive to variations in welding procedure. It has a higher ratio of proof stress to ultimate stress than the equivalent wrought material, so that failure in a transverse tensile test usually occurs outside the weld metal. However, austenitic alloys are used for applications operating at very low temperatures and also in the creep temperature

Austenitic and high-alloy steels

range, and in such cases the weld-metal properties may vary significantly with changes in composition and microstructure.

9.4.1 Subzero temperature

Austenitic stainless steel is used for equipment handling liquid gases, and is required to have good notch-ductility at the operating temperature, typically down to $-196\,°C$. A particularly exacting application is for superconducting magnets, where additionally the magnetic permeability must be maintained at the lowest practicable level, and where a fully austenitic weld metal is required.

Factors affecting the low-temperature toughness are the content of carbon, chromium, nickel, oxygen and ferrite, and, for welding with coated electrodes, the type of coating.

Carbon acts by strengthening the matrix and by increasing the content of second-phase particles. The degree of carbide precipitation is affected in turn by the thermal history of the weld and by the activity of carbide formers.

An increase in carbon content affects the microvoid coalescence mode of fracture and decreases the fracture energy. The Charpy impact energy at $-196\,°C$ is decreased by about 2.4 J for each 0.01 % C, and for preference carbon is kept below 0.04 %.

Chromium is the major carbide-forming element, and for this or other reasons has a negative effect, reducing the fracture toughness at $-196\,°C$ by about 1.2 J for each 1 % Cr. Nickel, however, increases the impact energy. Molybdenum, being a ferrite former, probably has an adverse effect.

Increasing oxygen and ferrite contents normally decrease the low-temperature toughness, and, for example, in submerged arc type 308 L weld metal, the Charpy V impacts at $+20\,°C$ and $-196\,°C$ were found to be (Ekstrom and Olsson, 1980)

$$C_{v+20} = 1491.1 - 386.2 O_2 - 3.3 FN \qquad (9.6)$$

$$C_{v-196} = 95.1 - 310.7 O_2 - 1.5 FN \qquad (9.7)$$

where O_2 means mass per cent of oxygen and FN is the ferrite number. It will be evident that oxygen has a much greater effect than ferrite content.

Figure 9.13 shows the Charpy impact energy at $-196\,°C$ as a function of oxygen for submerged arc welds (ignoring ferrite), basic-coated and rutile-coated electrodes. The results for coated electrodes (316LN) are for a 0.03C–(16–18)Cr–13Ni–2.2Mo–0.15N (316LN) deposit while the submerged arc composition was 0.02C–20Cr–10Ni (308L), the latter being more favourable for low-temperature notch-ductility. The lower impact figures for rutile-coated rods are associated with higher titanium and niobium contents in the weld metal.

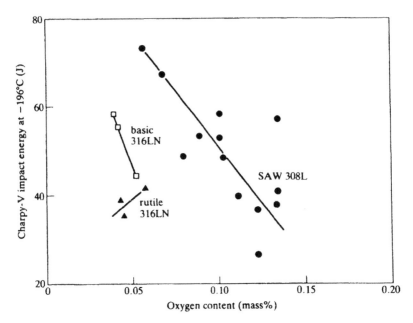

9.13 The Charpy V impact energy at −196°C for 308L submerged arc weld metal (Ekstrom and Olsson, 1980) and 316LN welds made with basic-coated and rutile-coated electrodes (Marshall and Farrar, 1984).

These elements are reduced from the rutile and are thought to form embrittling carbides and/or nitrides. (See also Fig. 9.17 which shows the effect of O_2/CO_2 in the shielding gas on low-temperature notch-ductility.)

9.4.2 Elevated temperature

The elevated-temperature tensile properties of austenitic Cr–Ni weld metal are slightly higher than or equal to those of the equivalent plate at temperatures below the creep range. The creep and rupture properties, however, may fall below the median line for wrought material, at least for short-time exposure. Weld metal is subject to various phase transformations at elevated temperature, so that the relative performance of weld and plate may change with time; this is illustrated for type 316 weld and base metal in Fig. 9.14. Carbon increases the rupture strength substantially. The ASTM 'H' grades of 18Cr–10Ni types have a specified carbon range of 0.04–0.10 %, while the heat-resistant casting HK 40, much used in petrochemical plant, has a carbon content of 0.4 %. Alloying elements that improve rupture strength include molybdenum, tungsten, titanium and niobium. Niobium is, however, associated with low rupture ductility and

Austenitic and high-alloy steels

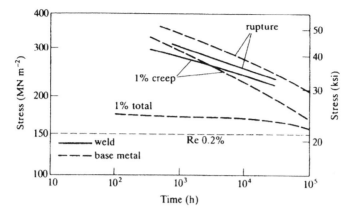

9.14 Stress for rupture and 1% creep for type 316 weld metal compared with those for wrought material of a similar composition (from Archer *et al.*, 1980).

with stress relief cracking. Types 304H and 316H are often selected for elevated-temperature applications.

Austenitic Cr–Ni weld metal deposited using basic-coated electrodes has a significantly lower rupture stress than that from rutile-coated types. This applies to E308, E316 and E310 compositions, and has been shown to be due to the same factor that causes embrittlement of rutile deposits in low-temperature testing: namely, reduction of niobium and titanium from the rutile. At elevated temperatures, niobium has the dominant effect. Figure 9.15 shows the time to rupture at 650 °C for E308 H weld deposit containing various amounts of niobium.

Ferrite content appears to have little effect on the rupture strength of weld metal, but the rupture ductility is lower with high (20%) ferrite content. Thus there is little penalty for maintaining a 3–10% ferrite range for elevated-temperature applications.

9.5 Transformation, embrittlement and cracking

Austenitic chromium–nickel steel weld metal may suffer a degree of embrittlement due to postweld heating, for example during stress-relief heat treatment. Such embrittlement is caused by carbide precipitation and by the transformation of ferrite to σ and χ phases. The degree of embrittlement so caused is variable, depending on the carbon, ferrite, chromium and molybdenum contents of the steel. Carbide precipitation may give rise to creep embrittlement and reheat cracking, and may promote intergranular corrosion (see Chapter 11). Austenitic stainless steel is subject to hydrogen embrittlement if the hydrogen content is

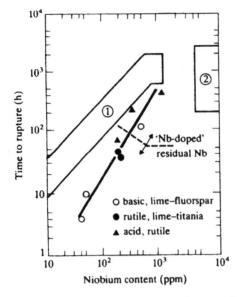

9.15 The effect of residual niobium on the stress rupture life of type E308L weld deposits at 650 °C and 200 MN m^{-2} stress. Results obtained by Oak Ridge National Laboratory on plate material are shown for comparison, Block 1: ORNL commercial and experimental 304, heats at 650 °C and 172 N mm^{-2}. Block 2: ORNL 347, typical values at 650 °C and 172 N mm^{-2} (from Marshall and Farrar, 1984).

high enough, and this has contributed to the disbonding of overlay cladding in hydrocracker reactors. It may also be subject to sulphide–phosphide precipitation on cooling through the overheating temperature range, resulting in elevated-temperature embrittlement.

9.5.1 Carbide precipitation

Figure 9.16 shows the time–temperature relationship for intergranular precipitation of carbides in austenitic chromium–nickel steels and weld metal as a function of carbon content. Precipitation occurs between 425 and 800 °C. Under welding conditions it is most rapid at about 650 °C and occurs preferentially at grain boundaries, within the ferrite phase (if present) or along slip planes in cold-worked material. The carbide so formed is normally chromium carbide, $Cr_{23}C_6$, and it may be redissolved by heating at between 1000 °C and 1100 °C for a short period. Quenching in water or, for thin sections, rapid cooling in air, will retain carbides in solution (**solution treatment**), provided that the carbon content does not exceed about 0.12%. Note that in ferritic chromium steels this heat

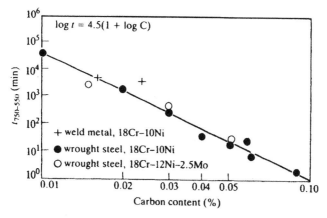

9.16 Sensitization time for austenitic chromium–nickel steels as a function of carbon content; t is the time (min) to cool from 750 to 550 °C (from Polgary, 1976).

treatment cycle results in carbide precipitation; this is because of the lower solubility of carbon in the ferritic alloy.

In fusion welding or brazing 18Cr–(10–12)Ni austenitic steel, there is a region of the heat-affected zone that is heated within the **carbide precipitation temperature range** and in which **intergranular carbide precipitation** will occur. Such precipitation is damaging to the corrosion resistance (see Section 11.4) and various means are used to circumvent it. The most common method is to add to the steel either niobium or titanium, both of which have a greater affinity for carbon than chromium and form carbides that do not go into solution at the normal hot working and annealing temperature of the steel. Thus the percentage of carbon in solution in a niobium- or titanium-bearing steel is normally extremely low and, when it is reheated within the precipitation temperature range, chromium carbide is not precipitated. **Niobium-** and **titanium-stabilized** 18/8 steels are resistant to intergranular corrosion in the heat-affected zone of welds except in special circumstances, which will be discussed later.

An alternative method of avoiding carbide precipitation is to reduce the carbon content of the steel to a low level, ideally below 0.03%. Such **extra-low-carbon** (ELC) steel is not subject to carbide precipitation during welding and is used to an increasing extent.

Electrodes and filler rod for welding stabilized austenitic steel must also be stabilized, since in multi-run welds the weld metal is heated within the precipitation range by subsequent runs. In coated-electrode welding and in other processes where there is metal transfer through the arc, titanium cannot be used as a stabilizing element, since too high a proportion is lost by oxidation. Niobium, on the other hand, transfers satisfactorily and is employed to stabilize coated electrodes and wire for metal inert-gas and submerged arc welding. The oxidation losses in oxyacetylene and tungsten inert-gas welding are small, and for these processes titanium-stabilized filler wire may be applied. For ELC steel either an extra-low-carbon or a stabilized electrode is used.

It is possible to remove the damaging effect of intergranular precipitation by heat treatment of the welded joint. Two alternative treatments have been used: either **carbide solution treatment**, which requires heating at between 1000 and 1100 °C for a short period followed by rapid cooling, or a **stabilizing anneal**, which consists of heating at 870–900 °C for 2 h. The stabilizing heat treatment may complete precipitation, remove microstresses and/or diffuse chromium into depleted regions. There are conflicting reports as to the effectiveness of a stabilizing anneal. In cither case it is necessary to heat the entire component, since local heat treatment develops a zone of carbide precipitation just outside the heated band. Generally it is more practicable and economic to use stabilized or ELC steel rather than apply postwelding heat treatment to an unstabilized steel.

Stabilization with niobium and titanium is not completely effective under severely corrosive conditions. If a stabilized steel is heated to very high temperature (1100 °C and over for Ti-bearing, and 1300 °C and over for Nb-bearing steel), some of the titanium or niobium carbide goes into solution. If it is subsequently cooled slowly or held within the carbide precipitation range, a steel so treated will suffer intergranular precipitation of chromium carbide. Now the parent metal close to a fusion weld boundary is heated through the temperature range in which titanium or niobium carbides are dissolved, while subsequent weld runs may heat all or part of the same region in the carbide precipitation range. Hence there may be a moderate intergranular chromium carbide precipitation close to the weld boundary even in stabilized steels, and this can be damaging if the joint is exposed to attack by (for example) a hot mineral acid. The degree of precipitation, and hence the risk of corrosion, is diminished by reducing the carbon content of the steel. See Section 11.4.2 for a discussion of the effect of intergranular carbide precipitation on corrosion resistance.

9.5.2 Phase transformations and segregation

It is sometimes necessary to stress-relieve heavy sections of austenitic stainless steel after welding, and a typical heat treatment requires holding the component at 850 °C for several hours. Operating temperatures may be up to, say, 750 °C for 18Cr–10Ni types and up to 1000 °C for heat-resisting grades. At such elevated temperatures, and particularly in the 600–900 °C range, transformations of ferrite to σ and χ phases may occur, together with precipitations of $M_{23}C_6$ carbides.

The composition of σ and χ phases in a type 316 weld metal after heat treatment at 850 °C, as determined by energy-dispersive X-ray (EDX) analysis,

Table 9.2. Composition of transformed phase in type 316 stainless-steel weld

Phase	Composition (mass %)			
	Cr	Ni	Mo	Fe
Carbide	65	1	5	25
σ	28	4	8	58
χ	23	4	13	57

Source: from Slattery *et al.* (1983).

is shown in Table 9.2. Formation of σ and χ phases is promoted by high chromium and molybdenum contents, and in type 316 weld metal the ferrite is enriched in these elements, as shown in Table 9.3 for two welds, one with 6% and the other 18% ferrite. Segregation of molybdenum is particularly high, resulting in the formation of substantial amounts of σ and χ in the higher ferrite weld metal after 850°C treatment.

25Cr–20Ni (type 310) and 23Cr–13Ni (type 309) heat-resisting steels may suffer a substantial loss of ductility after an 850°C stress-relief heat treatment or during service at similar temperatures. 25Cr–20Ni is usually fully austenitic, and embrittlement is due to carbide precipitation plus a γ–σ transformation. The second alloy has an austenite–ferrite structure, and in this case the ferrite transforms to σ.

Figure 9.17 shows the effect of stress relief at 850°C on the Charpy V impact strength at −196°C for three weld metals: E309, an extra-low-carbon fully austenitic steel approximating to the E347 composition and E310. As would be

Table 9.3. Composition of austenite and ferrite in type 316 stainless steel in the as-welded condition

Weld metal ferrite number	Element	Bulk analysis	Austenite	Ferrite	Change in composition (%)	
					Austenite	Ferrite
FN6	Cr	18.0	18	23	0	+27
	Ni	8.9	9	5	+6	−46
	Mo	1.85	2	4	−3	+82
	Fe	68.5	68	66	−1	−4
FN18	Cr	20.5	21	26	0	+27
	Ni	10.1	11	6	+5	−44
	Mo	3.1	3	5	−14	+56
	Fe	64.8	64	62	0	−4

Source: from Slattery *et al.* (1983).

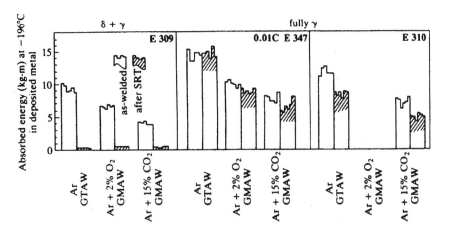

9.17 Charpy V impact test results at −196 °C for as-welded and stress-relief treated (SRT) weld metals (after Ogawa et al., 1982).

expected, the ferritic–austenitic steel is the most severely affected, followed by type 310, while the low-carbon fully austenitic type 347 is little changed. In all cases the low-temperature notch-ductility is reduced when O_2 or CO_2 is added to the shielding gas; this is due to an increase in the oxygen content from 0.01 % for argon shielding to 0.04 % for $Ar + 2\% O_2$ and 0.06 % for $Ar + 15\% CO_2$. The results shown in Fig. 9.17 are for heating at 850 °C for 2 h followed by a slow cool; after long-term exposure at this temperature, the type 310 weld metal would be more severely affected.

It will be evident that austenitic chromium–nickel steel weld metal is prone to microsegregation, firstly as a result of dendrite formation, secondly because of the partition of solute elements between ferrite and austenite, and thirdly due to the precipitation of carbides and intermetallic constituents. If this effect is severe enough, it may denude local regions of chromium and molybdenum to the extent that their corrosion resistance is significantly reduced. Leaving aside the question of carbide precipitation, which will be described in Chapter 11, segregation may reduce the passivity of weld metal to the extent that it becomes anodic to the surrounding area, thus causing preferential corrosion of the weld. Where such attack is possible, and for alloy compositions up to the molybdenum-bearing type 316, the effect of segregation may be compensated by a marginal increase in chromium and molybdenum contents, and some manufacturers of coated electrodes routinely make such adjustments.

The segregation problem is much more severe in the case of the austenitic super-alloys. These steels are intended to resist severe corrosive conditions where chloride ions are also present, and they typically contain about 6 % Mo. It is not practicable to compensate adequately for segregations with an iron-based alloy, so nickel-based alloys are used for welding consumables. Table 9.4 compares the

Table 9.4. Microsegregation of gas tungsten arc welds in superaustenitic chromium–nickel stainless steel

Filler	Element	Bulk	Composition (% by mass)	
			Dendrite core	Interdendritic
None	Mo	6.1	4.2	6.9
	Cr	20.2	19.3	20.5
	Ni	17.6	17.6	19.2
ERNiCrMo-3	Mo	7.6	5.8	10.5
	Cr	21.4	20.7	20.9
	Ni	47.2	47.7	43.6
ERNiCrMo-4	Mo	11.3	9.8	14.4
	Cr	17.3	17.2	18.1
	Ni	40.7	43.1	39.2

Source: Gooch (1996).

bulk alloy composition with that of dendrite core and interdendrite regions for weld deposits made in a superaustenitic steel without filler wire and with two nickel–base compositions, ER NiCrMo-3 and ER NiCrMo-4. The nickel–base alloys retain an adequate alloy content even in the dendrite core. There is evidence to show that intermetallic phases precipitate in the iron-based deposits during welding, and this is further disincentive to their use.

9.5.3 Reheat cracking

As well as being susceptible to heat-affected zone cracking during fusion welding, the 18C–13Ni–1Nb type steel may also crack adjacent to the weld during postweld heat treatment or during service at elevated temperature. Generally speaking, such behaviour has been shown by welds in thick material (over 20 mm) and when the service or stress-relieving temperature has exceeded 500 °C. The presence of notches produced by undercutting or other defects close to the weld boundary increases the risk of cracking. Cracks are intergranular and run through the heat-affected zone close to the weld boundary (Fig. 9.18).

It is possible that intergranular embrittlement may play some part in the reheat cracking of austenitic Cr–Ni steel, but the most important factor is strain-induced carbide precipitation. Fine carbides precipitate within the grains (**intragranular precipitation**) causing the grain to be harder and more resistant to creep relaxation than the grain boundaries. Thus any relaxation of strain due to welding or to externally applied loads is concentrated in the grain boundary regions and, if the strain so generated exceeds the rupture ductility, cracks may be initiated. Figure 9.18 shows a case where reheat cracking has been initiated

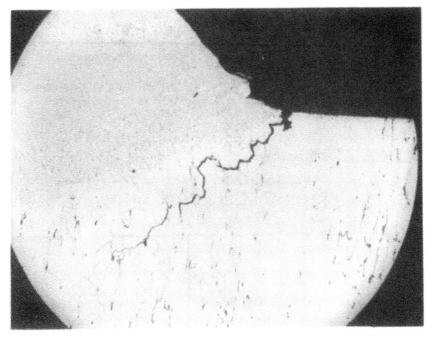

9.18 Reheat cracking of a niobium-stabilized stainless steel (×50, reduced by one-third in reproduction) (photograph courtesy of TWI).

by a small hot tear located at the toe of the weld, and one way of reducing the cracking risk is to grind the toes of all welds prior to heat treatment or service. However, a more effective solution (as with ferritic alloy steels) is to select a non-sensitive alloy.

Of the commonly available 18Cr–8Ni-type parent materials, the most crack-sensitive is the niobium-stabilized grade, followed by titanium-stabilized and unstabilized types. The molybdenum-bearing 18Cr–10Ni–2.5Mo grade has little, if any, tendency to crack in the heat-affected zone. The intragranular precipitate does not form in the straight molybdenum-bearing alloy. However, if niobium is present in a molybdenum-bearing steel, intragranular precipitation and reheat cracking are possible, the risk increasing with the amount of niobium. Therefore, when molybdenum-bearing steels are used for duties where reheat cracking is possible, the niobium and titanium contents of the alloy should be controlled.

9.5.4 Overheating and ductility-dip cracking

The problem of embrittlement due to intergranular sulphide and phosphide precipitation in the coarse-grain region of the heat-affected zone has already been discussed in relation to ferritic alloy steel in Chapter 8. The solubility of both these elements, and particularly that of sulphur, falls with decreasing temperature below the solidus, and this may result in intergranular precipitation and embrittlement. In austenitic Cr–Ni steels, this is manifested in ductility-dip cracking, which is a somewhat infrequent mode of failure in multipass 25Cr–20Ni fully austenitic weld deposits. Cracking is intergranular and is found in the earlier passes of the deposit. The cracks result from a combination of tensile stress and heating (due to later passes) in a low-ductility temperature range. Once ductility-dip cracking has initiated, it will usually propagate through successive passes but not through the final pass, so that it may not be detectable by normal means of inspection.

9.5.5 Hydrogen embrittlement

The possibility of hydrogen embrittlement and cracking of austenitic chromium–nickel steels has become evident in recent years from experience with hydrocracker reactors, where type 347 cladding and welds are exposed to hydrogen at pressures of 140 to 240 atm and temperatures up to 460 °C. The specific problems associated with such conditions are discussed in Chapter 11, but the main variables affecting hydrogen embrittlement are indicated in Fig.

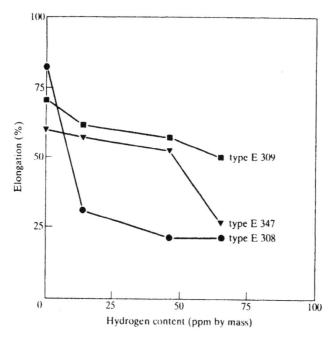

9.19 Tensile elongation of austenitic chromium–nickel steel weld metal as a function of hydrogen content (after Mima *et al.*, 1976).

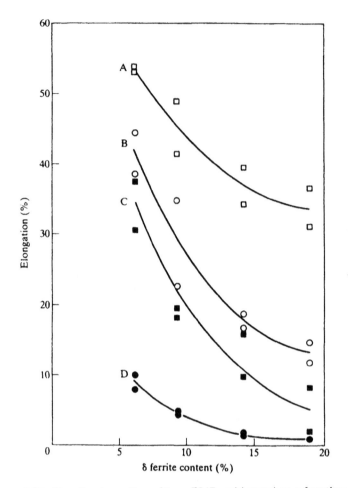

9.20 Tensile elongation of type E347 weld metal as a function of ferrite content in the hydrogen-free and hydrogen-charged conditions: A, as-welded, hydrogen-free; B, as-welded, hydrogen-treated; C, after postweld heat treatment at 750 °C for 70 h, hydrogen-free; D, after postweld heat treatment at 750 °C for 70 h, hydrogen-treated (after Kobe Steel Ltd, undated).

9.19 and 9.20. In some cases (e.g. the E308 weld metal in Fig. 9.19), embrittlement can be quite significant even with the level of hydrogen content obtained in weld metal deposited by rutile-coated electrodes. As a rule, however, serious embrittlement requires a combination of high hydrogen content (30–500 ppm) and the presence of substantial amounts of ferrite and/or σ phase. The unfavourable effect of postweld heat treatment temperatures of 700 °C and above will be evident from Fig. 9.19; unfortunately, this is the temperature range for postweld heat treatment of low-alloy steel pressure vessels.

9.6 The use of austenitic Cr–Ni alloys for repair welding, cladding and transition joints

One factor that is common to repair welding and to weld-deposit clad steel is the behaviour of hydrogen at an austenitic–ferritic junction, so it will be convenient to review this problem here.

The solubility of hydrogen in any particular alloy depends on factors such as thermal history and degree of plastic deformation, but reasonable values of the solubility of hydrogen and representative expressions for the diffusion coefficient may be calculated from the following.

1. Iron, ferritic

$$s_f = 47.7 p^{1/2} e^{-3272/T} \qquad (9.8)$$

2. Iron, austenitic

$$s_a = 7.5 p^{1/2} e^{-705/T} \qquad (9.9)$$

3. 2.25Cr–1Mo ferritic alloy steel

$$s_f = 21.4 p^{1/2} e^{-3275/T} \qquad (9.10)$$

$$D_f = 2.4 \times 10^{-7} e^{-2150/T} \qquad (9.11)$$

4. Types 347 and 309 austenitic steel

$$s_a = 7.1 p^{1/2} e^{-520/T} \qquad (9.12)$$

$$D_a = 1.6 \times 10^{-7} e^{-5130/T} \qquad (9.13)$$

In these formulae p is in atm, T is absolute temperature in K, s is in ppm and D is in m^2 s^{-1}. (The equations have the following sources: 9.8 Geller and Sun (1950); 9.9 Lothian (1974); and 9.10–9.13, Japan Pressure Vessel Research Council (undated).)

After the weld metal has solidified and starts cooling, there is a time (of the order of seconds) during which both weld metal and parent metal are austenitic, the diffusivity of hydrogen is relatively high, and hydrogen may penetrate the parent metal for a very short distance. With a ferritic weld metal the hydrogen spreads progressively into the parent metal on cooling to room temperature. With an austenitic weld metal, two factors influence the distribution: first, the solubility of hydrogen is greater in austenite (4000 times at 20 °C) than in ferrite, and second the diffusivity in austenite is lower (by a factor of 2.6×10^{-5} at 20 °C) than in ferrite. Then at the interface

$$\frac{c_a}{c_f} = \frac{s_a}{s_f} \qquad (9.14)$$

$$D_f \frac{\partial c_f}{\partial x} = D_a \frac{\partial c_a}{\partial x} \qquad (9.15)$$

where c_a and c_f are the hydrogen concentrations at the interface in austenite and ferrite respectively, s_a and s_f are the solubilities given by equations 9.8 to 9.10 and 9.12, and x is distance at right angles to the interface. The resulting distribution is shown (in part diagrammatically) in Fig. 9.21. Any hydrogen that has diffused across the interface into the ferritic material will be removed by

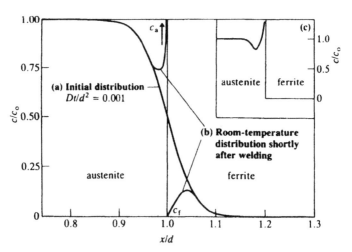

9.21 The hydrogen distribution across the interface of an austenite–ferrite weld: (a) during cooling above the γ–α transition; (b) after a short period at room temperature; (c) after a long period at room temperature. Distribution (a) has been calculated for the end of cooling through the austenite range, estimating that $D_{mean} = 3 \times 10^{-9}$ m^2 s^{-1}, $d = 3 \times 10^{-3}$, $t = 3$ s, giving $Dt/d^2 = 0.001$. The other two distributions are diagrammatic.

Austenitic and high-alloy steels 339

diffusion back into the austenite and outwards into the bulk of the parent metal. The eventual distribution after a substantial period of time is shown in Fig. 9.21(c). The hydrogen content of the ferritic material is maintained at a low level primarily because of the high ratio of solubility, s_a/s_f. This rather unexpected type of distribution has been calculated and confirmed by measurement of the analogous case of austenitic-clad plate charged with hydrogen from the clad side only at elevated temperature and then cooled to room temperature (Fig. 9.22).

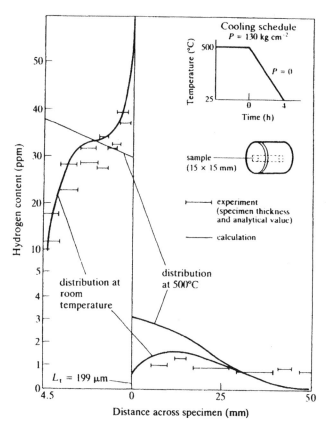

9.22 The hydrogen distribution across 50 mm 2.25Cr–1Mo steel clad with 4.5 mm type 347 austenitic Cr–Ni weld deposit, after exposing the clad surface to hydrogen at about 130 atm and then cooling at 500 °C h^{-1} with both surfaces exposed to air at atmospheric pressure. The ratio s_a/s_f increases exponentially as the temperature falls. L_t is the thickness of the transition layer between 2.25Cr–1Mo and type 347 stainless steel.

9.6.1 Repair welding

Problems may arise in the repair welding of machinery and parts due to high carbon content and high impurity levels, and such conditions demand the use of a weld metal that can accommodate dilution without solidification cracking and has a low hydrogen potential. The compositions frequently used for this purpose are 20Cr–9Ni–3Mo, 23Cr–12Ni (type 309), 22Cr–12Ni–3Mo and 29Cr–9Ni. The first-named alloy has a long history of use for welding hardenable alloy steel but does not have as much tolerance to dilution as the other types. Figure 9.23 shows dilution lines marked on the Schaeffler diagram. The upper line joins the 20Cr–9Ni–3Mo composition to 2.25Cr–1Mo; between 25% and 40% dilution the structure is predicted to be fully austenitic and therefore prone to solidification cracking. The lower line is for a type 309 weld and 0.3% C low-alloy parent metal. This also indicates a need to maintain the dilution below 24%, but in practice E309 electrodes have a greater dilution capacity than the 20Cr–9Ni–3Mo type. The 24Cr–9Ni is the most tolerant of dilution and is widely used for repair work, but undiluted it has a ferrite content of 40%, which may be undesirable.

The presence of hydrogen in the parent metal adjacent to the fusion boundary immediately after welding means that there is a risk of hydrogen cracking in sensitive materials. The fusion boundary itself may have a thickness of up to 100 μm, with a gradation of alloy content such that in part it is martensitic, and this may be subject to cracking. The 0.3–0.4% C low-alloy steels are intolerant of hydrogen and for such materials it may be necessary to preheat (say to 150 °C) and control the hydrogen content of the electrodes.

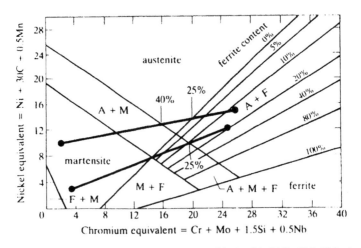

9.23 Dilution lines for 2.25Cr–1Mo welded with 20Cr–9Ni–3Mo (upper) and 0.3% C low-alloy steel welded with 23Cr–12Ni (lower) (from Gooch, 1980).

Austenitic and high-alloy steels

Wherever practicable, it is better to make repairs using weld metal of matching composition. However, there are cases where the required precautions, such as application of preheat, are undesirable or impossible, and where the use of austenitic electrodes is a useful alternative.

9.6.2 Cladding

In pressure vessel technology the cladding of steel with a corrosion-resistant overlay has a number of advantages. It is economic, since, except for thin sections, clad steel is cheaper than solid stainless steel. It minimizes the risk of catastrophic failure due, for example, to stress-corrosion cracking or hydrogen embrittlement and cracking of the austenitic material, while for high-pressure conditions, such as nuclear reactor vessels, hydrocrackers and coal conversion vessels, it provides the optimum combination of mechanical properties with corrosion resistance.

One method of applying a protective layer to a metal surface is by **thermal spraying**. In this process fine liquid drops of a suitable alloy are projected on to the surface, where they coalesce. Thermal spraying is not considered in this book, and interested readers should consult specialized publications such as the book listed in the further reading section at the end of this chapter.

Cladding techniques

Four different processes may be employed for cladding carbon or low-alloy steel with austenitic stainless steels; manual welding with coated electrodes; submerged arc welding with either a rod or strip electrode; and electroslag welding with a strip electrode. The strip is typically 0.4 mm thick and 25–75 mm wide. Figure 9.24 illustrates the principle of the submerged arc and electroslag cladding methods.

For the submerged arc process, the flux is similar to that used in submerged arc welding, and for cladding low-alloy steel a basic flux containing carbonates may

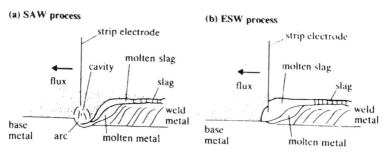

9.24 The strip cladding technique: (a) submerged arc welding; (b) electroslag welding.

be used to minimize hydrogen absorption. For electroslag cladding, however, gas-generating agents must be absent from the flux (otherwise an arc is generated) and it must be electrically conductive. Hydrogen absorption is not a problem with the electroslag process.

For both variants, magnetic control is usually required to improve bead profile and prevent undercut. Without such control, the electromagnetic force induced by the welding current acts inwards towards the centre of the weld pool, causing an inwardly directed flow and generating undercut. Permanent magnets or electromagnets are used to reverse the flow, thereby increasing the width of the bead and giving more uniform penetration.

Cladding with rod electrodes is traditional in the USA, but in Japan and Europe strip electrodes are most commonly used. The electroslag process has the advantage of reduced dilution as compared with submerged arc (typically 10% as compared with 20%). Thus it may be possible to obtain an acceptable cladding composition with one layer using electroslag, whereas with submerged arc two layers are usually required. On the other hand, the electroslag process generates a coarse-grained heat-affected zone, such that the sensitivity to **underclad cracking** and **hydrogen-induced disbonding** is increased. The automatic processes are used for cladding large areas but in pressure vessels there are certain places (e.g. around nozzles) that are not accessible to machine welding, and here the manual process with coated electrodes is employed.

Cracking during the cladding operation

Solidification cracking is usually avoided by ensuring that the microstructure of the deposit is austenite plus ferrite with 3–10% ferrite. In using the submerged arc process, dilution must be taken into account, and it is common practice to use type E309 (23Cr–12Ni) consumables for the first layer, since this composition has good resistance to the effects of dilution. For applications where a fully austenitic E316 type deposit is required to withstand the corrosive environment, the composition must be controlled in the manner discussed in Section 9.3.2.

Hydrogen-induced cold cracking in the boundary zone or heat-affected zone is avoided by controlling the moisture content of the flux and/or by using a carbonate-containing flux.

The mechanism of underclad cracking (UCC) is similar to that of stress-relief (reheat) cracking, but it occurs during welding. Location of the cracks is shown in Fig. 9.25. Successive weld passes reheat part of the coarse-grained region in the heat-affected zone of the previous pass, and at the same time this region is subject to a tensile strain in the welding direction. The risk of underclad cracking may be reduced by decreasing the heat input rate, and thereby reducing the grain size in the coarse-grained region. As a rule, however, there is not much scope for modifying the welding variables, and the susceptibility of the base material must

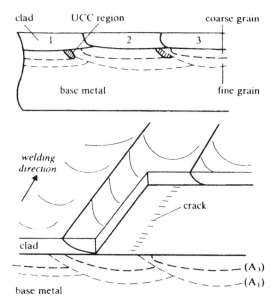

9.25 The location of underclad cracks (from Kobe Steel Ltd, undated).

be reduced either by selecting a less sensitive composition or by control of trace impurity elements.

Embrittlement caused by postweld heat treatment

The impact properties and ductility of the weld metal are reduced by postweld heat treatment, and this embrittlement increases with temperature (within the specified range) and time. In the fabrication of heavy pressure vessels the temperature is at the high end of the range (e.g. 680–720 °C) and the heating time 20–50 h. Figure 9.26 shows the results of bend tests on type E347 cladding as a function of the calculated ferrite content and the Larson-Miller parameter P. This figure combines temperature with time of exposure. It is commonly used as the independent variable in recording the results of creep tests, but it is used here as a measure of the combined effects of time and temperature during heat treatment.

Embrittlement of weld metal is due to a combination of σ phase formation and carbide precipitation, and is minimized by the use of low-carbon material and by keeping the ferrite content at the low end of the range.

During postweld heat treatment carbon migrates from the base metal into the overlay material to a depth of 20–50 µm, while in the same region the chromium and nickel content is reduced (Fig. 9.27). In the carbide migration zone, there is heavy carbide precipitation and the hardness is in the region of 350 VPN, while

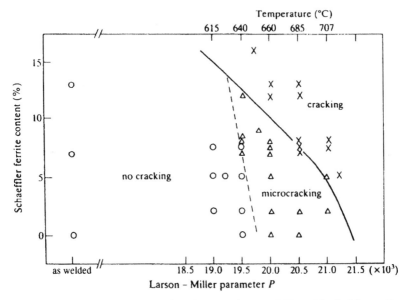

9.26 Cracking in a bend test of type 347 weld cladding after 25 h postweld heat treatment, as a function of ferrite content and Larson–Miller parameter, $P = T(K)[20 + \log t \text{ (h)}]$ (from Kobe Steel Ltd, undated).

on the ferritic steel side of the weld boundary the hardness is reduced. Beyond the carburized zone in the cladding there is a fully austenitic region with intergranular carbide precipitation, which grades into the normal austenite–ferrite structure. When the hardness of the carburized zone exceeds about 400 VPN, microcracks appear in the bend test as indicated in Fig. 9.26. This zone may be subject to hydrogen-induced cracking in service, as discussed in Chapter 11.

9.6.3 Transition joints

In large power stations, both ferritic and austenitic steels may be used for main steam piping and in superheaters, and there is a need for joints between these materials. Such joints may be made as separate elements welded to the ferritic steel (usually 2.25Cr–1Mo) on one side and the austenitic steel (usually type 316 H) on the other. Alternatively, a joint may be made directly by buttering the surface of the weld preparation on the ferritic side with an austenitic or nickel-base deposit, and then completing the joint with an austenitic or nickel-base filler.

Transition joint elements may be produced with a graded composition by vacuum arc remelting or electroslag melting, the ingots so obtained being punched and forged into tubular form. The intention of such a procedure is to avoid the sharp gradient of composition and properties inherent in a directly

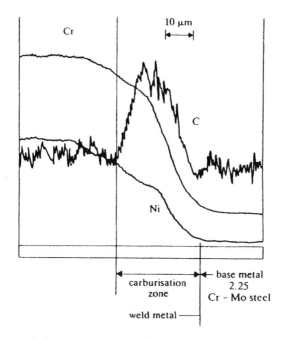

9.27 Element profiles close to the interface between type 347 weld overlay and 2.25Cr–1Mo base metal after postweld heat treatment (from Kobe Steel Ltd, undated).

welded ferritic–austenitic joint: for example, in an electroslag transition piece, the transition from ferritic to austenitic structure occurs over a distance of 0.5 mm whereas in a weld overlay the transition zone is 0.03–0.07 mm.

In main steam lines, the wall thickness of the pipe may be up to 85 mm and postweld heat treatment may be obligatory. Using austenitic chromium–nickel filler, the transition region is subject to carbon migration during such heat treatment. Nickel-base filler alloys, however, do not suffer carbon migration and are often preferred for this reason. Also, the differential of the coefficient of thermal expansion between weld metal and ferritic parent metal is much smaller with nickel-base than with austenitic fillers, so that the residual stress produced by postweld heat treatment is lower.

Metallurgical changes occur in the weld boundary zone of transition joints during service, and these are discussed in Chapter 11.

9.7 Corrosion-resistant steels: alloys and welding procedures

The more common grades of corrosion-resistant austenitic Cr–Ni steels are listed in Table 9.5. The AISI designations used here are internationally accepted, and

Table 9.5. Corrosion-resistant austenitic Cr–Ni steels

AISI no.	Composition (%)						Electrode type*
	C, max.	Cr	Ni	Mo	Ti	Nb	
304	0.08	18–20	8–11	–	–	–	E308
304L	0.035	18–20	8–13	–	–	–	E308L
304H	0.04–0.10	18–20	8–11	–	–	–	E308
316	0.08	16–18	11–14	2–3	–	–	E316 or E16–8†
316L	0.035	16–18	10–15	2–3	–	–	E316L
316H	0.04–0.10	16–38	11–14	2–3	–	–	E316
321	0.08	17–20	9–13	–	5 × C, 0.07	–	E347
347	0.08	17–20	9–13	–	–	10 × C, 0.10	E347
309	0.15	22–24	12–15	–	–	–	E309
310	0.15	24–26	19–22	–	–	–	E310

*To AWS A5.4–69.
† For high-temperature creep-resisting applications.

the group of steels shown are known as the **300 Series**. The corrosion problems mentioned below in connection with individual grades of steel are reviewed in Chapter 11.

Type 304 is normally specified in US practice for mildly corrosive conditions such as weak organic or inorganic acids, and it is welded with a matching filler, E308. Some carbide precipitation occurs in the heat-affected zone but intergranular corrosion will not take place except in thick plate (where the heat-affected zone is reheated by successive weld runs) and/or strong corrodents. It is not easy to determine the breakpoint between intergranular corrosion and no intergranular corrosion, however, and in European practice it is more common to specify stabilized steel, usually type 321, for all corrosive duties except those requiring more highly alloyed steel. Type 347 steel may be required for the containment of strong mineral acids where type 321 might be subject to knife-line attack. An alternative solution to such problems is to use 304L welded with 304L filler. Joints so made are not subject to any form of intergranular corrosion, and the absence of niobium decreases the risk of liquation cracking. A potential disadvantage of ELC steels is that carbon may be picked up during steel fabrication: for example, if a tube is annealed without thorough removal of drawing components it may be internally carburized. Check analysis of weldments may be desirable for critical applications. A second disadvantage is that ELC grades have low tensile strength, particularly at elevated temperature. Where a positive assurance of elevated-temperature properties is required (generally for temperatures over 400 °C), the H grade, which has a specified minimum carbon content, may be used.

The addition of molybdenum improves the corrosion resistance of austenitic Cr–Ni steel to almost all media except strong nitric acid. A titanium-stabilized type 321 steel with carbon content limited to 0.06 % would be preferable for both strong and weak nitric acid. The carbon content of Type 316 steel is also as a rule restricted to about 0.06 % maximum, and for most applications this is adequate. In more severe exposure, as in urea plants, the ELC grade with 0.03 % carbon maximum is specified.

Molybdenum has the capability of strengthening the protective oxide film, and of healing weak points where these occur. Nitrogen has a similar effect. Both these elements improve the resistance of stainless steels to pitting corrosion, and are beneficial in applications where chloride ions, which may penetrate the protective layer and initiate pits, are present. Such considerations have led to the development of the austenite superalloys, referred to earlier. Typically these contain 6 % Mo and 0.2 % N, as will be seen from Table 9.6.

For any given corrodent, the resistance to pitting depends upon composition, and may be expressed in terms of the **pitting resistance equivalent** or **pitting index**. One version of this quantity is given by Gooch (1996):

$$\text{Pitting resistance equivalent} = Cr + 3.3Mo + 16N \qquad (9.16)$$

where Cr, etc. are percent by mass of the element concerned. This figure is used for comparing different compositions, but is not used for quality control – in specifications, for example.

Amongst other applications, superaustenitic stainless steels are used for drums and pipework in the bleaching section of paper mills. In this section the wood fibres are treated with sodium hypochlorite solution, which would cause severe pitting and crevice corrosion of 300 series stainless steels. As already indicated,

Table 9.6. Typical compositions of proprietary superaustenitic corrosion-resistant steels and welding filler alloys

	Designation	Composition (%)							
		C	Cr	Ni	Mo	W	Mn	N	Nb
Base metals	25 6 Mo	0.02	20	25	6	–	–	0.2	–
	254S Mo	0.02	20	18	6	–	–	0.2	–
	AL 6XN	0.03	21	24	6	–	–	0.2	–
	654S Mo	0.02	22	22	7.3	–	3.5	0.5	–
Filler metals	ER NiCrMo–2	0.02	24	60	9	–	0.2	–	3.5
	ER NiCrMo–3	0.02	16	Bal	16	3.5	0.2	–	–
	ER NiCrMo–10	0.01	21	Bal	14	3.2	0.2	–	–

the filler material is a nickel-base alloy, either of proprietary composition, or to AWS ER NiCrMo–3.

Type 316H steel may be specified for heavy sections (say over 25 mm) as a precaution against reheat cracking. Type 316 was used successfully for the main steam lines in supercritical power boilers, the welds being made with 16Cr–8Ni–2Mo coated electrodes.

Austenitic Cr–Ni steels may be welded by all welding processes except some specialized types such as thermit welding. Preheat and postweld heat treatment are not normally required. Indeed, preheat may promote solidification cracking, and for the same reason it is advisable to keep the interpass temperature below 200 °C. Electrodes are usually basic-coated but lime-titania coatings are also used, and give a more uniform weld profile.

9.8 Heat-resisting steels: alloys and welding procedures

Table 9.7 lists a selection of alloys used for creep and oxidation resistance at temperatures up to 1000 °C. In addition, type 316 steel (Table 9.5) has been used for creep-resistant duties in the 400–600 °C temperature range. At higher temperatures type 316 may, under stagnant conditions, be subject to rapid oxidation (**catastrophic oxidation**), and the 25 Cr–20 Ni steel is more generally useful. Type 310 steel has relatively low strength and is mainly employed for tubing or for sheet-metal liners inside piping or vessels that are internally insulated. For high-temperature applications (800–1000 °C) and especially where there is a significant operating stress, the higher-carbon grades such as HK40 are employed. These steels are welded with matching electrodes or by automatic gas tungsten arc welding. HK40 weld deposits made with basic-coated electrodes have about half the rupture strength of the parent metal, and welds made by this process may suffer creep cracking in service. Gas metal arc

Table 9.7. Heat-resistant austenitic Cr–Ni alloys

Designation	Chemical composition (%)						Remarks
	C	Cr	Ni	Al	Ti	Nb	
HK40	0.35–0.45	23–27	19–22	–	–	–	Centrifugally cast
HU50	0.40–0.60	17–21	37–41	–	–	–	Centrifugally cast
Incoloy 800H	0.05–0.10	21	32.5	0.4	0.4	–	Extruded
Incoloy 802	0.4	21	32.5	0.6	0.8	–	Extruded
24 24 1.5	0.3	24	24	–	–	1.5	Centrifugally cast
50/50 Nb	0.05	48.5	50	–	–	1.5	Centrifugally cast

and gas tungsten arc deposits are much closer to the parent metal in creep strength, and coated electrode deposits may be improved by using high-purity material in combination with a lime-titania coating.

Centrifugally cast HK40 is widely used for tubes in steam–methane reformer furnaces and ethylene pyrolysis furnaces, for which it is the most economical material. Incoloy 800 is used in similar furnaces where greater ductility is required: matching electrodes cannot be used because of their susceptibility to solidification cracking, and nickel-base alloys are employed. Some of these alloys have a rupture strength equal to that of the parent material. Incoloy 802 is an extruded high-carbon Incoloy 800, and has a better resistance to carburization (which is a problem in ethylene furnaces) than HK40. The 802 alloy and the 24Cr–24Ni–1.5Nb alloy are more costly than HK40 and their use is correspondingly limited. The 24C–24Ni–1.5Nb alloy has improved rupture ductility, and welds with matching electrodes have a rupture strength equal to that of the parent metal. The 50Cr–50Ni–1.5Nb alloy is used for resistance to fuel-ash corrosion (elevated-temperature attack by deposits containing sodium and vanadium). It is welded by the gas tungsten arc and manual metal arc processes using matching electrodes. The higher-carbon heat-resistant alloys do not have much ductility at room temperature and, in making procedure tests, the bend test is omitted.

9.9 Hardenable high-alloy steels

The hardenable stainless steels, which are used for rocket parts and for corrosion-resistant pumps and other machinery, may be divided into three groups:

1   single-treatment martensitic age-hardening (17–4 PH);
2   double-treatment austenitic-martensitic age-hardening (17–7 PH and FV 520);
3   austenitic age-hardening.

The metallurgy of steels in group 1 is similar to that of 18% Ni maraging steels (see Section 8.7.4), and in particular the properties of welded joints may be largely restored by ageing after welding. The 17–4 PH type, which contains 17Cr–4Ni–4Cu and 0.3Nb, may, however, suffer cracking along the weld boundary during hardening when the restraint is severe, and under such conditions it is best to weld in the over-aged condition and fully heat-treat after welding. 17–4 PH castings are hot-short (i.e. brittle at elevated temperature), and crack in the heat-affected zone if the copper content is too high; a maximum of 3% copper is necessary for welding.

The steels in the second group are substantially austenitic after solution treatment at 1050 °C and cooling to room temperature. A small amount of δ

ferrite is normally present and sometimes some martensite, but the ductility is high so that normal forming and pressing operations are possible. Subsequent heating conditions the austenitic matrix, so that on cooling and holding at ambient or subzero temperature it transforms to martensite. There are two generally used transformation treatments: the first is to heat at 750 °C for 90 min, air cool to 20 °C and hold for 30 min; the second requires similar heating at 750 °C or alternatively short-time heating at 950 °C, followed by cooling to −75 °C and holding for 8 h to complete the martensite transformation. Ageing occurs in the temperature range 500–570 °C. Of this group, 17-7 PH type of steel (17Cr–7Ni–1.2Al) is usually welded in sheet form using the DC tungsten inert-gas process. It is not subject to cracking either in the weld or heat-affected zone. Various combinations of welding with heat treatment are possible but, in order to obtain 95–100 % joint efficiency, it is necessary to weld with the parent material in the solution-treated condition, and subsequently carry out one of the two transformation and ageing treatments.

Type 17–10 P is a fully austenitic chromium–nickel steel containing phosphorus, and hardens to a moderate degree owing to carbide and phosphide precipitation when heated at 700 °C. This alloy is used where a combination of non-magnetic properties and hardenability is required. High-phosphorus austenitic steel is extremely hot-short, so that fusion welding with a matching filler alloy is impossible. Crack-free weld metal may be obtained by using coated electrodes giving a 29Cr–9Ni deposit. Such deposits are partially ferritic and not hardenable, and the full mechanical properties are not obtainable in welded joints. Moreover, hot cracking may occur in the heat-affected zone, so that fusion welding of this alloy is difficult. Flash-butt welding, however, gives satisfactory joints.

None of the hardenable high-alloy steels discussed in this section requires any preheat before welding.

References

Allen, D.J. (1983) *Metals Technology*, **10**, 24.
Archer, J., Berge, Ph. and Weisz, M. (1980) Materials for the LMFBR, in *Alloys in the Eighties*, Climax Molybdenum Corp. Greenwich, Conn.
Bennett, W.S. and Mills, G.S. (1974) *Welding Journal*, **53**, 548–5.
Ekstrom, U. and Olsson, K. (1980) in *Weld Pool Chemistry and Metallurgy*, TWI, Cambridge.
Geller, W. and Sun, T. (1950) *Archiv für das Eisenhüttenwesen*, **21**, 437.
Gooch, T.G. (1980) *Metal Construction*, **12**, 622–631.
Gooch, T.G. (1996) *Welding Journal*, **75**, 135-s–154-s.
Japan Pressure Vessel Research Council (undated) *Hydrogen Embrittlement of Bond Structure Between Stainless Steel Overlay Weld and Base Metals*, Japan Pressure Vessel Research Council Report, no. MHE–10.

Kobe Steel Ltd (undated) *Overlay Welding with Strip Electrodes*, Kobe Steel Welding Technical Report, no. 539.
Kujampaa, V.P., Takalo, T.K. and Moisio, T.J.I. (1980) *Metal Construction*, **12**, 282–285.
Kuwana, T., Kokawa, H. and Naitoh, K. (1984) *Quarterly Journal of the Japanese Welding Society*, **2**, 669–675.
Lancaster, J.F. and Mills, K.C, (1991) *Recommendations for the Avoidance of Variable Penetration in Gas Tungsten Arc Welding*, IIW Document 212–796–91.
Lothian, M.R. (1974) in *Hydrogen in Metals*, American Society for Metals, Metals Park, Ohio.
Marshall, A.W. and Farrar, J.C.M. (1984) *Metal Constructions*, **16**, 347–353.
Mima, S., Watanabe, J., Nakayo, Y., Murakami, Y. and McGuire, W.J. (1976) *International Joint Pressure Vessels and Piping and Petroleum Mechanical Engineering Conference*, Mexico City, preprint PV 76-5-61.
Ogawa, T., Nakamura, H. and Tsunetomi, E. (1982) *Toughness at Cryogenic Temperature and Hot Cracking in Austenitic Stainless Steel Weld Metals*, International Institute of Welding document no. II–C–677–82.
Polgary, S. (1976) *Metal Construction*, **8**, 445.
Rozet, D., Campbell, H.C. and Thomas, R.D. (1948) *Welding Journal*, **27**, 484.
Schafmeister, P. and Ergang, R. (1939) *Archiv für das Eisenhüttenwesen*, **12**, 459–464.
Slattery, G.F., Keown, S.R. and Lambert, M.E. (1983) *Metals Technology*, **10**, 373–385.

Further reading

General

American Welding Society (1982) *Welding Handbook*, 7th edn, Section 4, AWS, Miami; Macmillan, London.
Folkhard, E. (1984) *Metallurgie der Schweissung nichtrostender Stähle*, Springer-Verlag, Vienna, New York.
Marshall, P. (1984) *Austenitic Stainless Steels*, Applied Science Publishers, London.
The Welding Institute (1979) *Trends in Steel and Consumables for Welding*, TWI, Cambridge.

Weld pool shape

Anon. *Thermal Spraying, Practice, Theory and Application*, American Welding Society, Miami.
Lancaster, J.F. and Mills, K.C. (1991) *Recommendations for the Avoidance of Variable Penetration in Gas Tungsten Arc Welding*; International Institute of Welding document no. 212–796–91.

Solidification cracking

Ogawa, T., Nakamura, H. and Tsunetomi, E. (1982) *Toughness at Cryogenic Temperature and Hot Cracking in Austenitic Stainless Steel Weld Metals*, IIW document, II–C–677–82.

Phase transformations

Gooch, T.C. (1996) loc. cit.
Slattery, G.F., Keown, S.R. and Lambert, M.E. (1983) *Metals Technology*, **10**, 373–385.

Austenitic–ferritic joints

American Welding Society (1982) *Proceedings of Conference on Joining Dissimilar Metals*, Pittsburgh, AWS/EPRT, Miami.

Repair welding

Gooch, T.G. (1980) *Metal Construction*, **12**, 622–631.

Cladding

Kobe Steel Ltd (undated) *Overlay Welding with Strip Electrodes*, Kobe Steel Welding Technical Report, no. 539.

Transition joints

Chilton, I.J., Price, A.T. and Wilshire, B. (1984) *Metals Technology*, **11**, 383–391.
Nicholson, R.D. (1985) *Materials, Science and Technology*, **1**, 227–233.
Yapp, D. and Bennett, A.P. (1972) in *Welding Research Related to Power Plant*, eds N.F. Eaton and L.M. Wyatt, Central Electricity Generating Board, London.

10
Non-ferrous metals

10.1 Aluminium and its alloys

Pure aluminium, aluminium–manganese and aluminium–magnesium alloys may be joined by most fusion welding processes, but the use of such processes for some of the higher-strength alloys is limited by a susceptibility to solidification cracking and by a reduction of tensile strength in the heat-affected zone. Porosity may also be a problem in shielded metal arc and gas metal arc welding.

A new process, friction stir welding (described in Chapter 2) is capable of making butt welds in sheet, plate and sections of aluminium and its alloys. Friction stir welding is a solid-phase process, and is not subject to the defects associated with fusion welds. Its potential value in the fabrication of aluminium alloys is considerable, but has yet to be fully evaluated.

10.1.1 Gas metal reactions

Hydrogen dissolves endothermically in liquid aluminium, but oxygen and nitrogen both combine with the metal. Aluminium oxide is one of the most stable of such compounds, and the concentration of free oxygen in the weld pool is likely to be very low. That of nitrogen is considered below.

Nitrogen

Figure 10.1 shows the nitrogen content of gas metal arc welds made with pure aluminium electrode wire in nitrogen–argon mixtures. Whereas with gas metal arc welding using a steel wire the nitrogen content of the weld metal increases with nitrogen partial pressure up to a saturation level, there is no clear evidence for such a limiting value for aluminium. As with steel, however, at low nitrogen partial pressure the nitrogen content of the weld decreases with increasing current. The relative porosity of these welds is shown in Fig. 10.2, from which it will be seen that small amounts of nitrogen in the arc atmosphere generate porosity at 300 A but not at 170 A or 240 A. Thus nitrogen contamination could result in porosity in high-current gas metal arc welding.

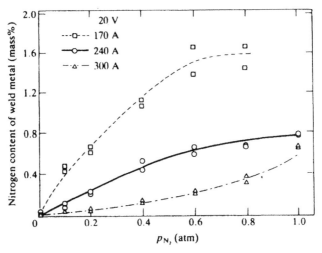

10.1 The effect of nitrogen partial pressure in nitrogen–argon mixtures of total pressure 1 atm on the nitrogen content of gas metal arc welds in pure aluminium (from Kobayashi *et al.*, 1970b).

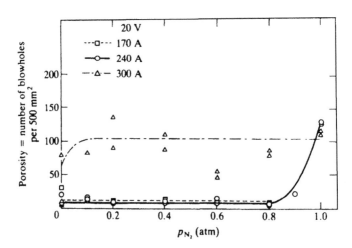

10.2 The effect of nitrogen partial pressure in nitrogen–argon mixtures of total pressure 1 atm on the porosity of gas metal arc welds in pure aluminium (from Kobayashi *et al.*, 1970b).

Oxygen

The dissociation pressure of Al_2O_3 increases with temperature, so that the solubility of oxygen in liquid aluminium would be expected likewise to increase with temperature. If this is the case, then particles of alumina will form in the weld pool on cooling and act as nuclei for gas pores. Such nuclei have indeed been found in hydrogen pores.

Prior to 1898, when the first effective gas welding flux was developed, the refractory oxide film that forms on aluminium prevented fusion welding. Fluxes are used for gas welding, brazing and soldering, and are typically a mixture of halide salts. They act by penetrating between the oxide and the liquid metal surface, and then dispersing and partially dissolving the film. In soldering a **reaction flux** containing stannous chloride or zinc chloride may be used; these salts are reduced by aluminium to tin and zinc, respectively, to provide all or part of the solder. An extended discussion of gas welding fluxes will be found in the book by West (1951).

A basically similar formulation is used for aluminium electrode coatings. The residue from such fluxes is corrosive to aluminium and they must be removed by water washing and, in extreme cases, by acid-pickling.

Fluxes are not required for the inert-gas-shielded processes. When operated with electrode positive, a non-thermionic cathode forms on the weld surface and this disperses surface oxides. In gas metal arc welding thin material, it may be desirable to add oxygen to the argon shield in order to limit movement of the cathode spots, but normally this addition is not required. Gas tungsten arc welding with electrode negative is possible using a helium shield. For top quality, the butting edges are cleaned by scraping, scratch-brushing or pickling prior to welding, but commercially acceptable welds have been made without these precautions. The surface of such welds is oxidized, but this oxide layer may be removed by scratch-brushing.

Hydrogen

The equilibrium solubility of hydrogen in liquid aluminium is

$$s = 625.2 p^{1/2} e^{-6355/T} \tag{10.1}$$

where s is in ml/100 g, p is partial pressure in atmospheres and T is absolute temperature. It may be calculated from equation 10.1 that the equilibrium solubility at 1 atm hydrogen pressure at 1300 °C is 11.0 ml/100 g, and at 660 °C it is 0.69 ml/100 g.

Figure 10.3 shows the effect of hydrogen partial pressure in a hydrogen–argon mixture on the porosity of non-arc-melted pure aluminium. These tests were made by levitation melting a sample of aluminium in the appropriate gas mixture and then dropping it into a copper mould so as to obtain a cooling rate similar to that of weld metal. In pure hydrogen the temperature of the molten sample was 1100 °C instead of 1230 °C for argon–hydrogen because of the higher thermal conductivity of hydrogen. The equilibrium hydrogen solubility at 1100 °C is 6.1 ml/100 g, as calculated from equation 10.1, while the solidified metal had a porosity of 30 %, corresponding to a gas content at NTP of 4.7 ml/100 g. Thus

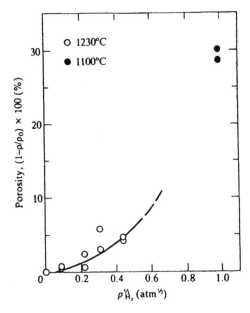

10.3 The effect of hydrogen partial pressure in hydrogen–argon mixtures on porosity of pure aluminium melted under non-arc (levitation) conditions and quenched at a rate equivalent to that experienced in arc welding (from Uda and Ohno, 1974).

the porosity in this case can be attributed to the fall in solubility of hydrogen from the liquid/metal temperature to the solidification temperature.

In arc welding, conditions are different in that, first, the solubility may be increased by some unknown factor, and second the hydrogen partial pressure may be quite low. In welding with coated electrodes, the hydrogen partial pressure is probably in the range 0.05–0.25 atm, but with gas metal arc welding it is not known. Most authors assume that there will be a significant amount of hydrogen dissolved in both cases and attribute porosity to the high ratio of solubility at the arc root temperature to that at the melting point. Thus, considering equilibrium conditions, if the hydrogen partial pressure at 1300 °C is p atm and the melt then cools to 660 °C, the pressure required to keep the gas in solution is now $(11/0.69)^2 p = 254p$ atm. To put this another way, if bubbles form when the amount of gas in solution is equal to the equilibrium solubility at 1 atm hydrogen pressure, then they will form provided that the hydrogen partial pressure at 1300 °C is higher than $1/254 = 3.9 \times 10^{-3}$ atm.

Hydrogen sources

There are many potential sources of hydrogen in aluminium welding. The first is the oxide film, which is hydrated to some degree, and on which an adsorbed film of moisture is invariably present when the metal is exposed to the atmosphere. In gas tungsten arc welding, evaporation of the oxide also evaporates any adsorbed or combined moisture that may be present, thus providing a mild source of contamination of the arc column. The effect of surface is more serious in gas metal arc welding, where a fine wire with high surface-to-volume ratio is fed continuously into the arc; for instance, lubricant residues on cold-drawn wire are a source of hydrogen. Cleanliness of the wire is therefore of prime importance in the gas metal arc process. Coated electrodes contain moisture in the coating, and under some conditions corrosion of the core wire underneath the coating may occur. The degree of contamination and, correspondingly, the weld porosity may be minimized by using electrodes as soon as possible after manufacture, by storing in a dry place, and by drying before welding. Regardless of the welding process, the material to be welded must be clean and dry. So also must be the welding equipment; condensation in gas leads, torch or welding gun of inert-gas processes may cause severe porosity. Porosity in the form of large discontinuous cavities or long continuous holes may occur from the use of excessive currents in gas metal arc welding with projected transfer. This defect (**tunnelling**) is due to high current causing turbulence in the weld pool. It may be overcome by limiting the current per pass, by using a gravitational transfer technique or by improving the inert-gas shielding. The risk of both hydrogen porosity and tunnelling in high-current welding may be reduced by operating with larger-diameter wires. The current density at the transition from large droplet to projected transfer is reduced with increasing diameter, from 150 A mm^{-2} for the smaller diameters to 20 A mm^{-2} for a 5.6 mm diameter electrode. Shaved wires, from which the surface layers have been physically removed, also reduce the incidence of porosity.

The main effect of porosity is to reduce the strength of the weld metal, and in designing welded aluminium structures or pressure vessels a joint efficiency factor corresponding to the anticipated weld quality must be used. In fully radiographed shop-fabricated work, it is practicable to set up a radiographic standard for porosity corresponding to a joint efficiency of 95 %.

Radiographically sound welds can be made in aluminium alloys by means of the inert-gas shielded processes, provided that adequate attention is paid to cleanliness of plate and welding materials, and that the work is done in a fabrication shop. In **site welds** some degree of porosity is to be expected because the necessary precautions are difficult to implement fully. Sound positional welds may, however, be made in piping where a backing bar is used.

Friction stir welds do not suffer from porosity, or from any other form of cavity formation, except that where the tool is removed at the end of a weld run, a hole is left where the pin was located; if it is not possible to use a run-off plate, it may be necessary to close this hole by welding or mechanically.

10.1.2 Cracking

Aluminium alloys may be subject to solidification or liquation cracking, and may suffer embrittlement and cracking at temperatures below the solidus. The higher-strength alloys may also fail by stress corrosion cracking.

Solidification cracking

The solidification cracking of aluminium and its alloys is associated in the main with intentionally added alloying elements rather than, as in steel, with the presence of low-melting impurities.

The cracking of aluminium fusion welds has been shown to follow the same pattern in its relation to alloy constitution as the cracking of castings, and crack-sensitivity relationships for castings may be taken as being applicable also to weld metal. The relationship between composition and the hot cracking of restrained castings in binary **aluminium–silicon** alloys is shown in Fig. 10.4 and the corresponding section of the aluminium–silicon phase diagram in Fig. 10.5. The degree of cracking rises to a maximum at about 0.5 % silicon and then decreases rapidly as the alloy content is increased.

The cracking curve follows the same trend as the solidification temperature range under equilibrium conditions, but is displaced to the left. Under continuous cooling conditions the constitution diagram is also displaced to the left, as shown by the broken line in Fig. 10.5, and the alloy content for maximum cracking is close to that for maximum freezing range in the modified diagram. Mechanical tests of cast aluminium–silicon alloys in the region of the melting point yield the type of results shown in Fig. 10.6 and 10.7. On cooling from the liquidus, the crack-sensitive 1 % silicon alloy acquires mechanical strength before it has any ductility, and there is a considerable temperature range over which it is very brittle. The 12 % silicon alloy is not crack-sensitive, and correspondingly has a very narrow brittle temperature range, as would indeed be expected from an alloy close to the eutectic composition. Pure aluminium also has a low crack susceptibility and narrow brittle range, but is much more sensitive to contamination than is the eutectic composition (see Fig. 10.4). The mechanism of this type of hot cracking in welds has already been discussed generally in Section 7.8.1. In a long-freezing-range alloy that is cooling from the liquidus, the growing crystals are at first completely separated by liquid and the alloy has no strength. As the temperature level falls, the volume of solid increases relative to that of the liquid, and at some point (the coherence temperature) the growing crystals meet and cohere. However, a limited volume of liquid still remains and persists down to the eutectic temperature, causing the metal to be brittle. At the same time, the solid portion contracts, and is therefore subject to a tensile stress, which may be high enough (depending on the degree of restraint) to cause failure of the weak, brittle matrix. The risk of cracking is greatest when a critically small

Non-ferrous metals 359

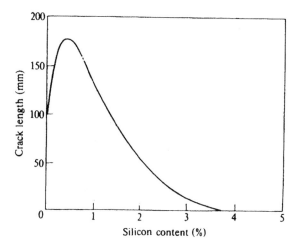

10.4 The length of cracking in restrained castings of binary aluminium–silicon alloys as a function of silicon content.

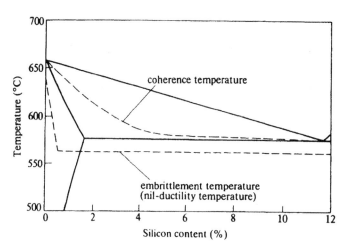

10.5 An equilibrium diagram for aluminium–silicon alloys (full lines) showing also coherence and nil-ductility temperatures on cooling (broken lines). The lower broken line represents the supposed solidus under cooling conditions.

volume of liquid metal is present below the coherence temperature. If the volume of eutectic present is relatively large, incipient cracks are healed by liquid that flows in from the weld pool. Such healed cracks may sometimes be seen in macrosections of aluminium welds.

Solidification cracking is severe with weld metal of the (3/4)Mg–1Si (Mg$_2$Si) type of composition and the fusion welding of this alloy with matching filler

360 Metallurgy of welding

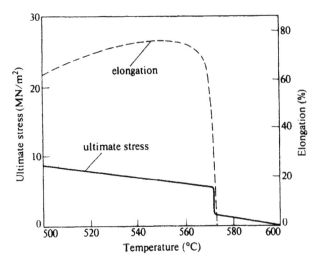

10.6 The mechanical properties of aluminium–1% silicon alloy on heating to temperatures close to the solidus (from Pumphrey and Jennings, 1948/49).

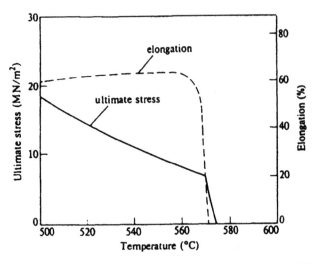

10.7 The mechanical properties of aluminium–12% silicon alloy on heating to temperatures close to the solidus (from Pumphrey and Jennings, 1948/49).

metal is only practicable under conditions of very low restraint. The higher-strength heat-treatable alloys are also prone to solidification cracking, particularly the Cu–Mg–Zn alloys used for aircraft applications such as 7075, 7079 and 7178. The Si–Cu–Mn–Mg types 2024 and 2219 are not quite as susceptible, while the

Mn–Mg–Zn alloys such as 7005, although subject to cracking under unfavourable conditions, are generally regarded as weldable.

Avoiding solidification cracking

Crack-sensitive alloys are usually welded with a filler metal that, when diluted by the parent metal, gives a weld-metal composition that is out of the cracking range. Filler wires containing 5 % Si or 12 % Si may be used, the latter being the eutectic composition for the Al–Si binary system. The Al–Mg and Al–Mg–Mn alloys such as 5554, 5356 and 5183 have been more frequently employed in recent years since they have an optimum combination of mechanical properties, corrosion resistance and crack resistance.

The degree of dilution depends on the type of joint, as well as on welding variables such as current and welding speed. Figure 10.8 is a dilution nomogram applicable to the inert gas-shielded welding of aluminium alloys under normal conditions. This enables the final composition of the weld metal to be estimated.

An additional means of increasing the crack resistance of aluminium-alloy weld metal is grain refinement. The total grain boundary area in any given volume of weld metal is, to a reasonable approximation, inversely proportional to the grain diameter. Therefore, when the grain size is reduced, any residual intergranular liquid film is also reduced and the probability of intergranular cohesion is increased. Figure 10.9 shows the results of a series of ring casting

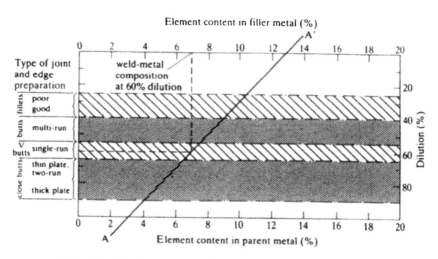

10.8 The dilution nomogram for fusion welds in aluminium. To find the weld composition at any dilution, a rule is laid between the appropriate compositions on the filler and plate scales (A–A'); the weld metal composition is then read off on either scale. In the example shown, the weld metal composition at 60 % dilution is 7 % (from Houldcroft, 1954).

tests on an Al–2Zn–2Mg alloy with various addition elements. This alloy is intended to represent the weld-metal composition obtained in welding an Al–Zn–Mg alloy with a 5 % Mg filler, assuming about 40 % dilution. There is a straight-line correlation between grain size and crack length in the restrained cracking test. The finest grain, and shortest crack length, was obtained by the addition of small amounts of Ti plus B. However, the grain refining effect of these additions is nullified by the presence of about 0.2 % Zr, and since zirconium is added to Al–Mg–Zn alloys to improve the resistance to stress corrosion cracking, Ti plus B additions cannot be used to reduce crack sensitivity.

A second means of grain refinement in aluminium alloys is to vibrate or stir the weld pool. This may be accomplished using an ultrasonic vibrator or by electromagnetic stirring, the latter technique being the more effective. Figure 10.10 shows the results of tests made using a modified Houldcroft-type specimen (illustrated in Fig. 10.11). An optimum reduction of crack sensitivity is obtained with an electromagnetic field alternating at 2 Hz. To obtain this benefit, it was necessary to add 0.16–0.24 % Zr to the basic Al–2Zn–3Mg alloy. In these tests it was demonstrated that in coarse-grained weld metal the residual liquid existed as

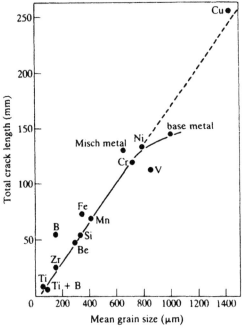

10.9 The relation between mean grain size and crack length in ring casting test using an Al–2Zn–2Mg alloy with and without addition elements. Pouring temperature, 750 °C; mould temperature 50 °C. Amount of element added is generally 0.5 wt% except for: Ti + B, 0.06 wt%; B, 0.2 wt%; V, 0.35 wt%; Misch metal, 0.3 wt% (from Matsuda *et al.*, 1983).

a continuous intergranular film, whereas in fine-grained material it was dispersed in globular form.

No melting takes place in friction stir welding, so there is no risk of solidification cracking. Stirring, however, has a similar effect to that produced by stirring the weld pool; namely it reduces the grain size as compared with that of the parent metal. The weld bead strength is correspondingly increased, and in tensile testing of such welds made in an annealed non-heat-treatable alloy failure occurs in the parent metal.

Other types of cracking

Liquation cracking in the heat-affected zone may occur when welding high-strength aluminium alloys of the Duralumin or aluminium–magnesium–zinc types. This type of cracking is caused by the presence of low-melting constituents in the structure and is associated with relatively low heat input rates. It may be overcome by using low-melting-point filler alloy or by increasing the welding speed. Cracking has also been observed well below the solidus (at about 200 °C) in high-strength aluminium alloys. This effect is also due to the formation of intergranular films at or near the solidus. These intergranular constituents cause embrittlement, which manifests itself as low-temperature cracking.

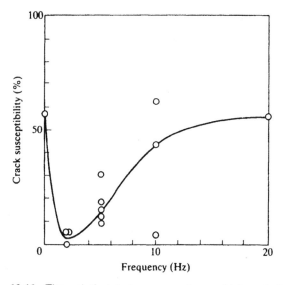

10.10 The relation between crack sensitivity of Al–2Zn–3Mg–0.24Zr alloy and frequency of applied magnetic field at a welding speed of 2.5 mm s^{-1}. Experimental conditions: gas tungsten arc welding with direct current, electrode negative, 55 A; electromagnetic stirring, 175 g (from Matsuda *et al.*, 1984).

364 Metallurgy of welding

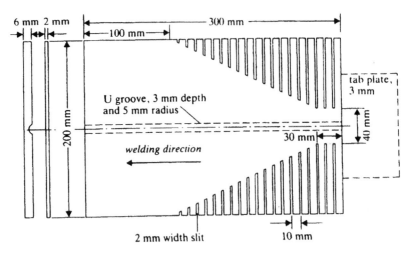

10.11. The modified Houldcroft test specimen for inert gas-shielded welding (from Matsuda *et al.*, 1982).

10.1.3 Mechanical properties

Sound fusion welds in annealed pure aluminium or any of the annealed non-heat-treatable alloys made with filler rod or electrodes of matching composition have a strength almost equal to that of the parent metal. When work-hardened plate is welded, the heat-affected zone may be fully or partly annealed, thus irreversibly reducing the tensile strength. This effect may be minimized by using the gas tungsten arc process with helium shielding and electrode negative, which reduces the width of the heat-affected zone.

The strength of heat-treatable aluminium alloys is obtained by solution treatment followed by quenching and then ageing, either at room temperature or, more usually, at elevated temperature. During the ageing treatment, the hardness and tensile strength increase because of precipitation and the formation of locally strained regions associated with the formation of clusters of solute atoms: Guinier–Preston zones. Heating at temperatures higher than the optimum value for age-hardening results in over-ageing; the Guinier–Preston zones dissolve, precipitates increase in size, and the matrix is softened.

Figure 10.12 shows the hardness distribution in the heat-affected zone of a gas tungsten arc welded Al–0.4Si–0.6Mg alloy 6063, initially in the fully hardened (T6) condition, together with the peak temperature reached during the weld thermal cycle. There is a fully softened zone adjacent to the fusion boundary, an intermediate zone where the hardness increases with distance from the fusion boundary, a third zone where the hardness is slightly higher than in the surrounding metal, and beyond this the unaffected parent metal. Figure 10.13 shows the same heat-affected zone after heating at 180 °C for 8 h. The relative

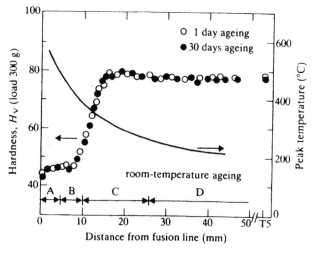

10.12 Hardness distribution in the heat-affected zone of a gas tungsten arc weld in Al–Mg–Si alloy 6063-T5. The peak temperature during the weld thermal cycle is also shown (from Enjo and Kuroda, 1982).

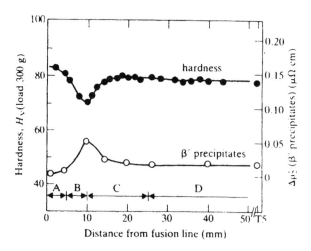

10.13 Hardness recovery due to ageing the weld shown in Fig. 10.12 for 8 h at 180 °C (from Enjo and Kuroda, 1982).

amount of β' precipitation, as determined by measurement of electrical conductivity, is also plotted. The β' phase consists of fine rod-like precipitates of Mg_2Si which normally contribute to the age-hardening effect. Comparing these two diagrams, it will be evident that, adjacent to the fusion boundary, the material has been fully solution-treated and has been rehardened by the ageing treatment. Farther out, there is a drop of up to 10 VPN in hardness, associated with precipitation. This may be an over-ageing effect or may be associated with some more complex mechanism. Beyond about 15 mm from the fusion boundary, the hardness is fully recovered. Note also from Fig. 10.12 that the 6063 alloy does not age-harden at room temperature.

The behaviour of the Al–Mg–Zn alloy such as 7005 is different in that, first, there is no over-ageing effect and, secondly, age-hardening occurs at room temperature. Thus the properties in the heat-affected zone may be almost completely recovered either by natural (room-temperature) or artificial (elevated-temperature) ageing of the joint.

Both Al–Mg–Si and Al–Mg–Zn type alloys may be welded with a 5.% Mg filler, and the weld metal, in addition to being resistant to solidification cracking, has tensile properties that match or are close to those of the parent metal. This is not the case with the higher-strength aircraft alloys of the Al–Cu–Mg–Zn type such as 7075, 7079 and 7178. It has not been possible to develop a filler alloy that combines matching strength with adequate resistance to cracking. These alloys also suffer from liquation cracking and reduction of strength in the heat-affected zone and are not normally welded. The older Al–Cu–Si–Mn–Mg aircraft alloys 2014 and 2024 are less sensitive and it is possible with these materials to obtain acceptable properties after welding. Indeed this type of alloy is fabricated by fusion welding in aerospace applications; outer rocket casings, for example.

In recent years there has been much interest in the development of aluminium–lithium alloys. These alloys contain additions of copper, magnesium, lithium, zirconium and, in some cases, silver. All the addition elements contribute to the age-hardening process, and in the fully heat-treated condition their tensile properties and fracture toughness equal, and sometimes exceed those of the 7XXX series. At the same time, their susceptibility to solidification cracking is similar to that of alloy 2024, so that fusion welding is possible.

Commercial alloys fall into two main classes, those with relatively high magnesium, and those with relatively high copper. Typical of the second category is type 2090, which has 2.7% Cu, 2.3% Li, 0.4% Mg, 0.4% Ag and 0.14% Zr. Fully heat treated it has a yield strength of 518 MN m^{-2} and an ultimate strength of 560 MN m^{-2}. As compared with 2024, its density is 9% lower, and its elastic modulus is 9% higher. This alloy is fusion welded using the gas tungsten arc process with direct current, electrode negative and helium shielding. When required, an aluminium–copper filler rod is used such as 2219 (6.3 Cu, 0.18 Zr, 0.06 Ti). Welded joints age-harden at room temperature in about 30 days.

Unfortunately, gas tungsten arc welds in aluminium–lithium alloys suffer from a very unusual and apparently intractable type of defect. In Section 7.7 it is stated that crystal growth in the fused zone of a weld is nucleated at the solid–liquid interface by crystals in the solid; the crystal growth is said to be epitaxial. In aluminium–lithium alloys this is not the case. In a narrow zone (thicknesses of from 20 to 600 micrometres have been measured) adjacent to the weld boundary equiaxed non-dendritic crystals form and are surrounded by a matrix of brittle aluminium–copper–magnesium–silicon eutectic. As a result, the strength, ductility and fracture toughness of welds may be reduced.

It has been suggested, on good evidence, that the equiaxed grains are nucleated heterogeneously in the weld boundary zone by lithium and zirconium-rich precipitates. There is no evident solution to the problem for gas tungsten arc welds. Other processes may not be so affected. Friction stir welding has been tried out for the welding of heat-treatable aluminium alloys with promising results. However, it must be borne in mind that welded structures almost always require manual welds for attachments and the like, whereas processes such as friction stir welding are primarily suited to main seam butt welds.

An important application for aluminium–lithium alloy is for the shell of the Space Shuttle external fuel tank. This is a large vessel, 28 feet (8.5 m) in diameter, and it was originally fabricated using the Al6Cu alloy 2219. Welds were made with the gas tungsten arc process and an Al–Cu alloy filler metal. In the early 1980s a plasma arc keyholing process was developed and this gave improved results.

The Al11Li0.4Mg0.4Ag alloy 2095 is 5% lower in density than 2219, and has 30% greater strength at the subzero operating temperature. The resultant weight saving per tank is about 7500 lbs (3400 kg). To use the new alloy for fuel tanks two changes to the weld procedure were required. Firstly, in order to avoid burn-out of lithium, it was necessary to apply inert gas shielding to the underside of the weld. Secondly, the filler alloy was changed to Al–5Si, type 4043. With these modifications, successful welds have been made. Non-destructive testing shows up occasional short cracks, which are subject to repair welding, but such cracking is to be expected with this type of alloy. There have been no reports of embrittlement due to the formation of a uniaxed zone, as reported earlier.

The high-strength heat-treatable aluminium alloys are in general susceptible to stress corrosion cracking in relatively mild environments such as weak chloride solutions or even in normal atmospheric exposure. Unwelded material is protected by cladding with Al–1Zn, which is anodic to the base material. Most stress corrosion cracking failures in Al–Zn–Mg weldments have been found in locations remote from welds. The Al–5Mg alloy may be susceptible to stress corrosion cracking due to precipitation of the β phase at slightly elevated temperatures. Al–Mg–Si heat-treatable alloys and the non-heat-treatable alloys other than the higher Al–Mg type do not suffer this type of attack.

10.1.4 Materials and applications

Welded aluminium products first became available early in the twentieth century. These were mainly vessels fabricated from commercially pure aluminium or the aluminium $1\frac{1}{4}\%$ manganese alloy, and they were fusion welded using the oxyacetylene torch with a halide flux, or were hammer welded. The hammer welding of aluminium is similar in principle to the force welding of iron. The plates to be joined were chamfered and overlapped, heated until a pine stick drawn across the surface left a brown mark, and then hammered until a sound joint was obtained. Aluminium vessels were used in the food industry, particularly in breweries. Between 1918 and 1939 aluminium–magnesium alloy plate was welded with coated electrodes to make some experimental structures, but there was no general application of this technique, in part because of concern about possible corrosion by flux residues.

The introduction of inert-gas shielded processes in the 1950s gave fresh impetus to aluminium welding, which found applications in the food, chemical and nuclear industries. More recently aluminium alloys, mostly aluminium–magnesium but also some heat-treatable magnesium–silicon tapes, have been employed on a large scale for ship superstructures and for accommodation modules on offshore rigs; also for helidecks on offshore structures. Aluminium is used here primarily to save weight, but the good resistance of the aluminium–magnesium alloys to marine atmospheres provides an additional incentive. In addition, aluminium–magnesium alloys may be specified because of their good low-temperature properties: for storage vessels containing liquefied natural gas on board ship, for example. Indeed, it may be said that welded aluminium has now become an essential element in marine structures.

The other major user of aluminium is the aircraft industry. The first successful all-metal aeroplane, the Junkers F13, flew in 1919, but this type of construction did not become widespread until the 1930s. In between these dates the fuselage of US aircraft consisted of a frame made of welded high-tensile steel tubing over which doped linen fabric was stretched. European machines used a variety of materials, but were mainly cloth mounted on a wooden framework.

The fuselage is now invariably of the all-metal type, with stress being carried in part by the airframe and part by the metal sheathing. The materials used are essentially the same as the Duralumin that was used to build the Junkers F13 in 1919, although improved in durability and strength. In particular, joints are still riveted, in spite of the weight penalty imposed by this type of construction. It must be accepted, of course, that the high-strength precipitation-hardened aluminium alloys are not ideal subjects for fusion welding, but the advent of the friction stir process opens up new avenues, and the all-metal, all-welded aircraft may not be such a remote possibility as was once the case.

Table 10.1 lists the composition and properties of some of the wrought aluminium alloys mentioned in this section.

10.1.5 Joining methods

Aluminium and aluminium alloys may be joined by fusion welding, resistance welding, solid-phase welding, brazing, soldering and adhesive bonding. Fusion

Table 10.1. Aluminium alloys

Alloy type	Designation	Nominal composition					Mechanical properties	
		Cu	Mn	Mg	Si	Zr	Yield strength (MN m^{-2})	Ultimate strength (MN m^{-2})
Non-heat-treatable	5083			4.3			125	275
Structural (marine) alloys	5154			3.5			85	215
Heat-treatable structural alloys	6061	0.3		1.0	0.6		225	295
	6063			0.7	0.4		180	200
Heat-treatable aircraft alloys	2024	4.35	0.6	1.5			395	475
	7075	2.5		2.5		5.5	450	520
	7178	2.0		2.8		6.8	480	550
Welding filler alloys	4043				5			
	4047				12			
	5350			5				
	5554			3				

welding was first accomplished by the oxyacetylene process but is now carried out using the inert-gas-shielded process. Gas tungsten arc welding with an AC power source is used for a wide variety of applications. During the electrode positive half-cycle the non-thermionic cathode removes oxide from the surface of the aluminium but the electrode overheats, while the electrode negative half-cycle allows the electrode to cool. With sinusoidal AC output there is a high restriking voltage as the workpiece becomes cathodic, with a delay in arc re-ignition which may result in extinction of the arc and, if the arc is maintained, a power imbalance between the two half-cycles that can overload the transformer. Older power sources therefore used a high-frequency arc igniter and some means of compensating the imbalance. Electronically controlled power sources that are now available produce a square wave with a sufficiently fast response to maintain the arc so this problem does not arise. Also, it is possible to change the balance between the two half-cycles. Increasing the duration of the electrode negative half-cycle reduces the heat load on the electrode, while increasing the electrode positive duration makes the cleaning action more effective.

Electrode-negative DC welding may be used with either argon or helium shielding, and has been applied to high-speed tube welding and for coil-joining welds in the production of aluminium strip. A helium shield is useful, as already indicated, for applications where deep penetration and a high heat input rate is required. The cost of helium, however, is a limitation to the use of this process, particularly outside the USA.

The gas metal arc process is operated with DC electrode positive, and the cathode has the same effect of cleaning up the surface as with gas tungsten arc welding. At medium current levels, the weld pool is roughly semicircular in cross-section, but as the current increases, a finger-shaped projection develops in the centre of the fusion bead. This finger-shaped penetration is undesirable in that it is intolerant to misalignment between the electrode and the centreline of the joint, and is thus more subject to lack of fusion along the centreline. This problem may be avoided by using helium or a helium–argon mixture as the shielding gas. With helium shielding the pool cross-section remains semicircular at high currents.

Various devices are used to overcome weld softening in medium-strength heat-treatable alloys. The first is to make allowance for this effect in the design stress used. The ASME code, for example, permits a design stress of 10 500 psi (72×10^6 N m^{-2}) for unwelded Al–Mg–Si alloy and 6000 psi (40×10^6 N m^{-2}) for the same alloy in the welded condition. It is sometimes possible in structural applications to place the weld in a region of low stress. In pipework, special pipes with thickened ends may be used for welding, the extra thickness being sufficient to compensate for softening.

In principle, it is possible to recover the full joint strength in the heat-affected zone by solution treatment and ageing, but for high-strength alloys the weld metal remains relatively weak, and since the whole component must be treated, this procedure is rarely practicable. Postweld ageing could, as indicated by Fig. 10.13, improve the heat-affected zone properties in some cases, but such treatment is rarely used.

Fusion welding is used mainly for pure aluminium, the non-heat-treatable Al–Mn and Al–Mg alloys, and the Al–Mg–Si and Al–Zn–Mg heat-treatable alloys. The higher-strength aluminium alloys of the Al–Cu–Mg and Duralumin types, which are specified mainly for aircraft structures, are difficult to fusion-weld effectively. Such alloys are normally joined by riveting, although a few fabricators employ spot welding as a joining method, and fusion welding is applied to space vehicle construction in certain cases.

Resistance spot welding has been employed for joining sheet aluminium, for example in the automobile industry. Aluminium is, however, an intrinsically difficult material to spot weld. The presence of a refractory oxide film on the metal surface may cause uncontrolled variations in the surface resistance, and standardized cleaning methods are necessary. The electrical and thermal conductivities are high, so that high-current welding transformers are required. The volume changes on solidification and cooling necessitate careful and rapid control of electrode loading. Low-inertia heads are also needed for projection welding because of the low elevated-temperature strength of the metal. Metal pick-up results in a relatively short electrode life. However, in spite of these disadvantages (as compared, say, with mild steel), spot welding has been used for a number of years as a means of joining sheet metal. Any of the aluminium alloys may be so joined, including the heat-treatable Al–Cu alloys.

Cold pressure welding is used in joining aluminium for certain special applications, notably cable sheathing. Ultrasonic welding has important applications in microjoining, as explained in Chapter 5. Electroslag welding is applicable to aluminium for the joining of thick sections such as may be used in the electrical industry. The industrial applications of friction stir welding are being explored, and it is expected to be a valuable addition to the wide range of joining processes currently used for aluminium.

There are a number of duties for which it is required to join aluminium to either carbon steel or austenitic chromium–nickel steel. These include cryogenic equipment, anode assemblies, busbars and cooking utensils. Fusion welding is not practicable because of the formation of brittle intermetallic compounds, and joints are made by soldering, brazing, diffusion bonding or friction welding. For parts of circular symmetry, friction welding is usually the best technique; sound joints can be made although brittle compounds may be present at the interface.

For certain applications of complex form (for example, automotive radiators and core-type plate heat exchangers), brazing is used for joining aluminium. Aluminium–silicon alloys may be employed as the brazing filler alloy and in some cases it is convenient to use sheet aluminium clad with the brazing alloy, the whole assembly being dipped in a bath of molten flux to make numerous joints simultaneously. Fluxes are normally corrosive and special procedures may be necessary for their removal. Potassium fluoroaluminate-based fluxes are, however, relatively non-corrosive.

Aluminium may be soldered to itself and to carbon steel, stainless steel, nickel alloys and copper alloys. Zinc or zinc–cadmium solders used with a zinc chloride flux give joints of relatively good corrosion resistance, but the melting range is high for solder, 335–415 °C. The tin–zinc eutectic mixture (91Sn–9Zn) melts at about 200 °C and gives joints suitable for atmospheric exposure provided that they are protected by paint. Lower-melting solders generally have low strength and poor corrosion resistance.

10.2 Magnesium and its alloys

10.2.1 Alloys and welding procedures

Magnesium finds its widest application in the aircraft industry and for launch vehicles and satellite structures, where its excellent strength/weight ratio can be used to full advantage. The pure metal has too low a strength for engineering use. The alloys may be divided into three main groups: **aluminium–zinc, zinc–zirconium** and **thorium** respectively. The Mg–Al–Zn alloys were the earliest in development and have the disadvantage of being susceptible to stress corrosion cracking; nevertheless, the 3Al–1Zn–0.4Mn alloy remains one of the most generally applicable and readily welded types. The Mg–Zn–Zr alloys with 2 % zinc or less also have good weldability, as do the thorium alloys, which are designed specifically for good strength at elevated temperatures. A 2.5Zn–1Mn alloy is used for aerospace duties.

There are a number of casting alloys that may be joined into structures by welding or may be repaired by welding. The principles outlined below apply to these alloys also.

10.2.2 Oxide film removal

Like aluminium, magnesium forms a refractory oxide that persists on the surface of the molten metal and tends to interfere with welding. However, magnesium oxide recrystallizes at high temperature and becomes flaky, so that the surface film breaks up more easily than that which forms on aluminium.

The mechanism of oxide removal by means of a flux in welding is probably similar to that in aluminium welding. Fluxes are typically mixtures of chlorides and fluorides of the alkali metals (e.g. 53 % KCl, 29 % $CaCl_2$, 12 % NaCl and 6 % NaF), and are highly corrosive to the base metal. For this and other reasons, gas welding is little used for magnesium and its alloys, the most important fusion welding process being gas tungsten arc welding with alternating current. Oxide is removed from the surface by arc action during the half-cycle when the workpiece is negative. Mechanical cleaning of the weld edges is essential for fusion welding. For spot welding, chemical pickling is necessary, combined with mechanical cleaning (with steel wool) immediately before welding.

10.2.3 Cracking

Zinc and calcium additions both increase the susceptibility of magnesium alloys to solidification cracking during welding. Zinc is a constituent of a substantial proportion of the alloys; in amounts of up to 2 % it is not deleterious, but alloys containing larger quantities, particularly those with 4–6 % Zn, have poor weldability. Aluminium, manganese and zirconium have little effect on this characteristic, but thorium and rare-earth elements are beneficial and tend to inhibit solidification cracking. Generally speaking, the most crack-sensitive magnesium alloys are the higher-strength high-alloy types, which suffer from cracking both in the weld and at the weld boundary.

10.2.4 Mechanical properties

Weld deposits of magnesium alloys solidify with fine grain and have a tensile strength frequently higher than that of the equivalent wrought material. Thus welded joints tested in tension commonly fail in the heat-affected zone, which may be embrittled by grain growth. Alloys that have been hardened by cold-working, and age-hardened material, soften in the heat-affected zone. Generally, however, the joint efficiency of fusion welds in magnesium alloys is good, and it is possible to use a relatively low-melting filler metal (e.g. the 6.5Al–1Zn type) for a wide range of alloys and achieve 80–100% joint efficiency 'as-welded'.

10.2.5 Corrosion resistance and fire risk

Magnesium alloys are commonly protected against atmospheric corrosion by means of a chromate dip. The green chromate layer must, of course, be removed from the vicinity of the joint before welding.

Aluminium-containing magnesium alloys are susceptible to stress corrosion cracking in the heat-affected zone of the welds and must be stress-relieved (generally at about 250 °C) after welding to prevent this type of attack. The zirconium- and thorium-bearing alloys are not susceptible to stress corrosion and do not require stress relief after welding.

There is a risk of fire if magnesium is allowed to accumulate in finely divided form, and proper attention must be paid to cleanliness in all operations involving cutting, machining and grinding. Except in the joining of foil, there is no direct risk of fire due to either fusion or resistance welding of magnesium.

10.3 Copper and its alloys

Gas welding was the first fusion process to be applied successfully to commercially pure copper, joints of acceptable strength being possible in **phosphorus-deoxidized copper**. More recently, inert-gas welding, using argon, helium or nitrogen as shielding gas, has greatly broadened the applicability of fusion welding to copper alloys. Coated electrode welding of pure copper, brass and cupronickel has not been successful in practice and, although satisfactory **tin bronze** and **aluminium bronze** electrodes are available, these are mainly used for weld overlays and dissimilar metal joints.

10.3.1 Gas–metal reactions

Measurements of the surface temperature of weld pools in the gas tungsten arc melting of copper range from 1350 °C at 100 A arc current to 1890 °C at 450 A. At such temperatures, the affinity of copper for oxygen is low, while it does not react with nitrogen at all. Both gases may dissolve in the liquid metal (although authorities differ on the solubility of nitrogen), and the solubility of hydrogen lies between about 10 ml/100 g and 20 ml/100 g. When hydrogen and oxygen are simultaneously present in the liquid metal or in the solid at elevated temperature, the steam reaction may occur:

$$Cu_2O + H_2 = 2CU + H_2O \qquad (10.2)$$

Much of the porosity to which fusion welds in copper are subject has been ascribed to the steam reaction. In addition, it is responsible for the intergranular embrittlement and cracking of the heat-affected zone in tough pitch copper welds made by the oxyacetylene process. Hydrogen generated by the combustion of acetylene diffuses into the heat-affected zone and reacts there with Cu_2O. The steam so formed precipitates at grain boundaries and causes the observed failures.

Nitrogen

The effect of nitrogen additions to argon gas on porosity for two materials is shown in Fig. 10.14. It is sometimes stated that nitrogen is insoluble in liquid copper, whereas these results clearly indicate a significant degree of nitrogen absorption. One possible reason is that the temperature of the liquid metal at the electrode tip, with a welding current of 300 A and a 1.2 mm diameter wire, is high, possibly close to boiling point. Secondly, it is possible that the solubility at the arc root is augmented, as in the arc melting of iron.

The use of nitrogen or nitrogen–argon mixtures may be desirable in the gas metal arc welding of copper in order to increase the heat input rate. Addition of

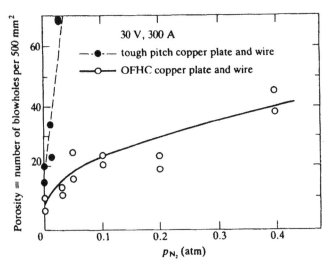

10.14 The effect of nitrogen content of nitrogen–argon atmosphere of total pressure 1 atm on the porosity of gas metal arc welds in copper. OFHC, oxygen-free high conductivity (from Kobayashi et al., 1970a).

titanium and aluminium, which are strong nitride formers, makes it possible to produce sound welds with such atmospheres.

Some tests that were made using gas metal arc welding in an enclosed chamber with mixtures of argon and other gases as the arc atmosphere are of interest. Two basic materials were used for both wire and plate, and the effect of the gas on the amount of porosity was assessed by radiography of bead-on-plate runs. The amount of porosity was recorded as the number of pores (blowholes) per 500 mm^2 of film. The gas content of the two materials is given in Table 10.2.

Figure 10.15 shows the effect of hydrogen additions for the same materials and welding method. The tough pitch copper is more severely affected than the oxygen-free high conductivity type, as would be expected from Equation 10.2. However, note the difference in scale of both porosity and gas additions; the slope of porosity vs partial pressure for oxygen-free high-conductivity copper is about the same as for tough pitch copper in Fig. 10.14. The effect of water vapour is shown in Fig. 10.16. In this case there is little difference between the two materials, suggesting that the water vapour is dissociating into hydrogen and oxygen in the arc and recombining in the liquid metal. Indeed, these results support the view that, in the absence of deoxidants, the steam reaction is a potent cause of porosity in copper welds.

Oxygen

The effect of oxygen is shown in Fig. 10.17. Up to a partial pressure of 0.1 atm, oxygen reduces the amount of porosity; above this level the porosity increases again. Carbon dioxide has a similar effect. The amount of oxygen present in the gas shield due to atmospheric contamination is certainly below 0.1 atm partial pressure, so this element is not likely to promote porosity in copper welds.

Avoiding porosity in fusion welding

Cracking in the heat-affected zone of copper welded by the oxyacetylene process may be prevented by the addition of between 0.02 and 0.10% phosphorus to the plate material. However, the weld metal itself is rarely free from porosity, and as a result the tensile strength of 'as-welded' oxyacetylene welds in phosphorus-

Table 10.2. Gas contents of tough pitch and OFHC copper

	O_2 (mass %)	H_2 (ml/100 g)
Tough pitch	0.03	0.008
OFHC copper	0.0007	1.80

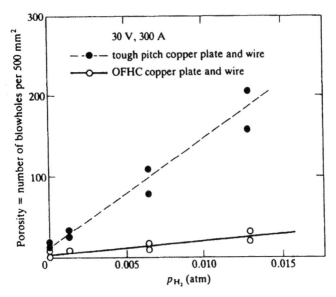

10.15 The effect of hydrogen content of hydrogen–argon atmosphere of total pressure 1 atm on the porosity of gas metal arc welds in copper (from Kobayashi et al., 1970a).

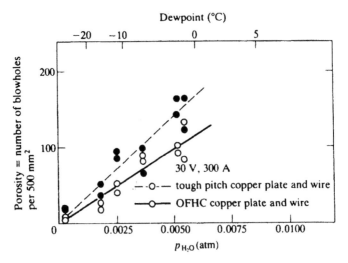

10.16 The effect of water vapour content of water vapour–argon mixtures at a total pressure of 1 atm on porosity of gas metal arc welds in copper (from Kobayashi et al., 1970a).

Non-ferrous metals

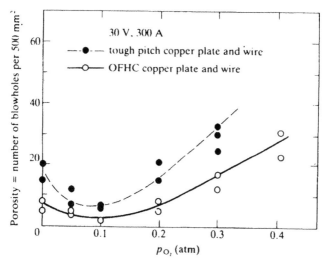

10.17 The effect of oxygen content of argon–oxygen mixtures at a total pressure of 1 atm on porosity of gas metal arc welds in copper (from Kobayashi *et al.*, 1970a).

deoxidized copper is substantially lower than that of the parent metal. Gas welds in copper are frequently **hammered** or **rolled** to remove distortion, and this treatment improves their tensile strength.

Inert-gas shielded welding does not cause gassing of the parent metal, whether or not it contains phosphorus. The weld metal itself, however, is grossly porous unless it is deoxidized. Phosphorus is not adequate for this purpose and a combination of either silicon and manganese or titanium and aluminium is added to filler wires and electrodes. Even using such powerful deoxidants, porosity may still occur in manual gas tungsten arc welding, particularly at restarts. Phosphorus-deoxidized plate is normally specified for gas tungsten arc or inert gas-shielded welding. Shielding gases are argon, helium or nitrogen, the latter being used when higher rates of heat input are necessary. There is little difference between argon and nitrogen shielding as regards the incidence of porosity, but helium-shielded welding appears to be somewhat less sensitive to this defect. Helium may not be available at an acceptable price outside the USA, however. Tunnelling porosity, which is associated with turbulence in the weld pool, may occur in gas metal arc welds if the current is too high: over about 350 A for nitrogen shielding and 450 A for argon shielding.

Copper alloys containing deoxidizing elements – aluminium bronze, tin bronze and silicon copper – are not subject to porosity, and may be welded without special additions to the filler metal. **Copper–nickel** alloys, however, suffer the same type of porosity as copper itself and require deoxidized filler rod. **Brasses** (both 70/30 and 60/40 types) are difficult to weld because of volatilization of zinc, and this effect, apart from interfering with visibility during welding, may also be associated with weld deposit porosity. In oxyacetylene welding, such porosity may be minimized by using an oxidizing flame and by adding a filler rod deoxidized with silicon, manganese or phosphorus. A copper with about 1.5 % zinc, known as **cap copper**, may, however, be welded by means of the gas tungsten arc process without deoxidized filler wire, and the resultant welds are sound.

10.3.2 The effect of high thermal conductivity

The high thermal conductivity of pure copper makes spot and seam welding impracticable, although resistance butt welding using high-capacity machines is possible and is practised for joining wire bar. The alloys have much lower conductivities, however, and normal resistance welding methods are possible.

Heat conductivity may give rise to difficulties such as cold shuts and lack of side-wall fusion in welding pure copper, particularly when the section of sidewall thickness is 25 mm or more. In order to increase the heat input per ampere, nitrogen is sometimes used as a shielding gas for both gas tungsten arc and gas metal arc welding. With nitrogen shielding and tungsten arc welding, 5 mm copper may be joined in one pass without preheat, as compared with 3 mm thickness under argon shielding. With pure nitrogen shielding the transfer characteristics of metal inert-gas welds are poor, but with argon–30 % nitrogen mixtures the good penetration of nitrogen shielding is combined with the good transfer characteristics of argon. Penetration per pass may be improved still further by using two gas metal arc welding guns mounted side by side along the weld seam. In this way high heat inputs may be obtained without the porosity that appears when the welding current is excessive. If the heat input obtained by such methods is still inadequate, it is necessary to preheat. Preheat temperatures of up to 600 °C have been used for welding thick copper. Such preheats necessitate insulation of the article to be welded and protective clothing for the welder. Alternatively, the use of electroslag welding could be considered.

10.3.3 Solidification cracking

Copper and copper alloys are rendered brittle and sensitive to hot cracking if excessive amounts of low-melting impurities, notably bismuth and lead, are present. A useful test for such embrittlement is a hot bend test conducted at 600 °C, the bending jig being heated if necessary. Alternatively, the ductility in a tensile test at 450 °C may be used as a measure of crack sensitivity. In general, it is necessary to restrict bismuth and lead contents to low levels in order to avoid hot shortness in the working of copper and copper-alloy products, such that these contaminants do not normally give rise to welding problems. However, single-phase aluminium bronze is notably sensitive to cracking during fusion welding (see Section 10.3.5).

10.3.4 Mechanical properties

The mechanical properties of unhammered oxyacetylene welds in copper are lower than those of the parent plate owing to the presence of porosity. Inert-gas shielded welds may also have lower strength than the parent in thick sections, but the reduction is less than for oxyacetylene welds.

The strength of work-hardened or age-hardened copper alloys is reduced in the heat-affected zone. However, the age-hardenable copper alloys, particularly **beryllium–copper** and **chromium–copper**, are rarely welded. In hard temper (work-hardened) copper, it must be assumed that the strength of a fusion butt weld is equal to that of the annealed material.

10.3.5 Alloys and welding procedures

There are two coppers that are mainly used for electrical transmission lines, **tough pitch copper** and **oxygen-free high-conductivity copper**. Arc welds in these materials are not very satisfactory; usually they contain defects and at best have undesirably high electrical resistance due to the use of deoxidized filler material, although in this respect boron-deoxidized wire produces welds of higher conductivity than other types. Cold pressure welding is applicable to rod and is used to make cold joints in electrical conductors. The process generally used for joining copper in electrical conductors is thermit welding, using a mixture of aluminium and copper oxide for the exothermic reaction. The mixture is held in a graphite crucible, with a steel disc covering the exit hole. It is ignited and in a very short time a pool of molten copper is formed, which melts the retaining disc and flows out into a mould surrounding the joint. The liquid copper has sufficient superheat to melt the joint faces and produce a sound weld. A collar of metal is left around the joint for strength and to ensure good electrical conductivity.

Phosphorus-deoxidized copper is the standard material for welded sheet and plate applications in copper. The copper–1.5 % zinc alloy (**cap copper**) has been employed for domestic hot water tanks, which are welded by the gas tungsten arc process without filler wire. This material gives sound ductile joints when so welded. **Silicon-bronze** may also be welded by the inert-gas processes without special additions, and gives sound ductile joints of mechanical strength equal to the parent metal. Resistance welding (and of course soldering) finds substantial application in sheet metal products made from brass.

Aluminium bronze is one of the more difficult materials to fabricate and weld owing to its susceptibility to hot cracking. The type most frequently specified is the single-phase Cu–7Al–2.5Fe alloy. Except for single-pass welds in thin material, the use of a matching filler material for welding this type of aluminium bronze is impracticable. Even if fissuring of the weld can be avoided, multi-pass joints may suffer embrittlement due to heat treatment of the weld deposit by subsequent runs. A duplex filler material containing about 10 % aluminium is, however, virtually free from any tendency to crack. A composition that has been successfully used for inert gas welding is nominally 10 % aluminium, 2.5 % iron and 5.5 % nickel. Duplex weld deposits may be subject to dealuminification in corrosive service, and this risk may be reduced by applying a single-phase capping run to the weld.

Although weld-metal cracking may be overcome by the proper choice of filler alloy, 93Cu–7Al plate material may sometimes crack during hot forming or welding. In welding, these cracks may extend for some distance away from the weld boundary. Intergranular cracking has also been observed close to the weld boundary. Such cracking is due to a deficiency of the plate material, the nature of which has not yet been explained.

Cracking of the weld metal appears to be associated with phase constitution in a manner analogous to that which occurs in fully austenitic steel: the single-phase alloy is subject to cracking, while two-phase alloys are not. A further similarity is that both materials have a narrow freezing range, so that cracking due to a wide freezing range inherent in the constitution of the alloy (as, for example, in aluminium alloys) is not possible. If cracking occurs at high temperature, it is probably due to the formation of liquid intergranular films of low melting constituents. Alternatively, cracking may take place at lower temperatures owing to the formation of brittle intergranular constituents.

Cupronickel is used mainly in sheet form for fabricated work, and the most suitable welding process is gas tungsten arc welding. Cracking is not a serious problem, and porosity is minimized by using a filler rod containing deoxidant. Cupronickel alloys vary in nickel content, typical compositions containing 5 %, 10 %, 20 % or 30 % nickel. If a matching filler is not available, the deoxidized 70/30 composition may be used. A suitable deoxidant is titanium.

Silicon bronzes may exhibit hot shortness at temperatures between 800 and 950 °C, and under conditions of restraint this may result in cracking of fusion welds. In general, however, these alloys present few welding difficulties; they have relatively low thermal conductivity ($54.4 \text{ W m}^{-1} \text{ K}^{-1}$) and are not subject to porosity in fusion welding. Silicon bronze may be welded using all the major welding processes. Generally, the speed of welding is high, and preheating rarely necessary, while the energy requirements for resistance welding are much lower than for other copper alloys: a reflection of their higher electrical resistance compared with copper.

10.4 Nickel and its alloys

Nickel has physical properties similar to those of iron, but differs metallurgically in that it does not undergo a γ–α transition, the face-centred cubic lattice structure being maintained down to room temperature. Metallurgical problems associated with the welding of Ni and its alloys include weld porosity, embrittlement by sulphur and other contaminants, and loss of corrosion resistance due to the formation of intergranular precipitates.

10.4.1 Gas–metal reactions

Nitrogen and hydrogen

The absorption of nitrogen in nickel follows a pattern similar to that for iron. For gas tungsten arc welding in an argon–nitrogen atmosphere, the solubility of nitrogen obeys Sievert's law up to a limiting value, above which the amount of gas absorbed is constant independent of the partial pressure of nitrogen. The solubility under non-arc melting conditions is very low: 0.0018 by mass % at 2000 °C and 1 atm pressure of nitrogen. In gas tungsten arc welding the solubility at partial pressures where Sievert's law is obeyed is 100 times the equilibrium value, while the saturation value (where the amount absorbed is constant) is about twice the equilibrium solubility at 1 atm. The welds become porous when the nitrogen content exceeds about 30 ppm, corresponding to a nitrogen partial pressure of 2.5×10^{-4} atm. Figure 10.18 shows nitrogen content as a function of partial pressure of nitrogen in the shielding gas.

Addition of hydrogen to the argon–nitrogen mixture increases the partial pressure of nitrogen at which porosity appears, so that with 10 % hydrogen in the mixture the critical partial pressure of nitrogen is increased to 5×10^{-4} atm. This is consistent with the absorption of nitrogen in the gas metal arc welding of iron, where, for example, a partial pressure of 0.04 atm nitrogen in argon gave a nitrogen content in the weld of 0.046 mass%, whereas the same partial pressure of nitrogen in hydrogen gave a nitrogen absorption of 0.012 %.

In gas tungsten arc welding nickel is also tolerant to hydrogen in argon–hydrogen atmospheres, such that porosity does not appear until the hydrogen content of the mixture is 50 %. In non-arc melting, however, porosity appears in solidified metal when the atmosphere contains 5 % hydrogen or more. These results are summarized in Table 10.3.

It is implied that hydrogen is absorbed at the arc root and bubbles out at the rear of the weld pool. In so doing it removes a proportion of the dissolved nitrogen. Thus it is possible to tolerate a higher partial pressure of nitrogen in the arc atmosphere before porosity occurs.

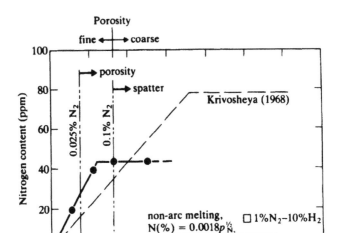

10.18 The nitrogen content of gas tungsten metal arc weld metal as a function of nitrogen partial pressure in an argon–nitrogen atmosphere (from Ohno and Uda, 1981).

Table 10.3. Critical nitrogen or hydrogen concentrations for the formation of porosity in solidified nickel for gas tungsten arc welding and levitation melting (from Ohno and Uda, 1981)

	Gas tungsten arc welding	Levitation melting at 2000 °C
Ni–H$_2$	50% H$_2$–Ar	5% H$_2$–Ar
Ni–N$_2$	0.025% N$_2$–Ar	100% N$_2$ (no porosity)
	1% H$_2$–0.025% N$_2$–Ar	–
Ni–H$_2$–N$_2$	10% H$_2$–0.05% N$_2$–Ar	–

Oxygen

Nickel has a higher solubility for oxygen than iron and a lower chemical affinity, such that the steam reaction is possible (the free energy of formation of nickel oxide is numerically smaller than that of H$_2$O) and the CO reaction

$$[O]_D + [C]_D = CO \tag{10.3}$$

could also occur. The equilibrium constant for reaction 10.3 in pure liquid nickel is

$$K = \frac{p_{CO}}{a_c a_O} = e^{4.851 + 7780/T} \tag{10.4}$$

Assuming that CO will be evolved from the melt when $p_{CO} > 1.0$, the maximum tolerable value for the solubility product at the melting point 1453 °C is

$$a_c a_O = e^{-(4.851+7780/T)} = 8.62 \times 10^{-5} \tag{10.5}$$

Now the oxygen content of gas tungsten arc welds in steel is typically 0.005–0.02 mass%. If the oxygen content of Ni welds is similar, the tolerable carbon content for avoiding porosity at 1453 °C would be 0.017–0.0043 %. The specified maximum carbon content for normal-quality commercially pure nickel is 0.08 % and for the low-carbon quality is 0.02 %. Thus, in the absence of deoxidants, porosity due to CO formation is possible.

Avoiding porosity

Pure nickel, the Ni–30Cu alloy Monel, and (to a lesser degree) the nickel–chromium–iron alloys are subject to porosity if the weld metal does not contain nitride-forming and deoxidizing elements. Filler rods and electrodes for the arc welding of these alloys are designed to give a weld deposit containing aluminium, titanium, or niobium, or a combination thereof.

It will be evident from the discussion of gas–metal reactions that porosity may be due to nitrogen evolution, to H_2O or CO formation, or to those reactions occurring in combination. In either event, such porosity may be minimized by the addition of elements such as aluminium and titanium, which form stable compounds with nitrogen and oxygen, and by avoiding atmospheric contamination, if necessary preventing the access of air to the underside of the weld.

Porosity in nickel welds may be eliminated in single-pass welds by the use of argon with up to 20 % hydrogen as a shielding gas, which is consistent with the experimental results described earlier. Addition of hydrogen to the shielding gas in gas metal arc welding, however, causes gross porosity. For steel welding by the gas metal arc process, oxygen or CO_2 is added to the argon shield to stabilize the arc; with nickel and its alloys, this causes undesirable oxidation of the weld and results in an irregular weld profile.

10.4.2 Solidification cracking

Hot-shortness of nickel and nickel alloys may be caused by contamination with sulphur, lead, phosphorus and a number of low-melting elements such as bismuth. These contaminants form intergranular films that cause severe embrittlement at elevated temperature. Hot cracking of weld metal may result from such contamination, but more frequently it occurs in the heat-affected zone

384 Metallurgy of welding

of the weld and is caused by intergranular penetration of contaminants from the metal surface (Fig. 10.19). Sulphur is a common constituent of cutting oils used in machining and therefore is frequently present on metal surfaces. Grease, oil, paint, marking crayons, temperature-indicating sticks or dirt may contain one of the harmful ingredients. Damaging elements may also be present on the surface of nickel that has been in service and requires weld repair.

Prior to any welding or brazing process where heat is applied, the metal surface must be cleaned. New material is scratch-brushed using a stainless-steel wire brush for a distance of at least 25 mm on either side of the joint, and then degreased with carbon tetrachloride, trichloroethylene or other solvent. Metal that has been in service needs more drastic treatment. It is ground, shot blasted or pickled in the region adjacent to the weld, and then degreased, as before. During fabrication, suitable steps must be taken to minimize the danger of contamination, in particular by establishing clean working conditions and by annealing either in electric furnaces or in furnaces fired with sulphur-free fuel. Cracking may occur in the weld or heat-affected zone of nickel and high-nickel alloys that have been work-hardened or age-hardened. Material should therefore be in the annealed or solution-treated condition before welding.

Welds made in the silicon-bearing 18Cr–38Ni alloy are likely to suffer from hot cracking if a matching filler metal is used. This alloy should be welded with a low-silicon 80Ni–20Cr type filler, avoiding dilution.

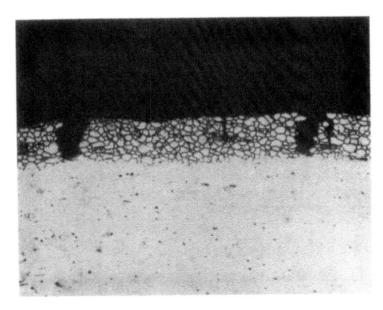

10.19 The hot cracking of nickel due to sulphur contamination of the surface (×100) (photograph courtesy of International Nickel Co. (Mond) Ltd).

Chromium-containing nickel-base alloys are less susceptible to damage by sulphur than commercially pure nickel and Monel. Nevertheless, it is essential with all these alloys to ensure that all sulphur contamination has been removed from the surface prior to welding.

10.4.3 Mechanical properties

The mechanical properties of properly made welds in annealed nickel and nickel alloys, other than the age-hardening types, are equal to those of the parent metal. Age-hardening alloys are normally welded in the solution-treated condition and age-hardened after welding. With an optimum combination of parent and filler metal compositions, joints of strength close to that of the fully heat-treated parent metal are obtained. Age-hardening treatments for nickel alloys are in the temperature range 580–700 °C. A number of filler alloys for the non-ageing materials contain sufficient titanium, aluminium or niobium to make the weld deposit harden if held within this temperature range, so that stress-relieving joints at 580–650 °C will result in some degree of weld hardening. Such hardening is not usually harmful in a non-corrosive environment, but its potential effect must be considered (see Section 10.4.4). If a non-ageing deposit is required for nickel–chromium–iron alloys, an 80Ni–20Cr filler alloy may be used. Welding fully heat-treated age-hardened alloys is only possible under conditions of minimum restraint and, to restore the full properties, solution treatment followed by ageing must be repeated after welding.

10.4.4 Corrosion resistance

Pure nickel and many of the nickel alloys are used for corrosion-resistant duties, as indicated in Table 10.4. Pure nickel has good resistance to caustic solutions, while the Ni–Mo and Ni–Cr–Mo alloys withstand some of the most severely corrosive environments encountered in chemical plant. These alloys are welded using the argon-shielded processes with electrodes that produce a deposit of matching composition, modified where necessary by the addition of deoxidizers. The corrosion resistance of such weld metal is generally adequate, but the base metal must, for exposure to severe conditions, be formulated to avoid intergranular precipitation due to the weld thermal cycle.

Considering Ni–Cr–Fe alloys in general, the effect of increasing the Ni content is to reduce the susceptibility to transgranular stress corrosion cracking and to increase the susceptibility to intergranular attack. Figure 10.20 shows this effect in terms of weight loss in a standard intergranular corrosion test as a function of nickel content. By the same token it is necessary, as the Ni content increases, to reduce the carbon content to lower values in order to maintain intergranular corrosion resistance. It may also be necessary to add vanadium as a stabilizing

Table 10.4. Nickel and nickel alloys

Chemical composition* (%)

Designation	Ni	Cr	Co	Cu	Fe	Mn	Mo	Si	W	C	S	P	Al	V	Ti	Mg	Typical use
Nickel 200	99.2 min.	–	–	0.25	0.40	0.35	–	0.15	–	0.10	0.005	–	–	–	–	0.10	Caustic service
Low-C nickel 201	99.0 min.	–	–	0.25	0.40	0.35	–	0.15	–	0.02	0.005	–	–	–	–	0.10	Caustic service over 316°C
Monel 400	63.0–70.0	–	–	28.0–34.0	1.0–2.5	1.25	–	0.5	–	0.15	0.02	–	0.5	–	–	–	Seawater, chlorides, HCl, HF
Hastelloy B	bal.	1.0	2.5	–	4.0–7.0	–	26.0–30.0	–	–	0.05	–	–	–	–	–	–	HCl service
Hastelloy B2	bal.	1.0	1.0	–	2.0	1.0	26.0–30.0	0.10	–	0.02	0.03	0.04	–	–	–	–	HCl service, as-welded
Hastelloy C	bal.	14.5–16.5	2.5	–	4.0–7.0	–	15.0–17.0	–	3.0–4.5	0.08	–	–	–	–	–	–	Strong oxidizers, mineral acids, wet chlorine gas
Hastelloy C276	bal.	14.5–16.5	2.5	–	4.0–7.0	1.0	15.0–17.0	0.08	3.0–4.5	0.02	0.03	0.04	–	0.35	–	–	As above, as-welded
Incoloy 825	40.0–45.0	20.0–24.0	–	1.5–2.5	bal.	–	2.5–3.5	–	–	0.05	–	–	0.20	–	0.7–1.1	–	Resistance to chlorides and mineral acids
Inconel 600	72.0 min.	14.0–17.0	–	0.50	6.0–10.0	1.0	–	0.50	–	0.08	0.015	–	–	–	0.30	–	Oxidation resistance to 1175°C; resistance to stress corrosion cracking; nuclear components
Incoloy 800	30.0–35.0	19.0–23.0	–	0.75	bal.	1.5	–	1.0	–	0.10	0.015	–	–	–	–	–	Petrochemical furnaces and piping; oxidation resistant

*Maximum except where a range is indicated.

element (note the compositions of Hastelloys B2 and C276 in Table 10.4). Sensitization is generally due to carbide precipitation, but, in the more complex alloys and in those containing molybdenum, intermetallic compounds or phases may also precipitate in the grain boundaries. Commercially pure nickel may also suffer weld decay and a low-carbon type (nickel 201) is used for the more severe environments. Intergranular attack of nickel-base alloys has been found in a number of acid media but it has also been observed in high-temperature water and in alkaline solutions. The standard test solution for detecting sensitization of these alloys is a mixture of sulphuric acid and ferric sulphate, as specified in ASTM A 262.

Nickel-base filler metal is used for joining austenitic Cr–Ni steels to carbon steel and low-alloy steel, particularly when the joint is to be given a postweld heat treatment. Where a Cr–Ni filler (say type 309 or 310) is used for such joints they are likely to embrittle by σ phase formation and by carbon migration from the ferritic steel into the weld deposit. Nickel-base weld metal does not suffer either of these two deficiencies. There is, however, the moderate hardening effect described earlier, and in most instances the weld metal will be sensitized, and may not be suitable for exposure to corrosive environments.

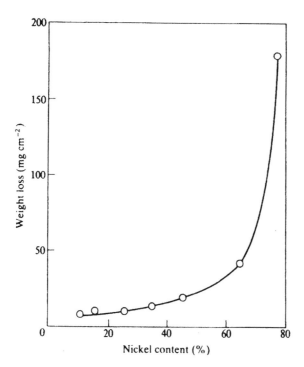

10.20 The influence of nickel content on intergranular corrosion on Ni–17Cr–Fe alloys. Tested 70 h in 5N $HNO_3 + 1 g\ l^{-1}$ Cr at 108 °C (from Heathorne, 1972).

10.4.5 Oxidation and creep resistance

A number of nickel-base and cobalt-base alloys have useful oxidation-resistance and creep-resistance properties, and at the same time retain adequate ductility after periods of ageing at elevated temperature. They are used for gas-turbine parts and in high-temperature furnaces for the petrochemical industry. Incoloy 800, for example, is specified for furnace tubes, headers and transfer lines in locations where there is a risk of thermal movement. Incoloy 800 behaves like a high-nickel steel when fusion welded, in that a matching-composition filler metal is too susceptible to solidification cracking for practical use. Nickel-base Ni–Cr–Fe filler alloys are therefore used to weld alloy 800, and the rupture strength of the weld metal is lower than that of the parent material. When this type of weldment is subject to strain in service, failure may occur in the form of creep cracking in the weld metal.

Problems associated with creep-resistant alloys are considered also in Section 9.8.

10.4.6 Alloys and welding procedures

Nickel and nickel-base alloys may be welded by nearly all the available welding processes. Those alloys that are employed for corrosion resistance (nickel, Monel, Hastelloy, for example) are normally joined using coated electrodes or gas tungsten arc welding. In gas metal arc and submerged arc welding it is necessary to restrict heat input rates to avoid solidification cracking, particularly with the complex nickel alloys and with cobalt-base alloys. In some instances, fusion welding is practicable in the flat position only. Gas-turbine alloys are welded by the gas tungsten arc and electron beam processes and also by resistance welding.

10.5 The reactive and refractory metals

The reactive metals (beryllium, niobium, molybdenum, tantalum, titanium, tungsten and zirconium) present some difficult welding problems. They have in common a high affinity for oxygen and other elements, and welding processes that employ a flux or permit exposure of heated metal to the atmosphere are inapplicable because of contamination and embrittlement of the joint. Likewise cleanliness is of special importance in welding reactive metals.

The processes that have been applied experimentally and in production are gas tungsten arc (DC electrode-negative), gas metal arc, electron beam, spot, seam and flash-butt, pressure, ultrasonic and explosive welding. Diffusion bonding is applicable to titanium alloys. Gas tungsten arc welding without filler wire addition is the most generally successful and widely used technique. The problems chiefly encountered are embrittlement due to contamination, embrittlement due to recrystallization, and porosity.

10.5.1 Embrittlement due to gas absorption

Beryllium, titanium, zirconium, niobium and tantalum react rapidly at temperatures well below the melting point with all the common gases except the inert gases. Contamination by dissolving oxygen and nitrogen from the atmosphere in the molten weld pool results in an increase in tensile strength and hardness and reduction in ductility (Fig. 10.21). As a result, even welds that have been effectively shielded show some hardness increase across the fused zone, and with increasing hardness the ductility of the joint is reduced (Fig. 10.22). Thus the ductility of the completed fusion weld depends upon the effectiveness of the gas shielding. For effective shielding it is necessary to provide a blanket of inert gas on both the torch side and the underside of the joint, and various devices, including welding jigs having an argon-filled groove below the joint line, are used for this purpose. Welding with a standard type of gas tungsten arc torch and with argon backing to the joint is practicable for titanium and zirconium. Better protection may be required for structural-quality joints, and for example this may be obtained by using a **trailing shield** (Fig. 10.23), which allows the welded joint to cool under an inert-gas shield. For gas metal arc welding, a leading as well as a trailing shield is required.

To reduce the risk of contamination still further, the welding may be done inside an argon-filled enclosure, such as a **plastic tent** that can be flushed with argon. Alternatively, a metal chamber fitted with glass ports for viewing the

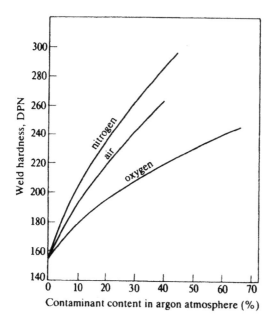

10.21 The effect of atmospheric contamination on the weld hardness in titanium (from Borland, 1961).

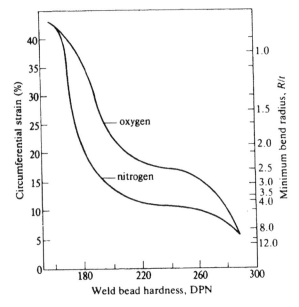

10.22 The relationship between weld hardness and ductility: titanium welds contaminated with oxygen or nitrogen (from Borland, 1961).

work and glove pockets for manipulation (**a glove box**) may be used. The most stringent precautions against contamination (such as may be necessary for fusion welding molybdenum or tungsten) require that the use of plastics and elastomers be kept to a minimum. The chamber is baked using built-in electric heaters to between 40 and 80 °C and is evacuated to a pressure of 1×10^{-2} N m^{-2} or less. It is then back-filled with purified argon, typically with an oxygen content of less than 1.5 ppm and water vapour less than 0.1 ppm (dewpoint −90 °C). Welding in such a chamber is carried out using a tungsten arc without any gas flow around the electrode.

Electron-beam welding avoids contamination problems entirely, since it is carried out in a vacuum. Resistance welds can normally be made in air because the time cycle is too short for any substantial degree of gas pick-up.

10.5.2 Porosity

Porosity in titanium fusion welds was at one time thought to be due to entrapment of argon gas between the butting faces of the joint, but it is now attributed to hydrogenation from inadequate precleaning. Metal produced by powder metallurgy often contains entrapped or dissolved gas that can be liberated on

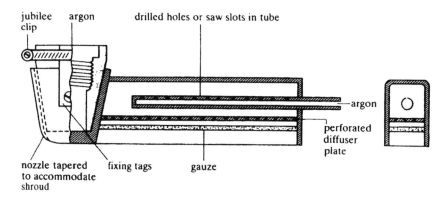

10.23 A typical design of a trailing shield for the argon-shielded gas tungsten arc welding of reactive metals (from Wilcock and Ruck, 1980).

fusion welding. Molybdenum and tungsten products usually derive from powder metallurgy compacts, and fusion welds in these metals may be so affected. Titanium and zirconium sheet and plate are normally rolled from cast ingots, and metal produced in this way is preferred for fusion welding. Porosity may also result (for example, in beryllium) from inadequate cleaning of the metal surface. Effective preweld cleaning is most important for all the reactive metals and is accomplished by machining, grit-blasting or grinding followed by degreasing, or by pickling. Given a clean metal surface, parent metal free from contamination, and defects and reasonable care in welding, porosity is not a serious problem with the metals under consideration.

10.5.3 Cracking

The unalloyed reactive metals are not prone to hot cracking, nor are the alloys that are commonly welded. The brittle metals beryllium, molybdenum and tungsten may crack at low temperature if welded under restraint, and joints are designed so as to minimize this hazard. Stress relieving also reduces the cracking risk and, for example, molybdenum may be heat treated at 900–925 °C for 1 h, and tungsten at 1400–1450 °C for 1 h.

10.5.4 Alloys and welding procedures

Titanium

Titanium is used on an increasing scale in gas-turbine engines, for air-frames, in seawater coolers and desalination plant, and in chemical plant. All these applications (with the exception of the plate heat exchanger) require the use of metal joining processes, particularly fusion welding.

Commercially pure titanium is used for corrosion-resistant applications, particularly for seawater exposure. Fusion welding is carried out by the inert gas shielded or plasma arc processes. Manual gas tungsten arc welding usually employs either a normal gas nozzle or a **gas lens**, which is a circular nozzle incorporating a gas distributor. The degree of contamination of weld metal can be measured by means of the hardness increase, which should not exceed 20 VPN for a single-pass weld. More simply it is assessed by the weld appearance. An acceptable weld is bright and silvery while contaminated welds are straw-coloured or, in the worst cases, a dull grey. For machine welding either a trailing shield or the type of jig illustrated in Fig. 10.24 may be used.

The metallurgy of titanium alloys is dominated by the phase change from hexagonal close-packed (α) to body-centred cubic (β), which in the pure metal occurs at 882 °C. Most alloying elements fall into one of two categories, α stabilizers or β stabilizers, depending on whether their presence raises (α stabilizers) or lowers (β stabilizers) the transformation temperature. The α-stabilized or near-α alloys, such as Ti–5Al–2.5Sn, are weldable more or less to the same degree as pure Ti. They are welded in the annealed condition.

The α–β alloys may be heat treated to high strength levels. The less heavily β-stabilized types, such as Ti–6Al–4V, may be welded in the annealed condition using the inert gas shielded processes and hardened by ageing during the postweld heat treatment. The higher-strength α–β alloys tend to be brittle in the

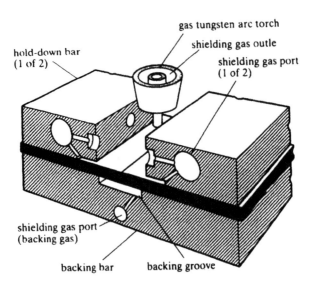

10.24 A gas-shielded jig for the gas tungsten arc welding of reactive metals (titanium in particular) (from Nippes, 1983).

heat-affected zone but acceptable welds may be obtained using electron-beam welding.

The ductility of a fusion weld may be improved by using a filler alloy of lower strength than the parent metal. Commercially pure titanium may be welded with an ELI (extra-low-interstitial) grade filler (AWS A.5.16 designation ER Ti–1), which has restricted carbon, oxygen, hydrogen and nitrogen contents. The alloys Ti–5Al–2.5Sn and Ti–6Al–4V are sometimes welded with pure titanium filler, there being sufficient dilution to ensure that the weld metal has the required strength.

Cleanliness of both parent metal and filler rod is most important. In addition to the contamination problems discussed earlier, titanium may be embrittled by hydrogen absorption, so that cleaning by degreasing or by pickling in nitric-hydrofluoric acid solution prior to welding may be required.

Zirconium, hafnium and tantalum

At room temperature, zirconium has a hexagonal close-packed structure (α-zirconium) and in the pure metal this transforms to body-centred cubic (β phase) at 860 °C. Hafnium has a similar structure but the transformation to the cubic β phase occurs at 176 °C. Zirconium and its alloys fall into two categories: commercial and nuclear grades. The commercial grades contain hafnium up to 4.5%, while the nuclear grades, although generally similar in alloy content, have a low hafnium content (maximum 0.01%). Pure zirconium is used for corrosion-resistant duties, for example as a lining for urea reactor vessels. In nuclear technology, Zircaloy is used for cladding fuel elements. Typical alloys are Zircaloy-2 (Zr–1.45Sn–0.14Fe 0.10Cr–0.60Ni) and Zr–2.5Nb.

Welding is by the gas tungsten arc process, direct-current electrode negative with matching filler rod, using the same precautions against contamination, including preweld cleansing, as for titanium. Weldments may be stress-relieved at 650–700 °C for 15–20 min.

Hafnium is used in the unalloyed form for control rods in nuclear reactors and for corrosion-resistant duties in the processing of spent nuclear fuel. It is welded using the same procedures as for zirconium.

Tantalum is employed primarily in chemical plant for its high resistance to corrosion, particularly against hydrochloric acid. It is welded using the gas tungsten arc process with the same precautions as those specified for other reactive metals, and also by the electron-beam process.

Beryllium

Because of its low neutron cross-section, beryllium has applications in the nuclear industry, while its mechanical properties make it attractive for use in space technology. In its optimum form as a hot extrusion, beryllium has a tensile strength of 500–700 MN m^{-2}, an elongation of 10–20%, a density about 70% of that of aluminium, and a modulus of elasticity four times greater (275–300 GN m^{-2}). Its disadvantages include a tendency to brittleness, toxicity and poor weldability.

Like aluminium, beryllium forms an adherent refractory oxide film and oxidized surfaces are cleaned prior to welding by an acid etch. If an edge preparation is required, the final machining cuts must be less than 0.075 mm to avoid twinning, which can result in welding cracks. Beryllium is also susceptible to solidification cracking owing to the presence of iron and aluminium, the content of which must be controlled. Gas metal arc welding with an Al–12Si (type 718) electrode has been used successfully for fusion welds. The heat input rate is kept low to minimize grain growth and embrittlement in the heat-affected zone. As with other reactive metals, it is essential to prevent atmospheric contamination. In addition, beryllium fume is toxic, so welding is carried out in a chamber (glove box) with a filtered exhaust system.

Molybdenum and tungsten

Molybdenum and tungsten are used mainly for high-temperature components in various industrial and domestic applications, and both metals may be joined by brazing, by the resistance, friction, laser and electron-beam welding processes as well as by gas tungsten arc welding. The crystal structure is body-centred cubic, as with iron, niobium and tantalum, and there is a temperature (the ductile-to-brittle transition temperature) below which the metal becomes brittle. However, the solubility of interstitial solute elements such as oxygen, carbon and nitrogen is much lower in molybdenum and tungsten than in the other three body-centred cubic elements, and the embrittling effect of these elements is correspondingly greater. Figure 10.25 shows the effect of oxygen, carbon and nitrogen on the ductile-to-brittle transition temperature of molybdenum. Other characteristics that affect weldability are the tendency to form coarse-grained structures on solidification and recrystallization, and strain-rate sensitivity. The strength, ductility and toughness of body-centred cubic metals falls with increasing grain size, so that if a fusion weld in tungsten or molybdenum is subject to tensile stress, the strain is concentrated in the coarse-grained heat-affected zone. Consequently, the strain rate is increased and the tendency to embrittlement accentuated. As a result, molybdenum and tungsten weldments may crack during cooling.

The risk of such cracking may be reduced by preheat, by postweld heat treatment, by selection of a suitable alloy–filler rod combination, and by avoiding contamination.

The required preheat temperature increases with the complexity and thickness of the item being welded, but in applying preheat it is also necessary to consider its possible effect on the extent of the coarse-grain heat-affected zone. Postweld heat treatment temperatures are given in Section 10.5.3.

Weldable materials include commercially pure molybdenum, either arc cast or power metallurgy, Mo–0.45Ti–0.10Zr (TZM), Mo–Zr–B and Mo–Al–B alloys. Filler alloys may be of matching composition or a molybdenum–rhenium alloy, such as Mo–50Re or Mo–20Re. TZM when gas tungsten arc welded with either a matching or a Mo–50Re filler has no significant ductility when tested at the strain rate normally used for steel ($10^{-2}-10^{-4}$ s^{-1}) but shows some elongation at low rates of strain ($10^{-6}-10^{-4}$ s^{-1}). The molybdenum–zirconium–boron alloy is more tolerant to recrystallization and has better fusion-welded properties.

The recognized precautions against contamination by oxygen and nitrogen are discussed in Section 10.5.1. In addition, carbon contamination due to the presence of lubricants must be avoided by degreasing or pickling the workpiece and by degreasing the welding chamber.

Brittleness may occur in spot-welded molybdenum. It may be minimized by inserting tantalum, zirconium, nickel or copper foil, or metal fibre between the surfaces to be spot welded.

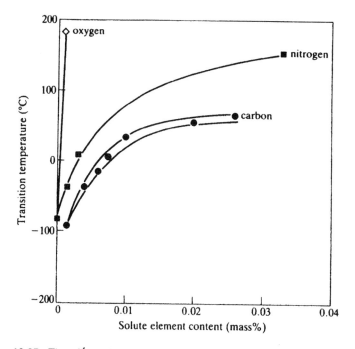

10.25 The effect of oxygen, nitrogen and carbon on the bend transition temperature of molybdenum (from Nippes, 1983).

10.6 The low-melting metals: lead and zinc

10.6.1 Lead

Although lead forms a notably protective oxide layer, it may readily be cleaned by scraping or scratch-brushing, and consequently is welded without flux. Gas welding is virtually the only method used for lead, and either the oxygen–coal gas, oxygen–hydrogen or oxyacetylene torches are applicable.

Metallurgical problems associated with the welding of lead are rare, most of the difficulties being manipulative. Carbon-steel chemical vessels handling corrosive fluids are sometimes homogeneously lined with lead. The metal surface is cleaned by shot blasting or other means and then tinned with a lead-tin solder using zinc chloride flux. Lead is then melted directly on to the tinned surface in two or three successive coats. Careful inspection for pinholes must be made between coats. Vessels may also be lined with sheet lead.

10.6.2 Zinc

Zinc is fusion welded either by the oxyacetylene welding process or by the gas tungsten arc process. In gas welding the oxide layer must be removed by means of a flux consisting of equal parts of ammonium chloride and zinc or lithium chloride. This is applied as an aqueous paste on either side of the joint, and welding is carried out with a neutral or slightly reducing flame, adding filler rod if necessary. Welds may be hammered to improve their strength, but such treatment must be carried out between 100 and 150 °C; outside this temperature range the metal is too brittle.

Zinc-coated sheet is successfully spot and seam welded in spite of reduced electrode life due to zinc contamination. Galvanized steel may be fusion welded using coated electrodes but there is a risk of porosity and cracking, and for stressed joints the galvanizing should be removed in the vicinity of the joint. Austenitic chromium–nickel weld metal will certainly crack if deposited on a galvanized or zinc-coated surface. Zinc-rich primers are often used for protecting steelwork before it is erected. Welds can be made through such primers if the paint thickness is not more than, say, 25 μm; with thicker layers there may be a reduction of impact properties of the weld metal.

Zinc-base die castings commonly contain 4 % aluminium and, because of the oxide skin that forms on heating, are most difficult to weld. Heavy fluxing and careful manipulation are the only available means to success in this case.

In fusion welding zinc and zinc-coated materials, there is a danger that toxic concentrations of zinc fume may build up. Good ventilation is therefore essential.

10.7 The precious metals: silver, gold and platinum

10.7.1 Silver

In its welding characteristics silver resembles copper: it has high thermal conductivity and low chemical affinity for atmospheric gases. It is, however, capable of dissolving oxygen, and rejection of this gas during welding can give rise to porosity.

Pure silver or the silver–30% copper alloy is used as a sheet metal lining or in the form of clad plate for vessels in the chemical industry. Liners were formerly welded using oxyacetylene torches (without flux), but current practice is to weld with the gas tungsten arc process, direct current, electrode negative. Welds free of porosity may be obtained by using lithium-deoxidized parent metal and filler wire. In welding clad plate, care must be taken to avoid iron contamination; this, however, is less difficult than with most clad material, because silver does not alloy with iron and has a much lower melting point.

Silver may be pressure welded, and spot welding is possible in spite of the high electrical conductivity.

10.7.2 Gold

Gold in fabricated form is used almost entirely for jewellery and the problem of welding in bulk form does not arise. Oxygen–coal gas welding of gold is practicable and is used by jewellers. Gold is also brazed and a series of gold brazing alloys are available. Gold may be readily pressure welded.

10.7.3 Platinum

Platinum is used for catalysts, insoluble anodes, thermocouples, laboratory ware, electrical contacts, and jewellery. Welding is accomplished by the oxyacetylene or gas tungsten arc processes (for example, for joining thermocouples), and it may be joined by resistance welding and pressure welding.

10.7.4 Other platinum-group metals

Palladium may be welded in the same way as platinum, but the use of reducing atmospheres must be avoided; in oxyacetylene welding a neutral or slightly oxidizing flame is used. Rhodium, iridium, ruthenium and osmium may be joined by tungsten inert-gas welding, the latter two in a controlled atmosphere chamber.

References

Borland, J.C. (1961) *British Welding Journal*, **8**, 64 ff.
Enjo, T. and Kuroda, T. (1982) *Transactions of the Japanese Welding Research Institute*, **11**, 61–66.

Heathorne, M. (1972) in *Localised Corrosion – Cause of Metal Failure*, ASTM Special Technical Publication STP 516.
Houldcroft, P.T. (1954) *British Welding Journal*, **1**, 470.
Kobayashi, T., Kuwana, T., Ando, M. and Fujita, I. (1970a) *Transactions of the Japanese Welding Society*, **1**, 1–11.
Kobayashi, T., Kuwana, T., Aoshima, I. and Nito, N. (1970b) *Transactions of the Japanese Welding Society*, **1**, 154–163.
Krivosheya, V.E. (1968) *Avt Svarka*, **4**, 13–14.
Matsuda, F., Nakata, K., Arai, K. and Tsukamoto, K. (1982) *Transactions of the Japanese Welding Research Institute*, **11**, 67–77.
Matsuda, F., Nakata. K., Shimokusi, Y., Tsukamoto, K. and Arai, K. (1983) *Transactions of the Japanese Welding Research Institute*, **12**, 81–187.
Matsuda, F., Nakata. K., Tsukamoto, K. and Uchiyama, T. (1984) *Transactions of the Japanese Welding Research Institute*, **13**, 57–66.
Nippes, E.F. (coord.) (1983) *Metals Handbook*, 9th edn, Vol. 6: Welding, brazing and soldering, American Society for Metals, Metals Park, Ohio.
Ohno, S. and Uda, M. (1981) *Transactions of the National Research Institute for Metals*, **23**, 243–248.
Pumphrey, W.I. and Jennings, P.H. (1948/49) *Journal of the Institute of Metals*, **75**, 203–233.
Uda, M. and Ohno, S. (1974) *Transactions of the National Research Institute for Metals*, **16**, 67–74.
West, E.G. (1951) *The Welding of Non-ferrous Metals*, Chapman and Hall, London.
Wilcock, A. and Ruck, R.J. (1980) *Metal Construction*, **12**, 219–221.

Further reading

American Society for Metals (1978) *Source Book on Selection and Fabrication of Aluminium Alloys*, American Society for Metals, Metals Park, Ohio.
American Society for Metals (1979) *Source Book on Copper and Copper Alloys*, American Society for Metals, Metals Park, Ohio.
American Welding Society (1982) *Welding Handbook*, 7th edn, Section 4, AWS, Miami, Macmillan, London.
Baker, R.G. (ed) (1970) *Proceedings of the Conference on Weldable Al–Zn–Mg Alloys*, TWI, Cambridge.
Dawson, R.J.C. (1973) *Fusion Welding and Brazing of Copper and Copper Alloys*, Newnes-Butterworth, London.
Hrivňák, I. (1992) *Theory of Weldability of Metals and Alloys*, Elsevier, Amsterdam.
Linnert, G.E. (1965) *Welding Metallurgy*, Vol. 1, American Welding Society, Miami.
Linnert, G.E. (1965) *Welding Metallurgy*, Vol. 2, American Welding Society, Miami.
Nippes, E.F. (coord.) (1983) *Metals Handbook*, 9th edn, Vol. 6: Welding, brazing and soldering, American Society for Metals, Metals Park, Ohio.
Schwartz, M.M. (1975) *The fabrication of dissimilar metal joints containing reactive and refractory metals*, Welding Research Council Bulletin, no. 210.
West, E.G. (1951) *The Welding of Non-ferrous Metals*, Chapman and Hall, London.

11
The behaviour of welds in service

11.1 General

Welding may generate both physical and metallurgical discontinuities in a structure and these in turn may adversely affect the service behaviour. The difference in properties between the weld and the parent metal is not very great with solid-phase welds made at elevated temperature, such as friction welds and hot pressure welds, but in fusion welds the weld metal and/or heat-affected zone may differ in various ways from the parent metal. Mechanical properties such as tensile and yield strength, hardness, impact properties, fatigue strength, creep-rupture strength and ductility, and resistance to slow or fast crack propagation may be affected. In addition, the resistance to various modes of corrosive attack may be modified; the weld or heat-affected zone may be subject to selective wastage, and if the conditions are such as to generate stress corrosion cracking then the residual stresses caused by fusion welding are normally sufficient to propagate such cracks. In the absence of metallurgical effects defects and discontinuities can give rise to various types of failure: for example, brittle fracture, accelerated fatigue, selective corrosion and stress corrosion cracking.

In this chapter an account is given of the way in which some of these metallurgical and physical discontinuities have affected, or may affect, the behaviour of welded structures in service.

11.2 The initiation and propagation of fast fractures
11.2.1 Brittle fracture

The term 'brittle fracture' is normally applied to fast unstable fractures that occur under plane-strain conditions; that is to say, there is little or no inward contraction of the metal at the fracture surface. The surface itself may be fibrous or matt in appearance, and show evidence of microvoid coalescence when viewed under the microscope. Alternatively it may exhibit shiny cleavage facets, or there may be a mixture of these types of fracture appearance. Brittle fracture is commonly associated with steel, but it must not be forgotten that other body-centred cubic metals also suffer a ductile–brittle transition, and that the mid-air explosion of the Comet I aircraft occurred as the result of the brittle unstable fracture of an age-hardened aluminium alloy. Theories that pertain to the conditions for initiating fast unstable fractures are described in Chapter 1.

The structures that have been most affected by brittle fracture in service are ships and pressure vessels, and some relevant cases are discussed below.

Ships and maritime structures

After the collapse of France during World War II, the US government decided to construct a large fleet of cargo ships and oil tankers. The only way to meet the production targets was to use welded instead of riveted construction. Work started at the Kaiser Swan Island shipyard in 1941, and ships were produced at a remarkable rate. There had been virtually no experience of this type of ship construction, however, and not surprisingly there were problems, notably because of brittle fracture.

Of the first 40 tankers built at the Kaiser yard, four broke in two as a result of brittle fracture. This was the most serious consequence; few of the observed fractures resulted in a total loss. Nevertheless the problem was a serious one, and at the time welded ships had a poor reputation.

Four factors contributed to the failures: design, material, workmanship and weather. Cracks originated at sharp hatch corners, at the ends of bilge keels and interrupted longitudinal stiffeners, and at other discontinuities. The steel had a transition temperature that was frequently above 0 °C. Weld quality was variable, and some of the failures originated at weld defects. And the North Atlantic weather was most unfavourable during winter. All these factors received attention; even the weather was improved by choosing more favourable routes. One expedient was to install up to five riveted seams in the hull. Riveted ships were not immune from brittle fractures, but the cracks rarely extended beyond a single plate. However, the ships containing these supposed crack-arrestors fared no better than the others.

By the late 1950s the loss rate of ships due to brittle fracture had ceased to be a significant problem. Design, steel quality and welding have all improved during the latter half of the century, and the type of failure experienced by the wartime ships is not likely to recur.

The 'Sea Gem'

Brittle fracture has not been a significant cause of loss in the case of offshore structures. In part this is because exploration in the North Sea, where winter temperatures are low enough to favour this mode of failure, did not start until the early 1960s, by which time the ship failures had given adequate warning of the potential danger. The *Sea Gem* accident was an exception to this rule.

This jack-up rig had been built in 1953, and had found various applications until in 1964 it was purchased by a British contractor for lease to British Petroleum as a drilling platform. To this end the deck was fitted with the necessary equipment and lengthened. The legs, of which there were ten, were cylindrical and 5 ft 11 inches (1.8 m) in diameter; these were also lengthened. A minimum impact strength of 25 ft-lbs (34 J) at 0 °C was specified for new plate, but the old structure was accepted without, so far as is known, any tests being made.

To move the platform, it was lowered until it floated with the legs projecting upwards, and it was then towed by a tug to the new drilling location. Drilling was carried out with the platform jacked up to about 18 m (60 ft) above sea level. When stationary and also during raising and lowering operations, the platform was suspended from grippers around the legs by tiebars, of which there were four to each leg.

On 27 December 1965, work being completed, preparations were made for a move. The air temperature was 3 °C, with a force 5 wind, and a wave height of up to 3 m (10 ft). Before starting to lower the platform it was decided to check the operation of the jacks by raising it a short distance. While this operation was under way, the platform suddenly heeled, and then, after stabilizing for a short time, fell into the sea about 15 m (50 ft) below. Only 1 man was injured by the fall, but 13 died subsequently of exposure or by drowning.

This catastrophe was undoubtedly due to the failure of the tiebars. Figure 11.1 shows the dimensions and profile of these items. They were flame cut from plate $2\frac{1}{2}$ inches (63 mm) thick, the specification being either ASTM A36 or a local equivalent. A36 is a plain structural carbon steel with no special requirements as to notch-ductility. A number of broken bars were recovered from the wreck, and the fractures were found to have originated at a sharp radius between the spade end of the bar and its shank (the left-hand end in Fig. 11.1). It was also found that surface weld runs had been made in some places, presumably to disguise scars due to the flame cutting. There were inclusions and cracks associated with these welds, and the hardness in the heat-affected zone averaged about 300 Vickers, with a maximum of 480. There were cracks, porosity and slag inclusions associated with these welds, and a brittle failure had originated from one of them. It was also found that in a number of cases the pin holes at the ends of the bars were elongated, indicating an overload. The design of the rig was such that jamming between legs and hull was possible, which could account for such an overload.

Impact tests were carried out on the recovered tiebar material. Averages from these tests are plotted as a transition curve in Fig. 11.2, and from this curve it may be determined that the 27 J transition temperature would be about 10 °C. Thus, at the time of the accident, some if not all of the tiebars were in a notch-brittle condition. On the night of 23–24 December a loud noise was heard and two tiebars were found to have fractured. The fact that these failures took place at night also suggests that low temperature was an important factor in the accident itself.

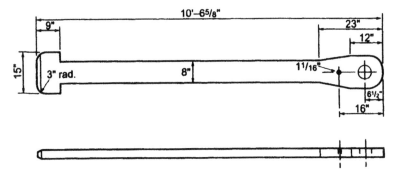

11.1 Dimensions of tiebars for jacks on the *Sea Gem*. 1 inch = 25.4 mm.

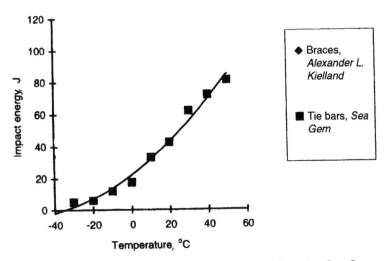

11.2 Charpy V energy transition curve for tiebars from the *Sea Gem*.

During the collapse and on impact with the water further brittle fractures took place, to the extent that a diver (possibly with some exaggeration) said that the wreck looked like a 'heap of broken glass'. The older part of the rig had spent most of its working life in the Middle East and Southern Europe, and was evidently not suited to North Sea conditions.

Ammonia storage tank failure

Liquid anhydrous ammonia is used as a fertilizer, and it is stored either at room temperature under pressure or at atmospheric pressure and maintained at subzero temperature. At a fertilizer plant in Potchefstroom, South Africa, the ammonia was stored under pressure in four 50-ton horizontal cylinders of the type known as bullets. On 3 July 1973 one of these bullets exploded while being filled from a rail car. An estimated 30 tons of liquid ammonia escaped from the vessel and a further 8 tons was lost from the rail car before the feed pump could be shut down. The air was still at the time of the spill and a large cloud formed, but within a few minutes a slight breeze arose, driving the gas over the perimeter fence and into a neighbouring township. Eighteen people died as a result of this spillage.

Local regulations required that pressure containers be given a hydrostatic test after a stated period of service, and the tank in question had been due for such a test shortly before the accident. There were some difficulties in carrying out the test, and the management of the works obtained permission to make an ultrasonic inspection of the vessel as an alternative. In the course of this examination one of the dished ends of the failed vessel was found to be laminated to an unacceptable degree in two places. One of these defects was repaired successfully by welding with coated electrodes, but the other required three repair attempts before it was cleared. The plate around this repair weld had low notch-ductility (its transition temperature was over 100 °C). A brittle fracture had initiated in this plate, and it then followed a roughly elliptical course around the repair weld, driven no doubt by its residual stress field. The cause of embrittlement of the plate material is not known.

The use of ultrasonic testing as a substitute for hydrostatic testing was fundamentally unsound. These two tests are different in character and serve quite different ends. The first hydrotest conditions the vessel, as discussed earlier, and subsequent ones demonstrate its continued integrity. Ultrasonic testing discloses cracks and laminations but it is unwise to assume that a vessel that has been in service and has passed such a test is safe. It could, as in the case in question, be brittle enough to fail in the absence of cracks.

11.2.2 Other modes of fast fracture

Fast fractures, that is to say, those that have a velocity of the order to $10^3 \, \text{m s}^{-1}$ may occur in the plane-strain mode, in which case they will be classified as brittle fractures, or they may be of the plane-stress type, when they will be classified as ductile. An example of the second type of failure is considered below.

The Alexander L. Kielland *accident*

The vessel in question was a semi-submersible platform that was used for accommodating offshore workers in part of the Norwegian sector of the North Sea. It was of a type known as 'Pentagone', and consisted of a lozenge-shaped platform supported by five columns that were themselves supported by submerged pontoons arranged more or less in an elliptical formation. The pontoons were restrained from spreading outwards by horizontal and diagonal tubular ties, and the deck was supported by compression members acting against the columns. The system was not unlike a set of five cranes, with their jibs pointing towards the centre of the platform and the bases prevented from outward movement by the horizontal stays, or braces. The design had no fail-safe element; on the contrary, failure of any of the tension members could throw unacceptable loads on its neighbours. Figure 11.3 shows a diagrammatic plan view of the arrangement.

A section of the design feature from which the failure originated is sketched in Fig. 11.4. This was an instrument housing, and it was made by boring a hole in one of the bottom horizontal braces, and fillet welding a flanged cylinder into this hole. The cylinder was made by rolling plate to the required diameter and making a closing longitudinal weld. The cylinder acted as a housing for the instrument and at the same time provided reinforcement for the opening.

Unfortunately the plate from which the fitting was made had exceptionally poor through-thickness ductility, as a result of which it suffered lamellar tearing

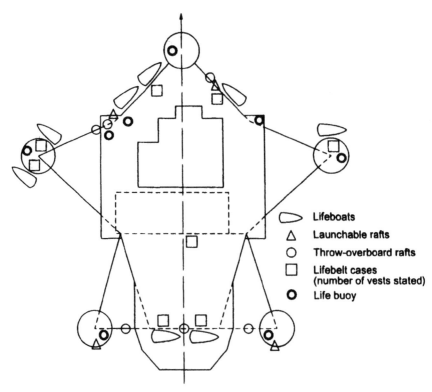

11.3 Diagrammatic plan view of the semi-submersible platform *Alexander L. Kielland.*

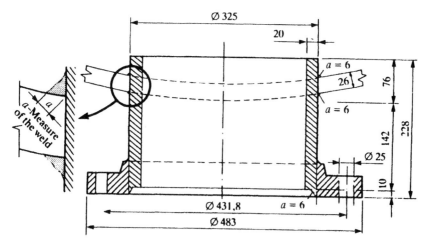

11.4 Section of the instrument housing which suffered lamellar tearing and which initiated the collapse of the *Alexander L. Kielland*: general arrangement and weld detail. The throat thickness of the weld was specified to be 6 mm.

during the welding operations (see Section 8.5.5). Reinforcement of the hole was therefore inadequate, and the brace into which the fitting was welded suffered a fatigue failure. Adjacent braces were then overloaded and failed progressively until the column broke away and the rig collapsed and capsized. Some 123 lives were lost in this tragic accident.

The braces that severed were recovered and the fracture surfaces examined. These were fibrous and lay at an angle to the pipe surface about 45° in one direction to 45° in the other. There was a variable amount of contraction, and in a few places the surface had arrow or chevron markings, similar to those found on plane-strain fractures. Such arrows point towards the origin of the fracture. In one instance this was clearly defined as being a small shiny flat area, but in all other cases the markings were too dispersed and the fracture appearance too uniform to identify the point of initiation. It seems reasonable, however, to interpret these failures as fast fractures that occurred under plane-stress conditions. The overload probably resulted in the formation of an internal cavity similar to that which appears in a cylindrical tensile testpiece after the onset of necking. Subsequently this cavity extended in both directions around the circumference of the brace.

The inertia of the structure was such as to preclude shock loading in this instance; however, the release of strain energy resulting from the initial failure resulted in severe vibration, which could have influenced the character of the fractures.

406 Metallurgy of welding

Earthquake fractures

As a result of two earthquakes, that at Northridge, California, in January 1994 and the great Hinshin earthquake that devastated the city of Kobe, Japan, in January 1995, steel-framed buildings suffered large numbers of brittle cracks. These events are a matter of concern on two counts; firstly because steel was expected to behave in a ductile manner when exposed to seismic shock, and secondly because in many cases there was no initiating crack. Fracture mechanics requires that a crack be present before it can extend in an unstable fashion. When the crack length is zero, the stress intensity factor is likewise zero.

The Richter scale is a measure of the amplitude of ground movement in a seismic disturbance, and is often taken as a measure of the intensity of the incident. The stability of buildings, however, is much more affected by accelerations produced by the seismic waves, which in turn relates to the frequency of the shock waves. So the Richter value is not always a good indicator for the amount of structural damage. Thus, at Northridge the Richter measurement was 6.8, a relatively modest value, but the ground accelerations were high, $1.2\,g$ and $1.8\,g$ in the horizontal and vertical directions respectively. Damage was extensive, although there were no fatalities. Over a hundred steel-framed buildings suffered plane-strain brittle cracks, mainly in the vicinity of beam-to-column welded joints. Figure 11.5 shows the general form of the cracks, which extended for a relatively short distance. In no case was a load-bearing member severed, and none of the steel-framed buildings collapsed. The brittle cracks appeared to initiate at the root of the weld, which had been made on to a backing bar. Subsequently an 'improved' joint was made by back-gouging and rewelding the root, and providing a fillet weld to minimize the stress concentration at the junction between beam and column. In testing, however, the improved joint performed no better than the original. It may be concluded, therefore, that any pre-existing cracks that might have been present were not the primary cause of the brittle cracks.

The Hinshin Japanese earthquake had similar characteristics to that at Northridge. It measured 7.2 on the Richter scale, and accelerations were high. The maximum ground speed was $1.04\,\mathrm{m\,s^{-1}}$ and the maximum displacement $0.27\,\mathrm{m}$. Damage was severe and extensive. Over 5000 people were killed, 27,000 injured, and 150,000 buildings were damaged. Some 90% of the casualties were due to the collapse of small wooden buildings. Some two-storey buildings also collapsed but taller buildings, and in particular multi-storey steel-framed structures, most of which had been built to earthquake-resistant standards, survived.

Against the good record for structural steelwork must be set the fact that brittle fractures were numerous. They were not confined to beam-to-column joints, but were found, for example, in column-to-column splices. They were not necessarily related to joints: some originated at access holes at the end of beams. Others

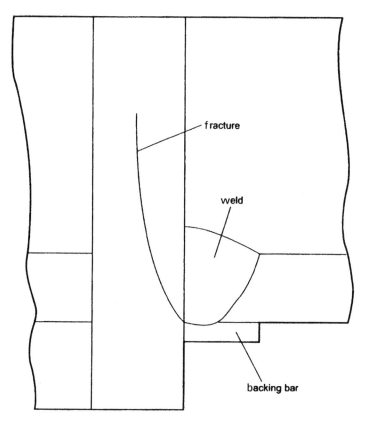

11.5 Typical brittle fracture in structural steelwork after Northridge earthquake.

occurred at the top of motorway bridge columns. Their incidence was widespread and varied.

In buildings fabricated to earlier codes, where partial penetration welds were allowed, the fractures showed no sign of plastic deformation. Fractures at joints incorporating full-penetration welds, however, were associated with a degree of plastic yielding.

Both earthquakes occurred in wintertime. However, the structures were mostly enclosed, so not exposed to excessively low temperatures. In addition, the Japanese steel had good notch-ductility; samples taken after the incident gave Charpy V-notch values of 50–80 J at 0 °C.

Fracture in the absence of cracks

The fractures described in the preceding two sub-sections have a number of features in common. Firstly, they occurred in the absence of a pre-existing crack.

Secondly, there did not appear to be any general yielding of the surrounding metal, although yielding local to the fracture was observed in a number of cases. Brittle, plane-strain fractures were more likely in the presence of a severe notch, such as that associated with a partial penetration weld. By contrast, where the stress concentration associated with the joint was small or (as in the case of the *Alexander L. Kielland* failures) entirely absent, yielding was likely to occur adjacent to the fracture.

The loading conditions likewise showed some similarities. In an earthquake the structural load imposed by the seismic tremor is roughly proportional to acceleration, and in both the Northridge and the Hinshin earthquakes accelerations were high. In the offshore incident the load was sufficient to fracture structural members and cause a catastrophic failure. It was calculated that the strain rate imposed on structural parts during the Hinshin earthquake was up to $0.1\,\text{s}^{-1}$. For the *Alexander L. Kielland* incident, the inertia of column and pontoon was such that full loading could have taken up to 1–2 s. Strain rates were therefore a few orders of magnitude higher than in standard tensile tests, but lower than that associated with impact loading.

Welding appeared to play a secondary role in this type of fracture. The failure could initiate at any point of stress concentration whether a weld was present or not; however, the most severe stress concentrations were associated with welded joints. These experiences underline the need to avoid stress concentration in the construction of earthquake-resistant steel structures, in so far as this is possible. Fortunately, earth tremors only rarely have accelerations of the magnitude noted in the two events recorded above, and only a minority of steel-framed buildings are erected in earthquake zones. There is no established theory relating to the phenomena described above, so that any discussion thereof must be somewhat speculative.

In Chapter 1 it was noted that a normally ductile metal could withstand very high rates of strain when exposed to uniaxial plastic strain, and that high strain rates were associated with a relatively modest increase in yield strength. It was also pointed out that such conditions would not occur in the case of multiaxial strain. In the extreme case of isotropic three-dimensional strain (equivalent to a negative hydrostatic pressure) no plastic deformation is possible, and if straining is continued to a sufficient degree, an internal fracture will occur, and its propagation will result in failure. This mechanism accounts for the fracture of a cylindrical tensile test bar, as described earlier. It may also be relevant to earthquake failures, where small volumes of metal near intersections (regions of stress concentration) may be subject to multiaxial strain. The evidence suggests that where strain rates are high and where displacements are sufficiently large, a fracture will initiate and develop into a running crack. Such cracks may be of the plane-strain type, with no bulk plastic deformation, or they may be accompanied by plastic strain adjacent to the crack. High stress concentration and low fracture toughness favour plane-strain fracture and vice versa.

11.3 Slow crack propagation

There are a number of modes of slow cracking that may affect welds, including hydrogen cracking, reheat cracking, creep cracking, stress corrosion cracking and fatigue. Cracks may also appear at the interface between ferritic and austenitic chromium–nickel steels and between ferritic and nickel–base alloys after service at elevated temperature. Cracking phenomena associated with corrosion, other than corrosion fatigue, are discussed in Section 11.4.

11.3.1 Fatigue

When a metal is subject to an alternating stress in the elastic range it may, after a given number of cycles, develop a surface crack. Further cycles of stress cause the crack to extend until the part fails in tension or shear across the reduced section. Carbon and carbon–manganese steels have a **fatigue limit**, which is an alternating stress range below which cracking does not initiate and fatigue failure does not occur. Other steels, and metals such as aluminium, do not necessarily have any fatigue limit, so that their fatigue strength is usually expressed as the limiting stress range for failure at an arbitrary number of cycles, say 10^6. Fatigue test results show a characteristic scatter, and an acceptable stress range is that appropriate to a given probability of survival.

The fatigue properties of fusion welds are reduced by two factors. Firstly, there is a stress concentration factor associated with the weld profile as noted in Section 8.4. Secondly, a small hot tear commonly forms adjacent to the fusion boundary in the case of fillet welds, particularly those made using coated electrodes. Figure 11.6 shows a diagram of such a weld after a period of fatigue loading, together with a micrograph showing how the fatigue crack extends from the base of the hot tear. For steel plate in the as-rolled condition a high proportion of the fatigue life is spent in the initial phase, where a surface crack is formed. For a plate on which there is a fillet weld this phase is short-circuited, and the fatigue life is correspondingly reduced. Figure 11.7 compares the fatigue properties of unwelded plate with one containing a drilled hole and a fillet welded specimen. It will be seen that not only has the life at any given stress level been reduced, but so has the fatigue limit. Reduction of the fatigue limit is due to stress concentration, for which the hot tear is also, in all probability, primarily responsible.

It is logical to suppose that the damaging effect of a fillet weld could be reduced by removing the hot tear, or by neutralizing its effect. This is indeed the case. Figure 11.8 compares fatigue curves for fillet welded steel in the as-welded, ground and peened (hammered) conditions. The toe of the weld is ground to a depth of about 1 mm using a disk grinder, while peening is carried out using a pneumatic hammer. This treatment puts the surface layers of the metal in compression, so reducing the damaging tensile stress. A still better result is

410 Metallurgy of welding

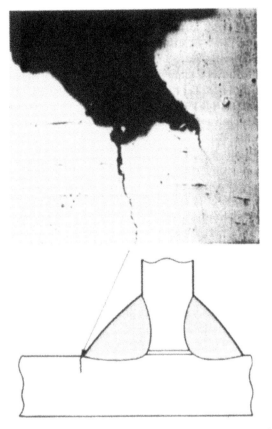

11.6 Cross-section of toe of fillet weld showing fatigue crack propagating from slag-filled discontinuity (Maddox, 1985).

obtained by melting the weld toe region using a gas tungsten arc welding torch. Grinding is usually the best choice for site welding; it may not give the best results but it is easy to apply and to inspect.

The amount by which fatigue strength of welded structural steelwork is reduced depends on the orientation of the weld relative to the loading direction; a transverse fillet welded attachment has a severe effect whereas that of a longitudinal butt weld is relatively small. In order to provide guidance for the design of steel bridges, welds are divided into categories according to stress orientation, as shown in Fig. 11.9, while the stress for a 97.5% probability of survival is given in Fig. 11.10. For the most recent fatigue design curves the latest edition of BS 5400 should be consulted.

Where an unwelded structural member is subject to cyclic loading, the resulting stress cycle will normally range more or less equally from compression

The behaviour of welds in service 411

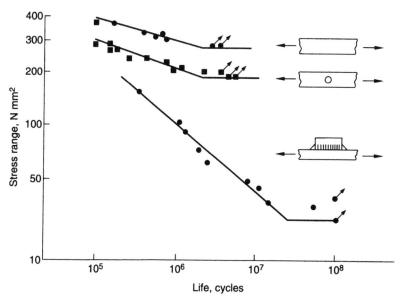

11.7 Comparison of the fatigue strengths of plain steel plate with a central hole and plate with a fillet-welded attachment. Stress zero to maximum (Maddox, 1985).

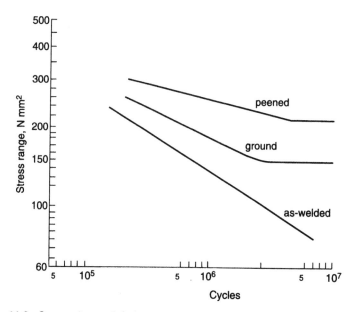

11.8 Comparison of fatigue strength of specimens with fillet-welded attachments in the peened, ground and as-welded conditions. Stress zero to maximum (Booth, 1986).

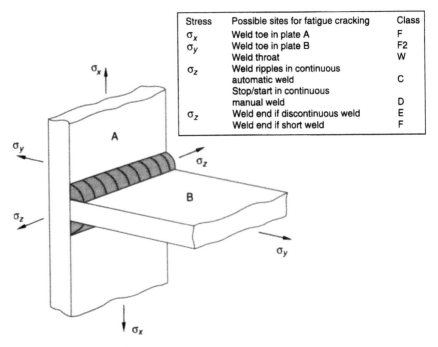

11.9 Welded joint classification for design against fatigue (Maddox, 1980).

to tension. Adjacent to a weld, however, there is a residual stress (in the as-welded condition, as will normally be the case for structural work) of yield point magnitude. In this region the compressive half of the stress cycle has the effect of reducing but not cancelling the tensile stress. Fatigue damage occurs only in the tensile part of the cycle. Thus, for a welded structure the stress range is, in effect, doubled, and this factor must be taken into account in the design.

The rate of growth of fatigue cracks, and of fatigue cracks in welds in particular, can be correlated with the stress intensity range ΔK to which the specimen or structure is exposed. The crack growth rate is given by

$$\frac{da}{dN} = C_0 (\Delta K)^n \tag{11.1}$$

where C_0 is a constant and n varies in the range 2–30. Figure 11.11 shows a typical plot. In fatigue there is no size effect and small laboratory test specimens can be used to assess growth rates in large sections. Fatigue cracking is primarily of concern in structures exposed to fluctuating loads, such as bridges, but may also be a factor in the design of pressure vessels.

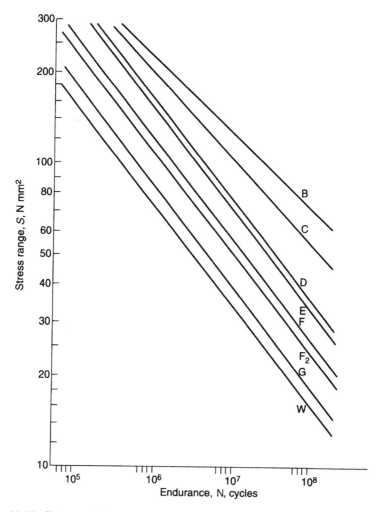

11.10 Fatigue design curves for the classes of joint illustrated in Fig. 11.9 (Maddox, 1980).

11.3.2 Reheat and creep cracking

Cracking in or adjacent to welds may occur in service at elevated temperature: for example, in steam power or petrochemical plant. Cracks may initiate during service at surface irregularities or may be initiated by pre-existing cracks formed during welding. In the latter case crack growth is usually by a creep mechanism, in which cavities nucleate and grow along grain boundaries. In elevated-temperature laboratory tests, it is found that where the test specimen contains a crack of length a, the crack growth rate is, in many instances, proportional to the

414 Metallurgy of welding

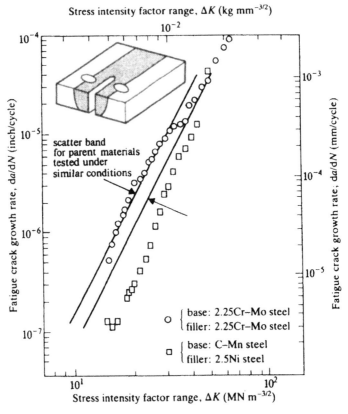

11.11 The fatigue crack growth behaviour of pressure vessel steel weldments tested in an air environment at 24 °C (from James, 1977).

nth power of the stress intensity factor:

$$\frac{\partial a}{\partial t} \propto K^n \tag{11.2}$$

The exponent n lies between 3 and 30, and for any given material has a value similar to the exponent for the rate of strain in secondary creep. In other cases the crack growth rate correlates better with the net section stress (load divided by uncracked sectional area). Justification for such correlations is possible using the techniques of non-linear fracture mechanics.

Reheat cracking occurred during service when 18Cr–12Ni–1Nb piping was used for supercritical steam power plant. The mechanism of such cracking is described in Section 9.5.3. Cracks initiated at the weld boundary in joints subjected to bending strain due to expansion of the piping system, and propagated through the heat-affected zone. The 18Cr–12Ni–1Nb alloy is no

The behaviour of welds in service   415

longer employed for steam service, but such cracking could occur owing to the inadvertent presence of niobium in other austenitic chromium–nickel grades.

Reheat cracking of ferritic alloy steel in service has been found in 0.5Cr–0.5Mo–0.25V piping, and the nature of the cracking is probably similar to that found after postweld heat treatment. Laboratory tests show that two cracking modes are possible, medium-temperature (350–600 °C) brittle intergranular cracking and a high-temperature (>600 °C) intergranular mode in which the fracture occurs by microvoid coalescence (see Section 8.5.6). For the medium-temperature type of cracking, the growth rate is an exponential function of the stress intensity factor K:

$$\frac{\partial a}{\partial t} = A e^{BK} \tag{11.3}$$

where A and B are constants. Figure 11.12 shows growth rates at 400–600 °C in a commercial 2.25Cr–1Mo steel. The tests were carried out by first subjecting the sample to a simulated weld thermal cycle such as to produce a microstructure similar to that of the coarse-grained heat-affected zone of a fusion weld. The specimens were notched and loaded in four-point bending to a given strain. They were then held at constant strain at the indicated temperature, and the crack growth determined by potential drop measurements. The medium-temperature

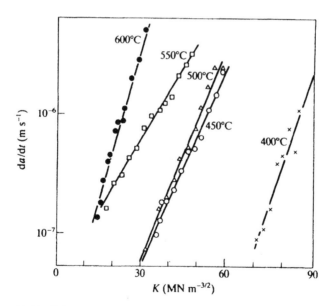

11.12 Brittle intergranular cracking of 2.25Cr–1Mo steel given a simulated weld heat-affected zone thermal cycle and subject to a stress relaxation test in the notched condition. The crack growth rate is plotted as a function of the stress intensity factor K (from Hippsley et al., 1974).

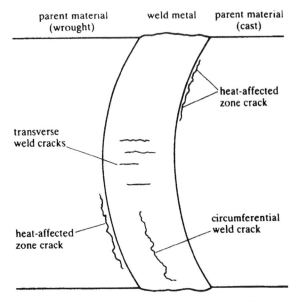

11.13 The typical form of cracking in 0.5Cr–0.5Mo–0.25V steam piping welded with 2.25Cr–1Mo electrodes after long-term service at elevated temperature (from Toft and Yeldhan, 1972).

brittle intergranular mode of cracking is associated with strain-induced segregation of sulphur to the crack tip. The degree of embrittlement due to sulphur segregation is augmented by the presence of phosphorus and tin (see Section 8.5.6).

The growth of reheat cracks at temperatures above 600 °C conforms to equation 11.2, and is similar to, possibly identical with, creep cracking. The embrittling elements phosphorus and arsenic also promote the high-temperature mode of cracking.

The morphology of cracks in welds made using 2.25 Cr–1Mo coated electrodes between 0.5Cr–0.5Mo–0.25V components in steam power piping is illustrated in Fig. 11.13. The operating temperature of such piping is between 550 and 580 °C, and the cracks result from a combination of stress relief cracking and creep cracking in service. The fractures result from the relaxation of strain by creep, and the mode of failure could reasonably be entitled **strain relaxation cracking**.

11.3.3 The cracking of ferritic–austenitic joints

Cracking in ferritic-austenitic interfacial regions may occur when the joint is held for long periods of time at elevated temperature, as in transition joints (see also Section 9.6.3) and hydrocracker reactor cladding (see also Section 9.6.2).

Transition joints

In large power-generating plant, austenitic chromium–nickel steels are used for the high-temperature parts of superheaters and reheat sections, and it is necessary to make welded joints between the austenitic steel and the 2.25Cr–1Mo ferritic alloy used at lower temperatures. Such joints are made in both small-bore tubes and large-bore pipes. Originally austenitic chromium–nickel steel weld metal was used, but more recently this has been superseded by nickel-based welds. Both types of joint fail after a period of service but nickel-based welds have a longer life. Operating temperatures are about 600 °C and the failures take the form of low-ductility creep cracking in the ferritic steel close to the interface.

When joints are made with an austenitic filler metal, there is a residual stress due to the mismatch of thermal expansion coefficients of austenitic and ferritic materials. This stress relaxes during service but is sufficient to promote failure. There is also a mismatch between the creep properties of the coarse-grained heat-affected zone and the parent metal on the ferritic side of the joint. Figure 11.14 shows time to failure versus temperature for cross-weld simulated transition joints where failure took place in the heat-affected zone, as compared with unwelded material. With low stress the heat-affected zone has a shorter rupture life than the parent metal, and therefore fails prematurely. Failures take the form of intergranular cracks about 100 μm from the fusion line. Carbon migration from the ferritic steel to the austenitic weld metal occurs at the operating temperature, and this may contribute to the heat-affected zone weakness. The cracks in nickel-based joints form in quite a different way. At the 2.25Cr–1Mo/Ni alloy interface there is a transition zone of width 0.5–50 μm, but generally about 1.0 μm, containing about 5 % Cr and 5 % Ni, and martensitic in structure. In long-term service, carbides precipitate in this region. When the martensitic band is narrow (~1.0 μm), the precipitates form a single line at the edge of the martensite bordering the ferritic steel. Where it is broader, the precipitates are finer and form in a band within the martensite. These two precipitation modes are known as type I and type II, respectively. Cracking is associated with type I carbides; microcracks initiate at the precipitates and grow to form a continuous crack.

Disbonding and cracking in hydrogenator reactors

Although this problem relates to a specialized field of process technology, its metallurgical features are of considerable interest. The reactors in question are used for the conversion of heavy oil distillates to lighter fractions by hydrogenation at elevated temperature (up to 460 °C) and high pressure (up to 250 atm). They are fabricated from 2.25Cr–1Mo steel up to 250 mm in thickness and weld deposit clad internally, normally with a first layer of type 309 followed by a surface layer of type 347 steel. After completion of all welding operations

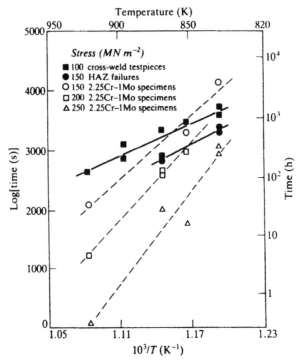

11.14 The variation of rupture life of cross-weld testpieces and parent 2.25Cr–1Mo steel as a function of temperature and applied stress (from Chilton *et al.*, 1984).

the vessels are given a heat treatment at 650–675 °C for about 12 h, which results in some carbon migration and precipitation at the austenite–ferrite interface.

Following a period of operation, which may be several years, cracks may form along the cladding interface and in severe cases these cracks join up to cause partial disbonding of the cladding. The location of cracks relative to the interface is shown diagrammatically in Fig. 11.15. Type A **disbonding** (as illustrated in Fig. 11.13) consists of intergranular cracks in the austenite adjacent to the interface. Type B cracking is also mainly intergranular and runs parallel to the interface. Type C cracks occur in the brown-etching martensitic zone, and type D in the black-etching carbon migration region. Laboratory tests indicate that type B cracks form most readily, and that disbonding is favoured by minimum dilution and a large austenite grain size close to the interface, such as is produced by electroslag cladding. Conversely, procedures that give a relatively wide transition zone between austenite and ferrite are less susceptible to cracking under test conditions. It is interesting to note that, as with transition joints made with nickel-base electrodes, better results are obtained with a wide rather than a narrow transition zone.

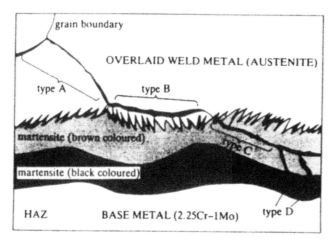

11.15 Types of crack found in disbonded weld overlays. Cracking is generally intergranular, types A and B (designated by some authors type II) in austenite and types C and D (otherwise designated type I) in martensite (from Matsuda and Nakagawa, 1984).

Disbonding is generally considered to be due to hydrogen-induced cracking. During operation, hydrogen diffuses through the vessel wall, and on cooling to room temperature the absorbed hydrogen is redistributed in the manner illustrated in Fig. 9.22, such that high concentrations of hydrogen occur in the austenite close to the interface. Calculation (with some support from measurement) indicates a hydrogen content of 350–700 ppm in the austenite and 1–5 ppm in the ferrite. Cracking in austenite is promoted by hydrogen above 500 ppm, and by sulphur, phosphorus and silicon; also, as stated earlier, by coarse grain. In laboratory tests, disbonding cracks start to form about 6 h after cooling to room temperature, and continue to propagate for several days.

In principle, disbonding could be prevented by holding the reactor at elevated temperature and low pressure prior to shut-down in order to outgas the shell, but this may present operational difficulties. Efforts at prevention are mainly directed towards modifying cladding procedures and minimizing the content of embrittling elements.

Cracking of fillet welds between internals and the austenitic cladding has also been found in hydrocrackers. This is due to a combination of σ phase embrittlement (σ phase being formed during postweld heat treatment) and hydrogen embrittlement.

11.4 Corrosion of welds

11.4.1 Localized wastage

Selective corrosion of welds occurs most frequently in media having high electrical conductivity such as seawater. It may be due to differences in structure, to segregation, or different alloy content. Figure 11.16 illustrates diagrammatically the different modes of attack. Selective corrosion of weld metal may be severe if the weld metal forms the anode of a galvanic couple. This has occurred for example in piping used for the injection of seawater into oil-bearing formations in the North Sea. The seawater so used is de-aerated and is nominally non-corrosive. However, process upsets or the addition of biocide may alter the balance; in some instances the weld metal is almost completely dissolved. Such attack may be prevented by the addition of small amounts of copper, chromium or nickel to the welding consumable. In aqueous solutions containing CO_2, attack may occur on either the weld metal or the heat-affected zone. In this case the attack on weld metal is not galvanic in character and alloying with copper, etc. is of little help. The preferential corrosion of the heat-affected zone of steel welds tends to be more severe with increased hardness. Warm seawater (on the shell side of a heat exchanger for example) may cause severe tramline corrosion on either side of the weld.

Austenitic chromium–nickel steels suffer preferential corrosion of the weld metal under various conditions. In acid chloride environments that are encountered in the food industry type 316 weld metal may be selectively attacked and one solution to this problem is to raise the molybdenum content of the weld metal, say, 1 % above that of the parent metal. This problem, which also occurs in urea plants, is due to microsegregation in the weld metal. It is discussed in Section 9.5.2, especially in relation to the use of high molybdenum superaustenitic steels in severely corrosive liquids.

Aluminium is a useful metal for handling strong nitric acid, but high-purity aluminium is required for the more severe conditions. In such conditions weld

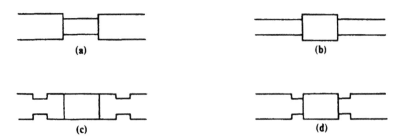

11.16 Selective corrosion of welds: (a) parent metal passive; (b) weld metal passive; (c) attack on the low-temperature region of the heat-affected zone; (d) attacks on the high-temperature region of the heat-affected zone.

metal is selectively attacked, probably because of the inevitable segregation in the cast structure of the weld.

Under special conditions exposure to elevated temperature may cause deterioration of welds. Hot, hydrogen-rich gas may attack carbon and low-alloy steels at temperatures above 250 °C, as described in Section 8.5.3. The operating conditions (partial pressure and temperature of hydrogen) below which carbon steel and Cr–Mo steels may safely be exposed to hydrogen are given by the **Nelson chart**, published by the American Petroleum Institute (API Publ. 941). Normally equipment is designed to conform to this chart, but in exceptional circumstances hydrogen corrosion may take place and it is not uncommon for the heat-affected zone of fusion welds to be preferentially attacked.

Under non-corrosive conditions carbon and carbon–manganese steel welds may lose strength owing to **graphitization**. If steel is held at temperatures above 450 °C for long periods of time, the cementite may decompose, eventually forming nodules of graphite in a low-carbon iron matrix. The breakdown of cementite appears to be promoted by the addition of aluminium to the steel. Carbon and C–0.5Mo steels are both susceptible to this defect; however, in carbon steel the graphite nodules tend to be scattered and do not form localized areas of weakness. In C–0.5Mo steel, the graphite forms 'eyebrows' in the heat-affected zone and there have been isolated failures in steam lines from this cause. Cr–Mo steels do not suffer graphitization since the carbides are more stable. For this reason, Cr–Mo material is often specified for elevated-temperature duties (particularly piping) where in other respects C–0.5Mo steel would be adequate. Alternatively, samples may be removed from the pipe from time to time and examined metallographically for evidence of graphitization.

11.4.2 Intergranular corrosion of welds in austenitic chromium–nickel steels

If an unstabilized 18Cr–10Ni type steel containing about 0.1 % carbon or more is brazed or welded, and is then exposed to certain corrosive solutions, intergranular corrosion will occur in the zone of carbide precipitation, which runs parallel to the weld (Fig. 11.17). The severity of this type of attack, which is known as **weld decay**, increases with the carbon content and with the corrosivity of the environment. Steels containing 0.08 % carbon or less may suffer weld decay attack in thick sections, but not when the thickness is 3 mm or less. Steels containing 0.03 % carbon and less and those stabilized by the addition of titanium or niobium are immune from weld decay in all thicknesses. The time–temperature relationships for sensitization of various grades of austenitic Cr–Ni steel are shown in Fig. 11.18.

The susceptibility of a steel to intergranular corrosion after welding is commonly tested by heating a sample at 650 °C for 30 min, and then immersing it in a standard test solution for the specified period. The most commonly used

422 Metallurgy of welding

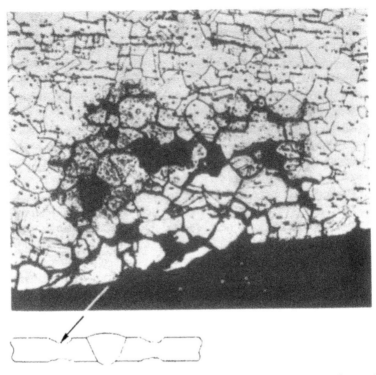

11.17 Intergranular corrosion in the heat-affected zone of a weld in unstabilized austenitic chromium–nickel steel.

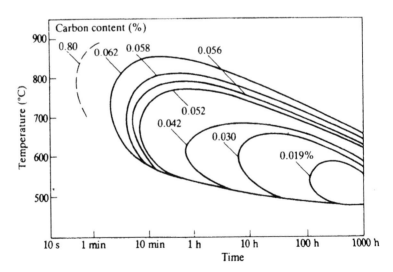

11.18 The effect of carbon on carbide precipitation: the time required for formation of harmful amounts of chromium carbide in stainless steels with various carbon contents.

tests are the copper sulphate–sulphuric acid and the boiling 65 % nitric acid tests. For detailed procedures the relevant ASTM or British Standard should be consulted. Tests may also be made in solutions representing the process fluid to which the welds will be exposed.

The addition of titanium in amounts exceeding four times the carbon content, or niobium in amounts exceeding ten times the carbon content, inhibits carbide precipitation in the normal weld decay zone but does permit some intergranular precipitation close to the fusion zone (see Section 9.5.1). Subsequent exposure to hot strong mineral acids may result in corrosion along a narrow zone adjacent to the weld boundary (Fig. 11.19). Such corrosion is known as **knife-line** attack. For any given carbon content, niobium-stabilized steel is more resistant to knife-line attack than the titanium-stabilized type because of the higher solution temperature of niobium carbide. The risk of this type of corrosion may be greatly minimized by limiting the carbon content of the titanium-stabilized steel to a maximum of 0.06 %. Steels containing less than 0.03 % C are not subject to knife-line attack.

The molybdenum-bearing grade with 0.08 % carbon maximum suffers weld decay only under severe conditions: for example, when immersed in hot acetic acid containing chlorides. For such environments the extra-low-carbon variety is used, or niobium or titanium is added in amounts sufficient to stabilize the carbon.

There are several theories concerning the mechanism of intergranular attack. The most widely held view is that the precipitation of chromium carbide at the grain boundary depletes the adjacent region of chromium, and this region then becomes anodic to the bulk of the metal and is preferentially attacked in corrosive media. Alternatively, it has been suggested that the precipitate is coherent, and that the coherency strains set up make the grain boundary region anodic and sensitive to preferential attack. Yet another theory is that the carbides are more noble than the surrounding matrix and cause galvanic corrosion of the intergranular regions. There is indeed some evidence that carbides are more noble than the base alloy, but in general it is not possible to find evidence that indicates positively which theory is most likely to be correct.

There are many non-corrosive or mildly corrosive conditions under which stabilization of austenitic stainless steel is not necessary – for example, in subzero temperature and decorative applications – and where for cost reasons unstabilized material is preferred. Such steels may have carbon contents up to 0.15 %, but in practice are best limited to 0.08 % maximum. Electrode and filler rod should in any event produce a weld deposit containing not more than 0.08 % carbon.

Because of awareness of the carbide precipitate problem, combined with the more ready availability of low-carbon grades, intergranular corrosion of austenitic stainless steel has become something of a rarity in process plant. However, intergranular stress corrosion cracking has occurred in the nuclear industry (Section 11.4.3).

424 Metallurgy of welding

11.19 Intergranular corrosion close to the fusion zone of a weld in austenitic stainless steel: knife-line attack (×50) (photograph courtesy of TWI).

Table 11.1 Alloy systems subject to stress corrosion cracking

Alloy	Environment
Aluminium base	Air
	Seawater
	Salt and chemical combinations
Magnesium base	Nitric acid
	Caustic
	HF solutions
	Salts
	Coastal atmospheres
Copper base	Primarily ammonia and ammonium hydroxide
	Amines
	Mercury
Carbon steel	Caustic
	Anhydrous ammonia
	Nitrate solutions
	$CO-CO_2$ solutions
	H_2S and cyanides in aqueous solutions
Martensitic and precipitation hardening stainless steels	Seawater
	Chlorides
	H_2S solutions
Austenitic stainless steels	Chlorides – inorganic and organic
	Caustic solutions
	Sulphurous and polythionic acids
Nickel base	Caustic above 315 °C (600 °F)
	Fused caustic
	Hydrofluoric acid
Titanium	Seawater
	Salt atmospheres
	Fused salt

Source: adapted from American Iron and Steel Institute (1977).

11.4.3 Stress corrosion cracking

Crack propagation associated with the combined effects of corrosion and stress is especially damaging; for example, when it occurs in a pipeline the whole line may be put out of action. Table 11.1 lists some of the alloys that may be affected by stress corrosion cracking. Of these, the most important are the ferritic and austenitic steels, and these two categories are considered below.

Stress corrosion cracking of ferritic steel welds

There is a group of chemical agents known collectively as 'poisons' that may cause cracking adjacent to welds in carbon steel. These are so called because they

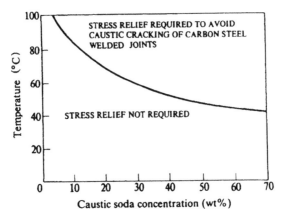

11.20 The conditions under which stress relief of welded joints in carbon steel is required in order to avoid stress corrosion cracking in caustic soda solution.

inhibit the action of catalysts, and they include H_2S and cyanides. In corrosion they have the effect of preventing or slowing down the recombination of hydrogen atoms, and thereby providing a source of hydrogen of high fugacity at the metal surface. This hydrogen dissolves in the metal to concentration levels that would be in equilibrium with very high pressures of molecular hydrogen. The gas may then precipitate at laminar discontinuities in the steel to form blisters or 'stepwise' cracks (cracks in which short laminar portions, associated with elongated inclusions, are joined by cracks at right angles to the laminar plane). Alternatively they may result in cracking adjacent to welds, of essentially the same type as the hydrogen-induced cracks that result from welding. Such cracking is possible in process lines carrying natural gas containing H_2S (sour gas) provided that moisture is present. It is avoided where the gas cannot be dried by limiting the hardness of welds to 200 BHN. A similar limit is often specified for water containing H_2S and/or cyanide. See also the next section.

Caustic alkalis, sodium and potassium carbonate, and nitrate solutions may cause intergranular stress corrosion cracking of carbon and low-alloy steel. Figure 11.20 shows the concentration–temperature limit above which it is necessary to stress-relieve carbon-steel welds in order to avoid caustic stress corrosion cracking. Stress corrosion cracking has also been observed where welds containing a pre-existing crack or crack-like defect are in contact with high-purity boiler feed water at elevated temperature.

Hydrogen pressure-induced cracking

This is a problem that arose in The Gulf in the 1970s. A spiral-welded pipe carrying sour gas ruptured only a few weeks after commissioning. The failures, which occurred preferentially adjacent to the welds, were due to H_2S corrosion, which saturated the metal with hydrogen and produced the step-like cracking described in the previous section, but not necessarily restricted to the weld region. Several other lines were affected.

The steel was strip containing a substantial amount of Type II manganese sulphide inclusions and was therefore particularly susceptible to delamination. Subsequently steelmakers have offered sulphur contents down to 0.002% (20 ppm) together with rare earth treatment.

However, a second problem arose. In continuous casting, manganese and phosphorus segregate to the central region, which results in a hard band in the rolled product. This may be subject to hydrogen cracking even in the absence of laminar inclusions. To avoid such segregation, manganese and phosphorus contents are reduced and the liquid steel is subject to electromagnetic stirring during the continuous casting operation.

Line pipe that may be employed in sour service is usually subject to testing in a simulated environment: either seawater saturated with H_2S or a similar solution acidified with acetic acid.

Stress corrosion cracking of austenitic chromium–nickel steels

Strong caustic alkalis may cause stress corrosion cracking in the austenitic stainless steels, but the agent most commonly responsible for this damage is the chloride ion. Chloride stress corrosion cracking is typically transgranular and branched (Fig. 11.21). Penetration may occur in a few seconds when sheet metal is exposed to hot concentrated chlorides, but at lower temperatures and lower chloride concentrations penetration may take many hours. Increased nickel content improves the resistance of a steel to this type of attack, while at the opposite extreme ferritic or substantially ferritic chromium steels are not susceptible to stress corrosion (Fig. 11.22). It should be noted that although ferritic chromium steels do not suffer stress corrosion cracking by chlorides, they may be so affected by wet hydrogen sulphide. The **threshold stress** below which this attack will take place is not well defined and must be assumed to be low; the residual stress developed by welding even thin sheet is sufficient to cause cracking if the corrosive environment is severe. Chloride stress corrosion cracking is usually caused either by accidental contamination of process steams by a chloride-containing aqueous solution, or by contamination of the metal surface by chlorides during manufacture or transport, followed by exposure to elevated temperature. The most severe condition is that in which the chloride-containing water is allowed to concentrate in contact with the metal surface. The design of stainless-steel equipment must be such as to avoid pockets or crevices where concentration can take place, and in manufacture, transport and storage,

428 Metallurgy of welding

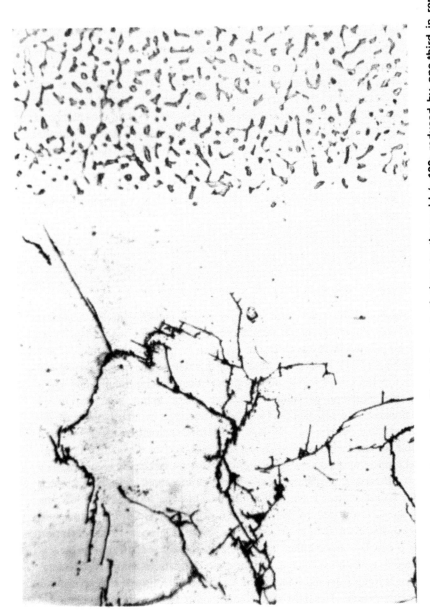

11.21 Stress corrosion in an austenitic stainless steel close to the weld (×100; reduced by one-third in reproduction) (photograph courtesy of TWI).

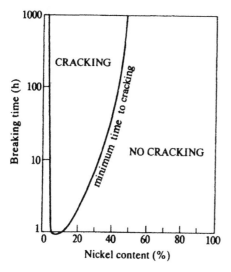

11.22 The effect of nickel in stress corrosion cracking (from Copson, 1959).

precautions must be taken against accidental contamination, for example by seawater.

Intergranular penetration and cracking of austenitic Cr–Ni steels may result from contamination of the surface by low-melting alloys followed by exposure at elevated temperature. Zinc from galvanized steel may cause such cracking in welds between stainless and galvanized steel. Spelter (60Cu–40Zn) will penetrate austenitic Cr–Ni steel at elevated temperature, so that brazing is carried out with low-melting silver brazing alloys. Generally speaking, it is advisable to avoid any contamination of austenitic stainless steel (e.g. by paint) if it is to be exposed to temperatures above, say, 400 °C.

Stress relief of welds at 900–1000 °C is sometimes employed as a means of avoiding stress corrosion cracking in piping. Where this is done, a stabilized or ELC steel must be used, the joint must be adequately supported to avoid hot cracking, and (for local postweld heat treatment) temperature gradients must be low enough to prevent the formation of residual stress at the edge of the heated zone. Lower temperatures are sometimes used, but in so doing the danger of σ phase formation and/or carbide precipitation at temperatures below 750 °C must be considered.

In practice it is exceptional to attempt such heat treatments. An alternative solution is to select a more resistant alloy. The molybdenum-bearing type 316 steel is significantly more resistant to stress corrosion cracking than types 304, 321 or 347. Reducing or increasing the nickel content may be beneficial. Figure 11.22 illustrates this possible approach. Higher nickel grades such as alloy 20 Cb 3 (with up to 35 % Ni) are more resistant but also more costly. A solution to this problem that has been adopted increasingly in recent years is to use a ferritic–austenitic stainless steel. The progenitor for such alloys was a 226Cr4Ni steel. This was prone to embrittlement by σ phase formation, but development has led to the production of a large family of alloys that are weldable and in some cases have a corrosion resistance similar to that of type 316. Fusion welds are made with consumables containing 2–3 % more nickel than the parent alloy, since otherwise the weld metal would be excessively ferritic.

11.5 Assessing the reliability of welded structures

Traditionally, the acceptability of fusion welds has been based on arbitrary standards. In general these standards exclude defects that are regarded as potentially hazardous, such as cracks and lack of fusion, while non-hazardous defects such as porosity and slag inclusions are rejected above a level known to be reasonably achievable. However, when the fracture toughnesses of the weld and parent material are known or can be measured, the application of fracture mechanics makes it possible to assess the risk of failure associated with any given defect, and in some cases such an analysis may avoid the unnecessary rejection of a weldment. For example, the radiographs of girth welds in a subsea pipeline were examined while still wet, and sentenced accordingly. Later examination of the same radiographs when dry showed that a number of welds contained defects outside the limits of the applicable code. However, by this time the pipe had been laid and repairs would have caused serious operational delays. A section of pipe was therefore recovered and CTOD tests carried out. Based on the results of these tests, together with a reliability analysis to the relevant standard, PD 6493, it was found possible to accept the welds.

Standard methods of calculating the risk associated with defects in nuclear pressure vessels are to be found in the ASME Code Section III Appendix G and Section XI, Appendix A. The Section XI technique has also been applied to non-nuclear pressure vessels. The original version of the British Standard PD6493: 1980 is based primarily on CTOD testing and incorporates a design curve from which it is possible to obtain the maximum allowable crack size for any given conditions. This curve has a built-in safety factor of 2. The 1991 revision of this standard provides for three levels of assessment. Level 1 is similar to the 1980 version. Level 2 does not include a built-in safety factor; users are required to apply their own factors. Level 3 is applicable to material of high strain-hardening capacity and would not normally be used for structural material.

The IIW document SST–1157–90 *Guidance on assessment of the fitness for purpose of welded structures* covers similar ground but is not a code and does not provide guidance as to the avoidance of brittle fracture. There are also Japanese and German standards for risk assessment.

The procedure used in PD 6493–1980 (which is similar to that of ASME XI) is generally as follows.

1. Determine the dimensions (widths and depth) of a crack that is equivalent to the observed defect, in accordance with specified procedures.
2. Calculate the crack size $\bar{a}_m$ that would cause failure by the following mechanisms:
   (a) brittle fracture;
   (b) ductile fracture or bulk yielding;
   (c) leakage;
   (d) corrosion;
   (e) creep.
3. Assess the size $\bar{a}$ to which the initial defect will grow during the design life of the vessel or structure due to:
   (a) fatigue;
   (b) corrosion;
   (c) creep crack growth;
   (d) creep fatigue interaction.

Then if $\alpha \bar{a}_m > \bar{a}$ the defect is acceptable. α is a safety factor: for PD 6493–1991 level 1 this is embodied in the design curve and here the defect is acceptable if $\bar{a}_m > \bar{a}$.

In practice, although the documents in question urge the user to consider the effects of corrosion, stress-corrosion cracking, creep crack growth and growth due to combined creep fatigue, quantitative guidance is given only for brittle fracture and fatigue. PD 6493 uses

$$\bar{a}_m = \frac{\delta_e E}{2\pi\sigma^2} \tag{11.4}$$

for $\sigma/\sigma_{yp} < 0.5$ and

$$\bar{a}_m = \frac{\delta_e E}{2\pi(\sigma - 0.25\sigma_{yp})} \tag{11.5}$$

for higher values of σ/σ_{yp}. In these expressions σ is the sum of all contributing stresses.

In principle it would appear that estimating the growth of an existing flaw due to fatigue would be simple, since

$$\frac{da}{dN} = C(\Delta K)^m \tag{11.6}$$

According to the IIW document values for the constants are, for da/dN in mm cycle^{-1} and ΔK in N mm$^{-3/2}$

$m = 3$

$C = 3 \times 10^{-13}$

in air and

$m = 3$

$C = 2.3 \times 10^{-12}$

in a marine environment. However, the shape factors for buried and surface cracks, and also the secondary stresses, vary with crack size. Therefore graphical methods are used to obtain the final value of $\bar{a}$.

A major evaluation of risk assessment has been undertaken in the Netherlands, using wide plate tests and full-scale trials of pipes and pressure vessels. The results up to 1990 indicated that all techniques based on CTOD and J-integral testing gave safe predictions of the brittle fracture risk. Where a comparison was made between ASME XI, PD 6493: 1980 and the revised PD 6493 all methods proved to be conservative, with the ASME XI values being closest to actual behaviour.

References

American Iron and Steel Institute (1977) *The Role of Stainless Steel in Petroleum Refining*. AISI Publ. SS607-477-20M–HP.
Booth, G. (1986) *Metal Construction*, **18**, 432–437.
British Standards Institution 1980, 1991 *Guidance on some methods for the derivation of acceptance levels for defects in fusion welded joints*, BSI Document PD6493.
Chilton, I.J., Price, A.T. and Wilshire, B. (1984) *Metals Technology*, **11**, 383–391.
Copson, H.R. (1959) in *Physical Metallurgy of Stress Corrosion Fracture*, Interscience, New York.
Hippsley, C.A., Rauh, H. and Bullough, R. (1984) *Acta Metallurgica*, **32**, 1381–1394.
James, L.A. (1977) *Welding Journal*, **56**, 386–391.
Maddox, S.J. (1980) *Metal Construction*, **12**, 531–533.
Maddox, S.J. (1985) *Metal Construction*, **17**, 220–224.
Matsuda, F. and Nakagawa, H. (1984) *Transactions of the Japanese Welding Research Institute*, **13**, 159–161.
Toft, L.H. and Yeldhan, D.E. (1972) in *Welding Research Related to Power Plant*, eds N.F. Eaton and L.M. Wyatt, CEGB, London.

Further reading

British Standards Institution (1980, 1991) BS PD6493: 1980, 1991.
Fracture toughness testing, 1965, ASTM STP 381.
Hertzberg, R.W. (1983) *Deformation and Fracture Mechanics of Engineering Materials*, 2nd edn, Wiley, New York.
Lancaster, J.F. (1996) *Engineering Catastrophes*, Abington Publishing, Cambridge.

Latzko, D.G.H. (ed.) (1979) *Post-yield Fracture Mechanics*, Applied Science Publishers, London.

Proceedings of conference on performance of pressure vessels with clad and overlayed stainless steel linings, 1981, Denver, ASME, Colorado.

Progress in flaw growth and fracture toughness testing, 1973, ASTM STP 536.

Appendix 1
Symbols

Symbols

		Units (nd = non-dimensional)
a	crack length, spherical radius	m
	acceleration	m s^{-2}
	activity	nd
B	basicity index	nd
	specimen width	m
c	concentration, specific heat	nd
C	constant	nd
d	lateral separation, width, grain size	m
D	coefficient of diffusion	m^2 s^{-1}
e	interaction parameter	nd
erf	error function	nd
E	modulus of elasticity	MN m^{-2}
	activity coefficient	nd
F	force	N
g	acceleration due to gravity	m s^{-2}
G	specific fracture energy	J m^{-2}
	free energy of formation	J mol^{-1}
	shear modulus	MN m^{-2}
	solidification temperature gradient	K m^{-1}
h, H	vertical height	m
I	current	A
J	current density	A m^{-2}
k	any proportional factor	nd
	Boltzmann's constant	J K^{-1}
K	absolute temperature	K
	equilibrium constant of a chemical reaction	nd
	stress intensity factor	MN m$^{-1/2}$, N mm$^{-3/2}$
	thermal conductivity	W m^{-1} K^{-1}
K_I	stress intensity factor in the opening mode (I)	MN m$^{-1/2}$, N mm$^{-3/2}$

Symbol	Description	Units
K_{IC}	critical value of K_I for unstable crack extension	MN m$^{-1/2}$, N mm$^{-3/2}$
K_n	modified Bessel function of the second kind of order n	nd
ln	logarithm to base e	nd
log	logarithm to base 10	nd
l, L	length, distance	m
m	any integer	nd
M	any metal	nd
n	any numerical factor	nd
	operating parameter	nd
N_0	Avogadro's number	mol^{-1}
p	pressure	N m^{-2}
q	rate of heat or energy flow	W
	electric charge	C
Q	quantity of heat or energy	J
r	radius	m
R	gas constant	J mol^{-1} K^{-1}
R, R_o	outer radius of a cylinder	m
R	solidification rate	m s^{-1}
s	solubility (gas in metal)	ppm, g tonne^{-1}
t	time	s
T	temperature	K, °C
U	strain energy	J
v	velocity	m s^{-1}
V	Voltage	V
V_o	dislocation velocity	m s^{-1}
w	thickness	m
W	work of cohesion	J m^{-2}
	rate of heat or energy flow	W
α	diffusivity of heat	m^2 s^{-1}
	coefficient of thermal expansion	K^{-1}
	degree of ionization	nd
γ	surface tension	N m^{-1}
	surface energy	J m^{-2}
δ	longitudinal displacement	m
	crack-tip opening displacement	m
δ_c	critical crack-tip opening displacement	m
Δ	increment	nd
ε	dielectric constant (relative permittivity)	nd
η	viscosity	kg s^{-1} m^{-1}
	arc efficiency	nd
θ	angle	nd
	absolute temperature	K
λ	wavelength	m
υ	kinematic viscosity	m^2 s^{-1}
	Poisson's ratio	nd
ρ	mass density	kg m^{-3}
σ	stress	MN m^{-2}
σ_u	ultimate stress	MN m^{-2}
σ_{ys}	yield stress	MN m^{-2}

τ	shear stress	MN m^{-2}
ϕ	angle	nd
	electron work function	J
ω	frequency	s^{-1}

Appendix 2
Conversion factors

To convert B to A multiply by	A	B	To convert A to B multiply by
1×10^{10}	ångström (Å)	M	1×10^{-10}
9.8692×10^{-6}	atmosphere (atm)	$N\ m^{-2}$	$1.013\ 25 \times 10^5$
1×10^{-5}	bar	$N\ m^{-2}$	1×10^5
$9.478\ 17 \times 10^{-4}$	Btu	$J = N\ m$	$1.055\ 06 \times 10^3$
$5.265\ 65 \times 10^{-4}$	Btu/°F	$J\ K^{-1}$	$1.899\ 09 \times 10^3$
$0.056\ 869$	Btu/min	$W = J\ s^{-1}$	$17.584\ 26$
$4.299\ 23 \times 10^{-4}$	Btu/pound	$J\ kg^{-1}$	2.326×10^3
$0.316\ 998$	Btu/ft^2 h	$W\ m^{-2}$	$3.154\ 59$
0.5778	Btu/ft^2 h for a temperature gradient of 1°F/ft	$W\ m^{-1}\ K^{-1}$	$1.730\ 73$
$0.238\ 846$	calorie (cal)	J	4.1868
$2.388\ 46 \times 10^{-5}$	cal/cm^2 s	$W\ m^{-2}\ s^{-1}$	4.1868×10^4
$2.388\ 46 \times 10^{-4}$	cal/g	$J\ kg^{-1}$	4.1868×10^3
$2.388\ 46 \times 10^{-4}$	cal/g °C	$J\ kg^{-1}\ K^{-1}$	4.1868×10^3
2.3889×10^{-4}	kcal/g mol	$J\ mol^{-1}$	4.1868×10^3
35.3147	cubic foot	m^3	$0.028\ 316\ 8$
6.1024×10^4	cubic inch	m^3	$1.638\ 71 \times 10^{-5}$
1×10^5	dyn (dyne)	N	1×10^{-5}
1×10^7	dyn cm	$N\ m$	1×10^{-7}
1×10^3	dyn/cm	$N\ m^{-1}$	1×10^{-3}
10	dyn/cm^2	$N\ m^{-2}$	0.1
1×10^7	erg	J	1×10^{-7}
1×10^3	erg/cm^2	$J\ m^{-2}$	1×10^{-3}
1×10^7	erg/s	W	1×10^{-7}
$3.280\ 84$	foot (ft)	m	0.3048
23.7304	foot poundal	J	$0.042\ 14$
$0.737\ 56$	foot pound force	J	$1.355\ 82$
$2.199\ 69 \times 10^2$	gallon (UK)	m^3	$4.546\ 09 \times 10^{-3}$
$2.641\ 72 \times 10^2$	gallon (US)	m^3	$3.785\ 41 \times 10^{-3}$
$1.543\ 24 \times 10^4$	grain	kg	6.4799×10^{-5}
1.0×10^{-3}	g/cm^3	$kg\ m^{-3}$	1.0×10^3
1.0	g mol (c.g.s.)	mol	1.0
1×10^{-7}	hectobar	$N\ m^{-2}$	1×10^7
$1.341\ 02 \times 10^{-3}$	horsepower	W	745.7
39.3701	inch	m	2.54×10^{-2}

Appendix 2

2.362 21	inch/min	mm s^{-1}	0.423 33
0.101 972	kilogram force	N	9.806 65
1.019 72 × 10^{-5}	kilogram force/cm^2	N m^{-2}	9.806 65 × 10^4
0.101 972	kilogram force/mm^2	MN m^{-2}	9.806 65
3.224 62	kilogram force/mm$^{3/2}$	MN m$^{-3/2}$	0.310 11
0.145 04	ksi (thousand pounds force/square inch)	MN m^{-2}	6.894 76
0.910 05	ksi√inch	MN m$^{-3/2}$	1.098 84
		N mm$^{-3/2}$	34.7498
9.999 72 × 10^2	litre	m^3	1.000 028 × 10^{-3}
0.039 370 1	mil (thou)	μm (micron)	25.4
0.621 371	mile	km	1.609 344
0.031 622 78	MN/m$^{3/2}$	N mm$^{-3/2}$	31.622 78
10	poise (dyn s cm^{-2})	N s m^{-2}	0.1
2.204 59	pound (lb)	kg	0.4536
7.233 01	poundal	N	0.138 255
0.204 809	pound force	N	4.448 22
1.450 38 × 10^{-4}	pound force/square inch	N m^{-2}	6.894 76 × 10^3
3.612 73 × 10^{-5}	pound/cubic inch	kg m^{-3}	2.767 99 × 10^4
1.55 × 10^3	square inch	m^2	6.4516 × 10^{-4}
10.7639	square foot	m^2	0.092 903
0.386 100 6	square mile	km^2	2.589 998
1 × 10^4	stoke, c.g.s. unit of diffusivity or thermal diffusivity (cm^2 s^{-1})	m^2 s^{-1}	1 × 10^{-4}
9.842 07 × 10^{-4}	ton (UK)	kg	1.016 05 × 10^3
1.102 31 × 10^{-3}	ton (short ton, US)	kg	907.185
0.100 361	UK ton force	kN	9.964 02
0.064 749	UK ton force/square inch	MN m^{-2}	15.4443
7.500 62 × 10^{-3}	torr (1 mm mercury)	N m^{-2}	133.322

Index

Acicular ferrite, 228–232, 298
Acrylic adhesives, 64–65
Active metal brazing, 114–116
Activity coefficient, 174
Adherend
  definition, 54
  surface preparation, 72–75
Adhesive types
  natural 54
  primer, 54–55
  thermosetting, 54
  thermoplastic, 54
Adhesive
  dispensers, 75–76
  mechanical properties, 79–80
Adhesively bonded joints
  applications, 81–83
  bonding procedures, 72–76
  design, 54, 80–81
  design stress, 56
  effect of water, 61, 68–71, 72, 73, 74
  in timber, 65–66
  mechanical properties, 66–67
  mechanical tests, 76–80
  quality control, 76
  upper service temperature, 66
Age-hardening
  aluminium alloys, 364–371
  nickel alloys, 385
  stainless steel, 316, 349–350
AISI 4130 and 4140 steels, 295
'Alexander L. Kielland', 403–405
Aluminium
  adhesive bonding, 72–74
  alloys, 367–371
  fracture toughness of steel, 246
  gas–metal reactions, 353–357
  as grain-refiner of steel, 234–235, 246
  graphitization of steel, and, 421
  mechanical properties, 361–371
  in self-shielded wire, 212
  solidification cracking, 358–361
  welding and brazing, 368–371
Aluminium bronze, 373, 377, 379–380
Aluminium–lithium alloys, 366–367

Ammonia tank failure, 402–403
Amorphous metal, 103
Anaerobic adhesives, 64
Anodizing, 72–74
Antimony in solder, 90, 91
Austenite
  forming elements, 318–319
  in carbon steel welds, 229–232
  solidification cracking and, 175–176
  transformation, 225–232

Bainite
  effect on HAZ toughness, 237
  in carbon steel weld metal, 231, 232
  in HAZ, 237
  structure, 226
Basicity index, 218–219
Basic oxygen process, 280–282
Bead-on-plate test, 293, 294
Beryllium, 379, 388, 393–394
Bessemer process, 280–281
Bismuth, 378, 383
Boeing wedge test, 78–79
Bond line, 54
Bonding
  adhesive to metal, 55–56
  covalent, 2
  in solid phase welds, 90–91
  ionic, 2
  mechanisms, 2
  metallic, 2
  of metals to non-metals, 3–5
  solid to liquid, 15–16
Boron, 97, 223, 232, 233
Brass, 377–378, 379
Braze welding, 101–102, 289
Brazing
  active metal, 114–115
  aluminium alloys, 371
  applications, 102–103
  atmospheres, 97–101
  capillary action, 85, 86–87
  ceramics, 114–116
  filler alloys, 95–97, 371
  fluxes, 97, 371

439

Brazing (*continued*)
  honeycomb structures, 102–103
  joint design, 94–95
  in microjoining, 121
Brittle
  fracture, 20–34, 399–403
  temperature range, 198–199, 223, 358–361
Butt welding, 41, 145
Buttering, 192–193, 272, 289

Calcium, 282
Cap copper, 378, 379
Capillary force, 85, 86–87
Carbide
  intergranular, 328–330, 346, 385–387
  intragranular, 333
  precipitation, 323, 328–330, 343
Carbon
  effect on solidification cracking, 223
  equivalent, 244–245, 267
  migration, 343, 345, 387
  porosity in welds, 180–184, 382–383
Cast iron
  welding, 287–289
Cast-to-cast variation, 313–316
Catastrophic oxidation, 348
Caustic soda, 426
Ceramics
  active metal brazing, 114–116
  ceramic–ceramic bonding, 130
  coatings, 114
  diffusion bonding, 116–118
  electrostatic bonding, 118–119
  field-assisted bonding, 118–119
  fracture toughness, 105, 106
  friction welding, 120
Cerium treatment, 272
Charpy impact test, 21, 28–29
Chevron cracking 269–271
Chevron markings, 21
Chi phase, 327–328, 330–333
Chromium–molybdenum steels, 301–303
Clad plate, 341–344
Clean steel, 282
Cleavage facets, 21
Cleavage strength, 2
Coated electrodes
  coating types, 130, 139
  moisture content, 258–260
Coherence temperature, 198
Cold pressure welding, 371
Continuous casting 283
Continuous cooling transformation diagram, 228, 232
Controlled rolling, 284–286
Controlled thermal severity (CTS) test, 263–264
Copper, 373–381
Corona discharge treatment, 56
Corrosion of welds
  austenitic Cr–Ni steel, 332–333, 347–348, 420

  carbon steel, 420, 421, 426
  caustic, 426
  hydrogen, 205, 250, 421
  intergranular, 303–304, 346, 385, 387, 421–425
  knife-line attack, 346, 423
  localized, 420–421
  modes, 420
  preferential, 420–421
  stress corrosion, 371, 373, 425–430
  weld decay, 421
Covalent bonds, 2
Crack
  extension force, 24
  growth
    creep, 416
    fatigue, 412, 414
    reheat, 274–278, 333–334, 342–343, 413–416
Crack velocity
  flexiglass, 32
  glass, 30–34
  steel, 32–33
  stress, 32–33
  temperature, 32
Cracking
  index, 224
  modes
    chevron, 269–271
    ductility tip, 204–205, 334–335
    hydrogen-induced, 166–167, 248–263, 419, 426
    intergranular, 201
    lamellar tearing, 271–274, 403–404
    liquation, 46, 201, 237, 366
    liquid metal, 329
    reheat, 274–278, 333–334, 342–343, 413–416
    shrinkage, 199, 200–201
    solidification see Solidification cracking
    stress-corrosion, 371, 373, 425–430
  tests
    compact tensile, 24–25
    CTOD, 25–26
    CTS, 263–264
    Houldcroft, 201, 202, 364
    implant, 264–265
    J-integral, 26–28
    Lehigh, 263, 264
    MISO, 201, 202, 225
    RPI, 266
    Rigid Restraint (RRC), 262, 265–266
    Tekken, 265–266
    Transvarestraint, 201, 202
    Y-groove, 265–266
Cupronickel, 377, 380
Cyanoacrylate adhesives, 56, 64–65, 81

Decohesion theory, 254–256
Delayed fracture, 257
DeLong diagram, 319

Index 441

Deoxidation, 175–176
Dew point, 100
Die bonding, 123, 124
Difussible hydrogen, 251
Diffusion
 bonding, 49–50, 116–118
 hydrogen, 191–192, 251
 general, 190–192, 193, 251
 treatment, 191
 water in adhesive, 69–71
 welding, 49–50, 116–118
Dilution, 192–193, 361
Dipole, 55–56
Disbonding, 417–420
Dislocation
 theory, 8
 velocity, 9–11
Dispersion forces, 8, 55
Dissimilar metal joints, 337–339, 344–345, 416–417
Dissociation
 gases, 187, 257
 metal oxides, 98
 pressure of oxides, 97–101
Dry joint, 15
Ductile behaviour
 crystal structure, 13–15
 fracture, 34–35
 structural integrity, 20
 and welding, 2, 3
Ductility dip cracking, 204–205, 271, 334–335

Earthquake fractures, 406–407
Edge preparation, 90–91
Electric resistance welding, 42–43
Electromagnetic stirring, 363, 427
Electron
 beam welding, 139, 390
Electroslag welding
 cooling rate, 157, 162
 grain size, 140, 161
 macrostructure, 236
 process, 131, 139, 140, 341–342
Electrostatic bonding, 70–71
Elevated temperature properties, 243, 326–327, 348–349
Embrittlement
 by PWHT, 208, 244, 247–248, 343
 ductility dip, 204–205
 gas absorption, 388–390
 HAZ, 245–248, 275–278
 hydrogen, 102, 248–263, 327–328, 335–336
 by liquid metal, 102
 recrystallization, 394–395
 sigma phase, 327–328, 330–333, 388, 419
 strain-age
  general, 213–214
  mechanism, 213–214
  in multi-run welds, 242–243
  sub-critical, 237

temper, 208, 278–279
transformation, 327–328
Epitaxial grain growth, 194–195
Epoxide adhesives, 63
Epoxide ring, 62
Energy release rate, 24
Explosion bulge test, 30
Explosive welding, 10, 12, 48–49

Fatigue properties
 carbon steel welds, 240, 409–413
 crack growth rate, 412, 414
 weld profile and, 240, 409
Ferrite
 acicular, 229–232, 237
 forming elements, 318–319
 grain boundary, 289–232, 237
 polygonal, 289–232, 237
 side plates, 289–232, 237
 Widmanstatten, 229, 237
Field-assisted bonding, 118–119
Fillet welds
 form, 146
 hydrogen cracking, 257
 profile, 146
Fish-eyes, 263
Flash-butt welding, 41, 43–44
Flux
 acid, 218–219
 active, 220
 basic, 218–219
 basicity, 218–220, 258, 261
 neutral, 220
 solder, 90–92
 types, 220, 258
Flux-cored electrodes, 132
Forge welding, 42, 44
Fracture
 appearance, 20, 28–29
 appearance transition temperature, 29
 ductile, 34–35
 energy, 24, 69, 78–79
 mechanics, 22–30
  measurement of K, 24–25
  theory, 22–25
 modes
  cleavage, 21
  microvoid coalescence, 21, 254
 toughness, 20
  ceramics, 105, 106
  definition, 5
  of HAZ in steel, 245–248
  measurement of, 24–30, 78–79
  properties, 37
  versus yield strength, 36–38
 velocity, 30–34
Free radicals, 62
Friction welding
 continuous, 44
 inertia, 44

Friction welding (continued)
　general, 41, 44–48
　metal-ceramic bonding, 120
　stored energy, 44
Friction stir welding, 46–48, 363, 367, 368
Full penetration welding, 146
Fully-austenitic steels, 306–307, 430
Fusion welding
　development, 128–129
　heat flow, 147–165
　joint design, 145–146
　power sources, 141–145
　processes, 131–138
　shielding methods, 130, 139

Galvanised steel, 396
Gamma loop, 303–304
Gas-metal reactions
　alloying additions, 174–175
　aluminium, 353–357
　carbon dioxide, 180–184
　in carbon steel, 211–215
　copper, 373–378
　desorption, 188
　diffusion, 190–192
　general, 169–191
　hydrogen, 170–172, 184–185, 211–212, 249–263
　interaction parameter, 174–175
　mechanism, 185–186
　nickel, 381–383
　nitrogen, 173–174, 177–179, 186–187, 212, 214, 353–354
　oxygen, 172–173, 179–180, 190, 214–216, 353, 397
　and porosity, 187–190, 397
　reactive metals, 388–390
　silver, 397
　tungsten and molybdenum, 394
Glass
　annealing temperature, 109
　bonding mechanism, 110
　ceramics, 111–114
　devitrification, 112
　fracture toughness, 105, 106
　fracture velocity, 30–34
　housekeeper joint, 108–109
　metal seals, 106–109
　sealing temperature, 111
　softening temperature, 107, 109
　strain temperature, 107
　viscosity, 107
　working temperature, 107, 109
Glass transition temperature, 66
Glove box, 389–390
Glue line, 54
Gold, 397
Grain
　coarsening temperature, 234

size
　aluminium, 196, 234, 235, 361–362
　austenitic, 234, 235, 237, 241
　cooling rate and, 236–237
　heat affected zone, 235–239
　solidification mode, 194–196
　titanium, 197, 234
　weld metal, 193–196, 240–241, 361–362
Grain growth region, 234–235
Grain refined region, 234–235
Graphitization, 421
Griffiths equation, 4, 21–22

Homogeneous linings, 396
Honeycomb structure, 82, 83, 103
Hydrocrackers, 417–420
Hydrogen
　absorption of, 170–172, 184–185, 211–212, 220–221, 354–357
　attack, 250, 421
　blistering, 250
　cold cracking and, 248–263, 419
　content of weld metal, 204, 210–215, 260–262
　diffusible, 251
　diffusion treatment, 191
　diffusivity in steel, 191–192, 251, 337
　embrittlement, 102, 248–263, 335–336
　notched tensile strength and, 252–253
　porosity, 187–188, 221–222, 354–357
　reactions in liquid metal, 170–172, 381–382, 390, 184–185, 211–212, 249–263, 355–357
　solubility, 171–172, 250–251, 337–353
　total, 251
　traps, 251
　water content of flux and, 258–261
Hydrogen-controlled electrodes, 260
Hydrogen pressure induced cracking, 285, 426–427

Impact properties
　cap and root of weld, 242–243
　heat affected zone, 245–248
　weld metal, 241–243
Impact test
　adhesives, 80
　Charpy, 28
Implant test, 264–265
Intensity of restraint, 262
Interaction parameter, 174
Interfacial energy, 56–59, 86–87
Intergranular
　corrosion, 303–304, 346, 385–387, 421–425
　precipitation, 328–330, 346, 385–387
Interpass temperature, 268–269
Intragranular precipitation, 333
Iron-carbon constitution diagram, 227

J-factor, 278–279
J-integral, 26–27

Joint
  clearance, 87, 95, 103
  type, 145–146

Keyhole weld, 139, 140
Knife-line attack, 346, 423
Kovar, 110, 111, 116

Lack of fusion, 15, 139, 166, 167
Lamellar tearing, 271–274, 403–404
Lamination, 283, 284
Lap-shear test, 76–77
Larson-Miller parameter, 343
Laser welding, 137
Lattice structure, 2, 6
L-D steelmaking process, 280–281
Lead
  homogeneous linings, 396
  in copper alloys, 378
  in solder, 87–89
  welding, 396
Lehigh slit groove test, 263, 264
Line pipe, 297–300
Liquation
  cracking, 46, 201, 237, 366
  in heat affected zone, 237
Liquid metal embrittlement, 102, 249
Lithium, 366–367
Low-hydrogen consumables, 260
Lower critical temperature, 225

Magnesium, 371–373
Magnetically impelled arc butt welding, 43
Manganese
  as deoxidant, 218–220
  cracking and, 222–225
  and overheating, 236
  slag-metal reactions, 216–220, 237
Manganese-molybdenum process, 114–115, 120
Manganese/sulphur ratio, 222–223
Marangoni flow, 19, 185–186
Maraging steel, 295–297
Martensite, 226–227, 229–232, 237–246
Metallic bond, 2
Metallizing, 114
Microalloyed steel, 287
Microjoining
  ball bonding, 125, 126
  diffusion soldering, 120, 126
  package, 120, 121, 122
  power devices, 126
  processes, 121
  silicon die bonding, 123–124
  wedge bonding, 125, 126
Microstructure, carbon steel welds
  heat affected zone, 233–239
  nomenclature, 230–232
  weld metal, 228–232
Microvoid coalescence, 21

Modified adhesives, 78
Moisture content of flux, control of, 258–260
Molybdenum, 390, 394–395
Monel, 379–380
Monomar, 61
Multipass welds, 140, 145–146

Napkin ring test, 80
Narrow gap welding, 140–141
Nelson chart, 421
Nickel
  fracture toughness of welds and, 241–246
  solidification cracking of steel and, 222–224
  welding of, 381–388
Nickel-iron electrodes, 288–289
Nil-ductility temperature, 29, 30, 198, 358
Niobium
  as carbide stabilizer, 303, 328–330
  as grain refiner, 234, 246, 287
  in steel, 234, 246, 287
  reheat cracking and, 414–415
Nitrogen
  as shielding gas, 374–379
  content of steel, 212
  embrittlement by, 212, 213
  porosity, 212, 353–354, 374–379, 381–382
No-gas welding, 139, 212, 214, 241, 243

Offshore structures, 400–402, 403–405
OFHC copper, 379
Open-hearth process, 280, 281
Operating parameter, 159
Overheating, 235–236, 334–335
Overlay weld cladding, 341–344
Overpressure test, 208–209
Oxide films, 40, 41
Oxide inclusions, 3, 21, 190, 215
Oxyacetylene welding, 128, 355
Oxygen
  content of weld metal, 214–216, 232–233, 353
  effect of microstructure, 232–233
  embrittlement by, 232
  effect on notch-ductility, 232–233
  porosity in silver, 397
  potential, 99
  reaction with liquid metals, 172–175, 179–180, 190, 214–216, 353–354, 382–383

Package, 120, 121, 122
Partial penetration weld, 145, 146
Partition coefficient 189
Pearlite, 225–226, 237
Peel forces, 76
Peel ply, 75
Peel test, 77, 78
Pellini drop-weight test, 29–30, 293
Peretectic reaction, 222–223, 316
Phenol-formaldehyde adhesives, 63–64
Phosphorous

Phosphorous (*continued*)
  as deoxidant, 373
  embrittlement, 95
  solidification cracking, 223, 224
  subsolidus cracking, 204–205, 334–335
Pitting corrosion, 347
Plane strain, 23
Plane stress, 23
Plasma
  cutting, 135
  welding, 135
Plastic flow, 7
Plastic tent, 389
Platinum metals, 397
Polar bonding, 55
Polymerisation
  addition, 62
  condensation, 62
  hardener, 62
  initiator, 62
  mechanism, 61–62
Polyurethane adhesives, 65
Porosity
  aluminium, 188, 356–357
  copper, 373–381
  general, 139, 187–190
  hydrogen, 187–188, 221–222, 356–357, 390
  mechanism, globular, 188, 356–357
  Mechanism, tunnel, 188–189, 357
  nickel, 187, 188, 381–383
  nitrogen, 187–188
  nucleation, 141–144
  reactive metals, 333, 390–391
  steel, 161, 167, 187, 189, 212
Positional welding, 146–147
Post welding heat treatment
  cast iron, 288–289
  dissimilar metal joints, 345
  embrittlement by, 208, 244, 247–248
  fracture toughness and, 244, 247–248
  nickel alloy steels, 300–301
  reheat cracking and, 417
  residual stress, 207–208, 244, 247–248
Power sources, 141–145, 369
Precipitation hardening
  aluminium, 360–371
  nickel welds, 385
  stainless steels, 316, 349–350
Preheat
  ANSI B31.3, 268
  AWS D1.1, 368
  BS 5135, 268, 290
  carbon steel, 268, 290
  cast iron, 287–289
  copper, 378
  general, 267–269
  hydrogen cracking and, 267
  lamellar tearing and, 273
  reinforcing bar, 289
Pressure vessels, 140–141

Pressure welding, 41, 42
Printed circuit boards, 81, 87–88, 92–94
Proof stress, 15
Pro-eutectoid ferrite, 229–232, 237, 241
Pulse-arc welding, 141–145

Quality control
  adhesively-bonded joints, 76
  spot welding, 76
Quasi-stationary state, 154

Rail steel, 289
Reaction bonding, 115
Radiographic inspection, 167–168
Rare-earth treatment, 272–282
Recrystallisation, 41–42
Reheat cracking, 274–278, 301, 333–334, 342–343, 413–416
Reinforcing bar, 289
Repair rate, welds, 167
Repair welding, 340–341
Residence time, 203
Residual stress, 205–209, 244, 247–248
Resistance welding
  aluminium, 370–371
  copper alloys, 379–380
  nickel alloys, 388
  processes, 128
  reactive metals, 390
Restraint
  hydrogen cracking and, 262–263
  intensity factor, 262
  RRC test, 262
  solidification cracking and, 199
Risk analysis, 430–432
Robots, 141
Roll bonding, 44
RPI test, 266

Schaeffer diagram, 318, 340
'Sea Gem', 400–401
Seam welding, 137
Self-shielded welding, 139, 212, 214, 241, 243
Sessile drop, 16–17, 56
Shielded metal arc welding, 130, 139
Shielding methods, 130, 131–137
Side plates, 229–232
Sigma phase 327–328, 330–333, 388, 419
Silicon
  as deoxidant, 218–220
  cracking and, 322–323
  slag-metal reactions, 218–220
Silicon bronze, 377, 379, 380
Silver, 397
Silver solder, 95, 96
Single-pass welds, 140–141
Slag-metal reactions, 216–220
Slip plane, 6, 7
Soldering
  condensation, 93

drag, 92
fluxes, 90–92
quality control, 94
reflow, 94
techniques, 92–94
wave, 92
Solders, 88–90, 371
Solid phase welding
 aluminium, 46, 48
 iron and steel, 42–48
 process, 42–48
Solidification
 cracking
  aluminium alloys, 46, 358–363
  austenitic Cr-Ni steel, 319–323
  carbon steel, 222–225
  copper alloys, 378–379
  general, 198–201
  nickel, 383–384
  solder, 88
  tests, 201–202
 modes, 193–196
 parameter, 194–195
Specific fracture energy, 5
Spot welding, 130, 136, 149–153, 370, 372, 395
Steam reaction, 204, 3773–374, 382
Steel, austenitic, Cr-Ni
 brazing, 95, 96, 101–102
 carbide precipitation, 328–330, 346
 cast-to-cast variation, 313–316
 chi phase, 327–328, 330–333
 cladding, 341–344
 constitution, 316–318
 corrosion, 346–348, 420–430
 cryogenic properties, 325–326
 ductility dip cracking, 334–335
 elevated temperature properties, 326–327, 348–349
 embrittlement, 327–328
 extra-low carbon, 329–330
 ferrite-free, 321–323
 gas-metal reactions, 310–316
 grades, 346–348
 hardenable, 316, 349–350
 heat-resisting, 348–350
 hydrogen embrittlement, 327–328, 335, 336
 mechanical properties, 324–327
 reheat cracking, 333–334
 repair welding, 340–341
 segration, 331–333
 sigma phase 327–328, 330–333
 slag-metal reactions, 310–313
 soldering, 90
 solidification cracking, 319–323
 stabilization, 328–330, 421–425
 stress corrosion cracking, 427–430
 subsolidus cracking, 334–335
 transformations, 316–318, 327–328, 330–333
 transition joints, 290–291
 weld pool shape, 313–316
Steel, ferritic
 acicular ferrite, 298
 balanced, 283
 calcium-treated, 282
 carbon-manganese, 289–290
 continuous casting, 282–284
 controlled-rolled, 284–285
 corrosion-resistant, 289, 303–307
 cryogenic, 300–301
 ferritic stainless, 303–307
 heat resistant, 301–303
 killed, 283
 line pipe, 297–300
 low-alloy, 301–302
 low temperature, 300–301
 maraging, 295–297
 microalloyed 246, 287
 microstructure, 225–228
 normalised, 284
 pearlite-reduced, 298
 quenched and tempered, 293–297
 rail, 289
 reinforcing bar, 289
 rimming, 283
 semi-killed, 283
 structural, 289–293
 transformations, 180–193
 transition joints, 290–291, 358–360
 weathering-resistant, 289
 ultra-high-strength, 246–247
Steel, ferritic-austenitic, 306–307, 430
Steelmaking
 casting, 282–284
  controlled rolling, 284–287
  developments, 280–282
Step cooling, 278
Stovepipe welding, 295, 298
Strain-age embrittlement, 213–214, 237–238
Strain rate, 9–12
Stress concentration factor, 248–249
Stress corrosion cracking, 371, 373, 425–430
Stress, intensity factor K, 23
Stress-strain curve, 13–15
Stud welding, 135
Submerged arc welding
 chevron cracking in, 220
 flux
  basicity, 218–220, 258, 261
  moisture in, 220, 258–259
  reactions, 216–220
  types, 220, 258
 mechanical properties, 297, 298
 oxygen in weld metal, 214–220
 process, 131, 140, 341–342
Sulphur, effect on
 hardenability of steel, 238–239
 hydrogen embrittlement and cracking, 238–239
 lamellar tearing, 271–274
 line pipe steel, 285, 426–427

Sulphur, effect on (*continued*)
    nickel, 383-384
    reheat cracking, 416
    solidification cracking, 222-225
    through thickness ductility, 271-274
    weld pool flow, 313-316
Superaustenitic steels, 332-333, 347-348
Superglue, 56, 64-65
Superplastic forming, 52-53
Surface energy
    dispersion and polar components, 56-58
    general, 16-17, 56-61
    measurement, 56-58
    values, 59-60
Surface tension
    solid-liquid interactions, 16-20
    surface-active elements, 17-20
    undercut, 17
    weld bead profile, 17
    weld pool flow, 17-20, 313-316

Tantalum, 393
Temper embrittlement, 208, 278-279, 306
Tensile strength
    alloy steel, notched, 252-254
    aluminium welds, 364-367
    carbon steel welds, 240-241
    copper welds, 379-380
    high-alloy steel, 324-327
    nickel welds, 385
    perfect crystal, 4
    zirconia, 5, 6
Thermal cycle, calculation, 156-164
Thermal severity number, 263-264
Thermit welding
    copper, 379
    process, 128
    rails, 289
Thermocapillary flow, 20
Thermomechanically controlled rolling, 284-287
Through thickness ductility, 271, 273-274, 290, 403
Tin
    deposition from flux, 88
    in solder, 87-89
Tin bronze, 377
Titanium
    as carbide stabilizer, 303, 328-330
    as deoxidant, 174-176
    as grain refiner, 233-235, 287
    in microalloyed steel, 287

welding of, 391-392
Toe crack, 257
Tough pitch copper, 373-381
Transformation products, steel, 225-232
Transition joints, 344-345, 416-417
Transition temperature range, 20-21
Tungsten, 394-395
Tunnel porosity, 188-189, 357, 377

Ultrasonic testing, 43, 168
Ultrasonic welding, 41, 99
Underbead crack, 257
Underclad cracking, 342
Undercut, 17, 196, 197
Uniformity of weld, 192-193
Unit cell, 6
Upper critical temperature, 225

Van der Waal's forces, 55-56
Vanadium, 234, 235, 287, 385-387
Vibratory stress relief, 208

Weld bead profile, 240
Weld decay
    austenitic Cr-Ni steels, 421-424
    nickel-base alloys, 385-387
Weld defects, 166-168
Weld overlay
    bead profile, 341-342
    cladding, 341-344
    disbonding, 417-419
Weld procedures, 267
Weldability equivalent, 245
Welding positions, 146-147
Welding power sources
    general, 141-145
    inverter, 142-145
    pulsed arc, 141-142
Welding processes, 131-137
Wetting
    angle, 200
    brazing and soldering, 86-87
    solidification cracking and, 200
Wood adhesives, 65-66
Work of adhesion, 59-61

Yield strength and strain rate, 9-15
Young-Dupré equation, 16-17, 56

Zinc, 396
Zirconium, 235, 393